VOLUME SEVENTY SIX

ADVANCES IN QUANTUM CHEMISTRY

Novel Electronic Structure Theory: General Innovations and Strongly Correlated Systems

Edited by

PHILIP E. HOGGAN

Institute Pascal, UMR 6602 CNRS,
Aubière Cedex, France

ELSEVIER

ACADEMIC PRESS
An imprint of Elsevier

Academic Press is an imprint of Elsevier
50 Hampshire Street, 5th Floor, Cambridge, MA 02139, United States
525 B Street, Suite 1800, San Diego, CA 92101-4495, United States
The Boulevard, Langford Lane, Kidlington, Oxford OX5 1GB, United Kingdom
125 London Wall, London, EC2Y 5AS, United Kingdom

First edition 2018

ISBN: 978-0-12-813002-5
ISSN: 0065-3276

For information on all Academic Press publications
visit our website at https://www.elsevier.com/books-and-journals

Publisher: Zoe Kruze
Acquisition Editor: Sam Mahfoudh
Editorial Project Manager: Shellie Bryant
Production Project Manager: Vignesh Tamil
Cover Designer: Greg Harris

Typeset by SPi Global, India

CONTENTS

Contributors xi
Preface xv

Section I
Molecular Electronic Structure Methodology and Reviews

1. Present Status of Selected Configuration Interaction With Truncation Energy Error **3**

Carlos F. Bunge

1. Introduction 4
2. Atomic Self-Consistent Fields 6
3. Molecular Self-Consistent Fields 9
4. CI and General Tools 10
5. Energy Contribution of a Single Configuration 17
6. Summary of SCI 21
7. Mechanics of the Selection Process 24
8. SCI on Large Systems 26
9. Final Remarks 27

Acknowledgments 28
References 29

2. Recent Developments in Asymptotic Expansions From Numerical Analysis and Approximation Theory **35**

Avram Sidi

1. Introduction 36
2. Trapezoidal Rule Approximations and Generalizations of the E–M Expansion 38
3. Error Expansions for Gauss–Legendre Quadrature 46
4. Asymptotics of Legendre Polynomials and Legendre Series Coefficients 49

References 56

3. The Kinetic Energy Pauli Enhancement Factor and Its Role in Determining the Shell Structure of Atoms and Molecules **59**

Eduardo V. Ludeña, Darío Arroyo, Edison X. Salazar, and Jorge Vallejo

1. Introduction 60
2. The Concepts of Shell Structure and the Pauli Potential 61

3. The Pauli Enhancement Factor and Shell Structure 64
4. Discussion 74
Acknowledgments 75
References 75

4. On the Laguerre Representation of Coulomb Functions and the Relation to Orthogonal Polynomials 79

Jessica A. Del Punta, Gustavo Gasaneo, and Lorenzo U. Ancarani

1. Introduction 80
2. Definitions and Notations of Coulomb Wave Functions 82
3. Series Representation of Coulomb Wave Functions in Terms of Laguerre-Type Functions 84
4. Connection With Orthogonal Polynomials 91
5. Summary 96
Appendix: Integral Involving Whittaker Function $W_{\chi,\frac{\mu}{2}}(z)$ 96
References 100

5. Numerical Hartree–Fock and Many-Body Calculations for Diatomic Molecules 103

John C. Morrison and Jacek Kobus

1. Introduction 104
2. The Domain Decomposition Algorithm 105
3. A Model Problem 107
4. Many-Body Calculations of the Correlation Energy 108
5. Numerical Solutions of the Orbital and Pair Equations 112
References 116

6. Local Effective Hartree–Fock Potentials Obtained by the Depurated Inversion Method 117

Alejandra M.P. Mendez, Dario M. Mitnik, and Jorge E. Miraglia

1. Introduction 118
2. Theory 119
3. Results 127
4. Conclusions 128
Acknowledgment 130
Appendix 130
References 131

7. 4-Center STO Interelectron Repulsion Integrals With Coulomb Sturmians **133**

James E. Avery and John S. Avery

1. Introduction 133
2. Coulomb Sturmian Expansions of STO 134
3. STO Electron Repulsion Integrals 139
4. Checks of Convergence and Accuracy 141
5. Discussion 145

References 146

8. Recurrence Relations for Four-Electron Integrals Over Gaussian Basis Functions **147**

Giuseppe M.J. Barca and Pierre-François Loos

1. Introduction 147
2. Four-Electron Integrals 149
3. Fundamental Integrals 152
4. Recurrence Relations 154
5. Algorithm 158
6. Concluding Remarks 162

Acknowledgments 162

References 162

9. Ab Initio Molecular Local Nuclear Magnetic Shielding Tensors **167**

Nabil Joudieh, Ali Bağcı, and Philip E. Hoggan

1. Introduction 168
2. HFR Formalism for Field and Moments 169
3. HFR With Field and Moments: NDDO Level 171
4. HFR/NDDO London Approximation for Fields 172
5. HFR INDO Formalism for Field and Moments Within the London Approximation 173
6. Second-Order Energy Correction $E^2\left(\vec{B}, \vec{\mu}\right)$ Using CPHF 174
7. Double Perturbed HF 176
8. Exchange of Perturbation in HF 177
9. HFR Exchange of Perturbations 177
10. Perturbation Exchange for Molecular Diamagnetism 178
11. Theorem of Exchange of Perturbation Directions 179
12. The Tensors χ_{he}, $\Phi_{\kappa, he}$, and $\sigma_{\kappa, he}$ 180
13. Discussion and Conclusion 181

References 181

Section II
Electron Correlation in Atomic Systems

10. Atomic Electronic Structure Computations With Hylleraas-CI Wave Functions **187**

Frank E. Harris

1. Introduction 188
2. Wave Functions and Potential-Energy Matrix Elements 189
3. General Methods for Exponential Correlation 196
4. Hylleraas Integrals 197
5. Unlinked Three- and Four-Electron Integrals 198
6. The Auxiliary Two-Body Integrals g_i 201
7. Check of $\mathcal{K}_2$ 203
8. Kinetic Energy 204
9. Kinetic Energy in Hylleraas-CI 206

Appendix. Angular Momentum Formulas 207
References 208

11. One-Particle Effective Potential for Helium Atom **211**

Daniel Gebremedhin and Charles Weatherford

1. Introduction 211
2. Preliminary 212
3. Construction of Basis Set 214
4. Eigen Equation and the Iterative Scheme 216
5. Results and Discussion 217
6. Conclusion 221

Acknowledgments 221
References 222

12. Configuration Interaction Study of the ^3P Ground State of the Carbon Atom **223**

María Belén Ruiz and Robert Tröger

1. Introduction 223
2. Configuration Interaction 224
3. Selection of Configurations 227
4. Calculations 229
5. Conclusions and Perspectives 234

Acknowledgments 236
References 236
Further Reading 238

Section III
Strong Correlation in Electronic Systems

13. Extension of the Configuration Interaction Monte Carlo Method to Atoms and Molecules **241**

Alessandro Roggero and Francesco Pederiva

1. Introduction 242
2. Method 242
3. Results 247
4. Conclusions 252

Acknowledgment 252
References 252

14. Time-Dependent Linear-Response Variational Monte Carlo **255**

Bastien Mussard, Emanuele Coccia, Roland Assaraf, Matthew Otten, Cyrus J. Umrigar, and Julien Toulouse

1. Introduction 256
2. Theory 257
3. Results 264
4. Conclusions and Perspectives 268

Acknowledgments 268
References 269

15. Quantum Monte Carlo Calculations for Industrial Catalysts: Accurately Evaluating the H_2 Dissociation Reaction Barrier on Pt(111) **271**

Philip E. Hoggan

1. Introduction 272
2. Setting Up the Model Geometries 273
3. QMC Methods 274
4. Perspectives and Conclusions 276

Acknowledgments 277
References 277

16. Density-Based Analysis of Spin-Resolved MP2 Method **279**

Mateusz Witkowski, Szymon Śmiga, and Ireneusz Grabowski

1. Introduction 279
2. Computational Details 281

3. Methodology and Results 282
4. Final Remarks 288
Acknowledgments 290
References 290

17. Time-Dependent Configuration Interaction Using the Graphical Unitary Group Approach: Nonlinear Electric Properties 295
Patrick J. Lestrange, Mark R. Hoffmann, and Xiaosong Li

1. Introduction 296
2. Theory 297
3. Results 302
4. Conclusions 308
Acknowledgments 309
References 309

18. Hybrid Treatments Based on Determinant Seniority Numbers and Spatial Excitation Levels in the Configuration Interaction Framework 315
Diego R. Alcoba, Alicia Torre, Luis Lain, Ofelia B. Oña, Gustavo E. Massaccesi, and Pablo Capuzzi

1. Introduction 316
2. Theory 317
3. Results and Discussion 321
4. Concluding Remarks 330
Acknowledgments 331
References 331

19. Potential Energy Curves for the LiK^+ and NaK^+ Molecular Ions With the Coupled Cluster Method 333
Monika Musiał, Anna Bewicz, Patrycja Skupin, and Stanisław A. Kucharski

1. Introduction 334
2. Synopsis of the Theory 335
3. Results and Discussion 337
4. Conclusions 346
Acknowledgment 347
References 347

Index *351*

CONTRIBUTORS

Diego R. Alcoba
Facultad de Ciencias Exactas y Naturales, Universidad de Buenos Aires; Instituto de Física de Buenos Aires, Consejo Nacional de Investigaciones Científicas y Técnicas, Buenos Aires, Argentina

Lorenzo U. Ancarani
Théorie, Modélisation, Simulation, SRSMC, UMR CNRS 7565, Université de Lorraine, Metz, France

Darío Arroyo
Center of Nanotechnology Research and Development, CIDNA, Escuela Superior Politécnica del Litoral, ESPOL, Guayaquil, Ecuador

Roland Assaraf
Laboratoire de Chimie Théorique, Université Pierre et Marie Curie, Sorbonne Universités, CNRS, Paris, France

James E. Avery
Niels Bohr Institute, University of Copenhagen, Copenhagen, Denmark

John S. Avery
H.C. Ørsted Institute, University of Copenhagen, Copenhagen, Denmark

Ali Bağcı
University of Warsaw, Warsaw, Poland

Giuseppe M.J. Barca
Research School of Chemistry, Australian National University, Canberra, ACT, Australia

Anna Bewicz
Institute of Chemistry, University of Silesia in Katowice, Katowice, Poland

Carlos F. Bunge
Instituto de Física, Universidad Nacional Autónoma de México, Mexico City, Mexico

Pablo Capuzzi
Facultad de Ciencias Exactas y Naturales, Universidad de Buenos Aires; Instituto de Física de Buenos Aires, Consejo Nacional de Investigaciones Científicas y Técnicas, Buenos Aires, Argentina

Emanuele Coccia
Laboratoire de Chimie Théorique, Université Pierre et Marie Curie, Sorbonne Universités, CNRS, Paris, France; Dipartimento di Scienze Fisiche e Chimiche, Universitá degli Studi dell'Aquila, L'Aquila, Italy

Jessica A. Del Punta
Universidad Nacional del Sur—IFISUR; Universidad Nacional del Sur, Bahía Blanca, Argentina; Théorie, Modélisation, Simulation, SRSMC, UMR CNRS 7565, Université de Lorraine, Metz, France

Gustavo Gasaneo
Universidad Nacional del Sur—IFISUR, Bahía Blanca, Argentina

Daniel Gebremedhin
Florida A&M University, Tallahassee, FL, United States

Ireneusz Grabowski
Institute of Physics, Faculty of Physics, Astronomy and Informatics, Nicolaus Copernicus University, Torun, Poland

Frank E. Harris
University of Utah, Salt Lake City, UT; Quantum Theory Project, University of Florida, Gainesville, FL, United States

Mark R. Hoffmann
University of North Dakota, Grand Forks, ND, United States

Philip E. Hoggan
Institute Pascal, UMR 6602 CNRS, Aubière Cedex, France

Nabil Joudieh
University of Damascus, Damascus, Syrian Arab Republic

Jacek Kobus
Institute of Physics, Faculty of Physics, Astronomy, and Informatics, Nicolaus Copernicus University, Torun, Poland

Stanisław A. Kucharski
Institute of Chemistry, University of Silesia in Katowice, Katowice, Poland

Luis Lain
Facultad de Ciencia y Tecnología, Universidad del País Vasco, Bilbao, Spain

Patrick J. Lestrange
University of Washington, Seattle, WA, United States

Xiaosong Li
University of Washington, Seattle, WA, United States

Pierre-François Loos
Research School of Chemistry, Australian National University, Canberra, ACT, Australia; Laboratoire de Chimie et Physique Quantiques, Université de Toulouse, CNRS, UPS, Toulouse, France

Eduardo V. Ludeña
Center of Nanotechnology Research and Development, CIDNA, Escuela Superior Politécnica del Litoral, ESPOL, Guayaquil, Ecuador; Centro de Química, Instituto Venezolano de Investigaciones Científicas, IVIC, Caracas, Venezuela

Gustavo E. Massaccesi
Ciclo Básico Común, Universidad de Buenos Aires, Buenos Aires, Argentina

Alejandra M.P. Mendez
Instituto de Astronomía y Física del Espacio (IAFE), CONICET-Universidad de Buenos Aires, Buenos Aires, Argentina

Jorge E. Miraglia
Instituto de Astronomía y Física del Espacio (IAFE), CONICET-Universidad de Buenos Aires, Buenos Aires, Argentina

Dario M. Mitnik
Instituto de Astronomía y Física del Espacio (IAFE), CONICET-Universidad de Buenos Aires, Buenos Aires, Argentina

John C. Morrison
University of Louisville, Louisville, KY, United States

Monika Musiał
Institute of Chemistry, University of Silesia in Katowice, Katowice, Poland

Bastien Mussard
Laboratoire de Chimie Théorique, Université Pierre et Marie Curie, Sorbonne Universités, CNRS; Institut des Sciences du Calcul et des Données, Université Pierre et Marie Curie, Sorbonne Universités, Paris, France

Ofelia B. Oña
Instituto de Investigaciones Fisicoquímicas Teóricas y Aplicadas, Universidad Nacional de La Plata, CCT La Plata, Consejo Nacional de Investigaciones Científicas y Técnicas, La Plata, Argentina

Matthew Otten
Laboratory of Atomic and Solid-State Physics, Cornell University, Ithaca, NY, United States

Francesco Pederiva
University of Trento; INFN-TIFPA, Trento Institute for Fundamental Physics and Applications, Trento, Italy

Alessandro Roggero
Institute for Nuclear Theory, University of Washington, Seattle, WA, United States

María Belén Ruiz
Friedrich-Alexander-University Erlangen-Nürnberg, Erlangen, Germany

Edison X. Salazar
Theoretical Chemistry, Zernike Institute for Advanced Materials, University of Groningen, Groningen, The Netherlands

Avram Sidi
Technion – Israel Institute of Technology, Haifa, Israel

Patrycja Skupin
Institute of Chemistry, University of Silesia in Katowice, Katowice, Poland

Szymon Śmiga
Institute of Physics, Faculty of Physics, Astronomy and Informatics, Nicolaus Copernicus University, Torun, Poland

Alicia Torre
Facultad de Ciencia y Tecnología, Universidad del País Vasco, Bilbao, Spain

Julien Toulouse
Laboratoire de Chimie Théorique, Université Pierre et Marie Curie, Sorbonne Universités, CNRS, Paris, France

Robert Tröger
Friedrich-Alexander-University Erlangen-Nürnberg, Erlangen, Germany

Cyrus J. Umrigar
Laboratory of Atomic and Solid-State Physics, Cornell University, Ithaca, NY, United States

Jorge Vallejo
Center of Nanotechnology Research and Development, CIDNA, Escuela Superior Politécnica del Litoral, ESPOL, Guayaquil, Ecuador

Charles Weatherford
Florida A&M University, Tallahassee, FL, United States

Mateusz Witkowski
Institute of Physics, Faculty of Physics, Astronomy and Informatics, Nicolaus Copernicus University, Torun, Poland

PREFACE

The present volume of *Advances in Quantum Chemistry* collects selected expert contributions on Novel Electronic Structure Theory From General Innovations to the Difficult Challenge of Strongly Correlated Systems.

The content focusses essentially on the rather small but vitally important energy contribution associated with many-electron behavior as compared to that of a system of (model) independent particles which can be used for convenience and as a reference.

This volume is organized in three sections:

I Molecular Electronic Structure Methodology and Reviews,

II Electron Correlation in Atomic Systems,

and

III Strong Correlation in Electronic Systems.

Many of the contributions are based on invited orals at MESBA (Molecular Electronic Structure in Buenos Aires) which was held in Buenos Aires, Argentina on September 19–23, 2016, www.iafe.uba.ar/mesba/program.html.

This international workshop drew about 100 participants from all five continents to Buenos Aires, Argentina.

The scientific themes covered in MESBA are reflected by the sections of this book.

Section I opens with the present status of Selected Configuration Interaction With Truncation Energy Error by Carlos F. Bunge. Avram Sidi reviews work often used in this field on asymptotic expansions and Eduardo Ludeña shows a theoretical view of the shell structure of atoms and molecules. Continuing with fundamental electronic structure theory, Ugo Ancarani describes the Laguerre representation of Coulomb functions and how they are related to orthogonal polynomials.

A frequent reference for atomic and molecular electronic structure theory is solving the Hartree–Fock equations which result in one-electron spin–orbitals that can be built up into antisymmetrised products to provide many-electron reference wave functions. Numerical Hartree–Fock and Many-Body Calculations for Diatomic Molecules are presented by John C. Morrison, whereas Dario Mitnik describes the associated local potentials.

James Avery presents 4-Center STO Interelectron Repulsion Integrals With Coulomb Sturmians and Pierre-François Loos gives Recurrence

Relations for Four-Electron Integrals Over Gaussian Basis Functions. Nabil Joudieh's Ab Initio Molecular Local Nuclear Magnetic Shielding Tensors are presented by Philip Hoggan.

Section II mostly treats configuration interaction (CI) work on correlated electron structure of atoms. It opens with review and new material on Atomic Electronic Structure Computations With Hylleraas-CI Wave Functions by Frank E. Harris and Charles Weatherford presents the One-Particle Effective Potential for Helium Atom. The section closes with María Belen Ruiz's Configuration Interaction Study of the ^3P Ground State of the Carbon Atom.

Section III contains Quantum Monte Carlo (QMC) and configuration interaction methods to account for strong electron correlation. The QMC contributions include an Extension of the Configuration Interaction Monte Carlo Method to Atoms and Molecules by Francesco Pederiva and Time-Dependent Linear-Response Variational Monte Carlo by Emanuele Coccia which are methodological and one application of QMC calculations for evaluating the hydrogen dissociation reaction barrier on Pt(111) by Philip Hoggan; the Density-Based Analysis of Spin-Resolved (Møller–Plesset) MP2 Method by Szymon Śmiga is a correlation method of perturbative form.

Xiaosong Li presents Time-Dependent Configuration Interaction Using the Graphical Unitary Group Approach and Diego Alcoba illustrates Hybrid Treatments Based on Determinant Seniority Numbers and Spatial Excitation Levels.

An application to potential energy curves for the KLi^+ and NaK^+ molecular ions with the coupled cluster method is presented by Monika Musiał to close this section.

It has been a pleasure editing this exciting collection of work which I am convinced the reader will appreciate.

Finally, the volume editor dedicates his contributions to the memory of his father, David G. Hoggan (15.10.1927–05.11.1986), since work on this volume began on the 30th anniversary of his death. David Hoggan specialized in the study of Medieval French Epic poems. He taught at the University of Aberystwyth, GB for 35 years, after completing a Doctorat de l'Université de Strasbourg.

Philip Hoggan,
July 21, 2017

SECTION I

Molecular Electronic Structure Methodology and Reviews

CHAPTER ONE

Present Status of Selected Configuration Interaction With Truncation Energy Error

Carlos F. Bunge[1]
Instituto de Física, Universidad Nacional Autónoma de México, Mexico City, Mexico
[1]Corresponding author: e-mail address: bunge@fisica.unam.mx

Contents

1. Introduction 4
2. Atomic Self-Consistent Fields 6
3. Molecular Self-Consistent Fields 9
4. CI and General Tools 10
   4.1 Projection Operators 11
   4.2 Subclasses 12
   4.3 Partition of Degenerate Spaces: Interacting Spaces 13
   4.4 Fast-Converging CI Expansions: NOs and Ergonals 13
   4.5 Multireference CI 15
   4.6 Atomic Orbital Bases With Prescribed Completeness Error 16
5. Energy Contribution of a Single Configuration 17
   5.1 Dick Brown's Original Formula 17
   5.2 A Priori Brown's Formulae 19
6. Summary of SCI 21
   6.1 Preselection of Configurations Without TEE 21
   6.2 Selection of Configurations With TEE 22
   6.3 SCI by Parts 22
   6.4 Truncation Energy Error 23
   6.5 Valence SCI and Related Models 23
7. Mechanics of the Selection Process 24
8. SCI on Large Systems 26
9. Final Remarks 27
Acknowledgments 28
References 29

Abstract

Configuration interaction (CI) starts from a matrix-eigenvalue equation involving an atomic or molecular electronic Hamiltonian represented by a complete set of Slater determinants made up of a given orbital basis. Full CI scales unfavorably with number

Advances in Quantum Chemistry, Volume 76
ISSN 0065-3276
https://doi.org/10.1016/bs.aiq.2017.05.001

of orbitals and number of electrons relative to all other orbital methods. Recent work on 10-electron systems (Ne and H_2O ground states, the latter at many internuclear distances), and using large orbital bases, shows that up to sextuply excited configurations can be selected a priori, quantitatively and very efficiently by means of Brown's formula, leading to unsurpassed accuracy and understanding. Selected CI (SCI) suggests an array of promising and unexplored models and calls for new vistas demanding new algorithms. Here I review SCI with truncation energy error in the light of new software suitable for considerably larger systems.

1. INTRODUCTION

The basic equation of atomic and molecular theory for the stationary states of a system of N electrons and M nuclei was introduced by Schrödinger[1–5]:

$$H\Psi_\mu = E_\mu \Psi_\mu, \tag{1}$$

where the nonrelativistic electronic Hamiltonian H is increasingly often enriched by relativistic and QED terms. Beyond one electron systems H is not separable thus exact solutions of (1) do not exist whereas approximate solutions are the only alternative.

Hylleraas[6] told us "it was argued, particularly by Max Born, that the crucial test of the correctness of wave mechanics was to be found in its application to the Helium atom, in particular to its ground state." In September 1929 at Det Skandinaviske Naturforskermoto, in Copenhagen, Hylleraas[7] presented a nonrelativistic calculation of the first ionization potential of He yielding a value of 198,211 cm^{-1}, in agreement (to within 400 parts per million) with the experimental value at the time of 198,298 cm^{-1}. By 1958, Pekeris[8] included corrections for nuclear motion, relativity and QED to obtain a theoretical value of 198,310.67(1) cm^{-1}, to be compared with the best (but less accurate) experimental value of 198,310.82(15) cm^{-1} of Herzberg.[9]

Pekeris' feat marked the beginning of a new epoch: for the first time, a many-electron quantum mechanical calculation had beaten experiment, adding an additional motivation for the new generations to get involved in accurate calculations. The latest value for the ^{4}He ionization potential is 198,310.66637(20) cm^{-1},[10] offering a new challenge for calculation and, possibly, for new theory as well.[11,12]

Pekeris-like accuracy requires explicit use of the inter-electronic distance r_{12} in the wave function. The r_{12} ingredient dominates all applications on two-electron atomic systems,[13,14] providing the best nonrelativistic energies, and also the best tests of QED itself. Comparative accuracy has been extended to three-electron atoms.[15,16]

Correlated Gaussians[17–19] are definitely superior for higher-electron atoms, and also for molecules.[20] Work on four-electron atoms[21–24] and more recently on five-electron atoms[25,26] anticipates widespread use and very high accuracy for these systems. A tour de force for a six-electron system was reported to be underway.[27]

Beyond five electrons, however, these calculations are expected to be exceptional rather than routine, due to the computational requirements involved, as r_{12} methods scale with no less than a noninteger power (greater than one) of the factorial of the number of electrons. Thus, although ab initio orbital methods are not expected to achieve the accuracy of r_{12} methods, their applicability to much larger systems makes them the chief theoretical tool for electronic structure.

We use a single set of programs for nonrelativistic and relativistic atomic calculations as well as for molecular nonrelativistic calculations. Program AUTOCL[28] generates fairly complete lists of atomic and molecular symmetry eigenfunctions, as is usual in ab initio electronic structure programs. Starting in 2006,[29,30] AUTOCL incorporated a unique feature in electronic structure calculations: truncated lists of symmetry configuration-state-functions selected with corresponding truncation energy error (TEE). Beginning in 2014,[11] a straightforward preselection of configurations allows practically unlimited control of configuration lists and TEE through a well-defined and general sensitivity analysis.

Program ATMOL evaluates variational energies, wave functions, and TEE. These may be used to calculate either natural orbitals (NOs) and data needed by both AUTOCL and ATMOL for the evaluation of TEE, or atomic multipolar electric and magnetic transition probabilities in nonrelativistic or relativistic regimes.

The use of AUTOCL is thought-provoking while that of ATMOL has become routine, except when it is used to feed AUTOCL with NOs and other data used to evaluate TEE. Since programs AUTOCL and ATMOL use one- and two-electron integrals from program NWchem,[31] plans for incorporating relativistic molecular calculations will proceed after finding an external source of these integrals.

All supportive numerical data for this chapter concern extensive calculations on well-known closed-shell systems (Ne and H_2O ground states, the latter at several internuclear distances) discussed at length in Refs. 11 and 12, so this chapter will be fully devoted to a self-contained exposition of the ideas, methodologies, and current limitations of SCI and related models.

The equations in this chapter serve one or other purpose: as illustrative material to support the text, or as new reference information for those interested in incorporating configuration selection with TEE into their own work.

The accuracy of orbital methods rests on good orbitals together with the ability to deal with the many-electron problem. Sections 2 and 3 briefly review self-consistent field (SCF) orbitals for atoms and molecules, respectively. Section 4 deals with the variational orbital approach to approximate Schrödinger's equation for stationary states. Configuration interaction (CI) is an essentially intractable problem: CI up to n-excitations (CI-nx) and full CI (FCI) when n equals the number of electrons. General issues and useful tools will be introduced at this stage.

All orbital methods, even Hartree–Fock, eventually become unmanageable with increasing number of electrons and size of the orbital basis, however, FCI has by far the worst scaling of them all.[11] Thus some kind of selected CI (SCI) is imperative. Section 5 is devoted to formulae used to select a single configuration.

Section 6 briefly summarizes SCI with an a priori selection scheme which also provides a TEE.[29] In general, the number of sufficiently large nonzero matrix elements of the eigen-problem may grow too large, so that a divide-and-conquer (CI by parts) strategy is unavoidable.[30]

Section 7 takes up the mechanics of the selection process itself, which requires intensive use of program AUTOCL, and Section 8 describes some salient features of our new computer program dealing efficiently with up to 100-electron systems. Conclusions are presented in Section 9.

2. ATOMIC SELF-CONSISTENT FIELDS

The first successful theory of atomic electronic structure was elaborated by Bohr before the advent of quantum mechanics. Starting with the hydrogen atom[32] and concluding with his remarkable "The Constitution of Atoms and the Physical and Chemical Properties of Elements,"[33] Bohr provided an explanation of the periodic table,[34] chemical valency, and chemical reactivity, as a direct consequence of atomic electronic structure. The path was set for a future exclusion principle for electrons.

In his exclusion principle,[35] Pauli introduced a new theoretical property of the electron: "being two-valued, not describable classically." Uhlenbeck and Goudsmit immediately recognized Pauli's "two-values" as a quantum number of a spin angular momentum.[36,37] The Dutch team received a letter from Heisenberg[38] the day after their Naturwiessenschaften publication, congratulating them for their "courageous note," together with a disturbing "What have you done with the factor 2?" Months later, Thomas[39] provided a justification for the factor of 2.

The present physics community strongly discourages "courageous notes" challenging accepted knowledge, such as the quantum theory of fields applied to the calculation of stationary states.[11,12,40] Thus we will here skip all concerning relativistic and QED calculations focusing just on variational methods applied to nonrelativistic electronic structures, for which new advances are sufficient to naturally debunk established prejudice.

The first general method to approximate the solutions of Schrödinger's equation was invented in 1928 by Hartree,[41] who introduced a nonrelativistic quantum mechanical self-consistent field potential for electron motion, without explicit consideration of antisymmetry of the wave functions upon permutations of electron coordinates. In Hartree's method, the individual Bohr orbits were replaced by one-electron wave functions which we now call orbitals. Hartree calculated a reasonable value for the ionization potential of He,[42] and he also further considered Rb^+. Hartree's predicted X-ray spectra for Rb^+[42] agreed rather well with experimental results.

In order to comply with Bohr's explanation of the periodic table (or with Pauli's exclusion principle) Hartree assigned no more than two electrons to each orbital. Soon afterward, Paul Dirac, who was working on a full agenda at the frontiers of physics,[43] attempted to incorporate antisymmetry[44] in Hartree's nonrelativistic self-consistent fields without real success.

The introduction of Dirac's paper[44] contains his time-honored dictum: "the underlying physical laws necessary for the mathematical theory of a large part of physics and the whole of chemistry are completely known and the difficulty is only that the exact application of these laws leads to equations [Schrödinger's equation] much too complicated to be soluble." Dirac's maxim remains undisputed, justifying our continued quest to find more reliable and efficient methods to approximate solutions of Schrödinger's equation.

Dirac had just advanced the idea of using determinants in connection with antisymmetry,[44] which would have been correct dealing with Dirac's four-component one-electron functions, however, Hartree's theory was a nonrelativistic one, and Dirac's immediate concern was with Hartree's self-consistent fields. After about half a year, Slater, who had already been close to the exclusion principle,[45] invented spinorbitals,[46] whereby nonrelativistic orbitals acquired an extra formal dimension. Determinant wave functions expressed as antisymmetrized products of spinorbitals, also known as Slater determinants, provided the needed theoretical ingredient for a satisfactory nonrelativistic theory of atomic and molecular electronic structure.

On December 19, 1929, in a very short note,[47] Slater also explained how Hartree's self-consistent fields with and without Pauli's exclusion principle could be derived from a variation principle, but he did not immediately write the equations explicitly. Ten days later, in Russian (first reference in Ref. 48) and two months later in German, Fock[48] published the explicit derivation of the variational equations for Slater determinants which have since been known as Hartree–Fock (HF) equations.

The early evolution of nonrelativistic self-consistent fields (SCF) has been superbly recounted by Fischer.[49] Also, and this is little known, Charlotte, as she is universally known, enriched Hartree's legacy with decisive contributions to the computational methodology of numerical nonrelativistic HF[50] at a time when almost no one believed in it.[51] Her first multiconfigurational Hartree–Fock (MCHF) program[52] allowed for HF calculations[53] that look complicated even today, however, energies were still obtained with no more than seven significant figures.

Although Hartree had used superposition of configurations with orbitals obtained within the SCF approximation,[54] it was Charlotte alone who pioneered: (i) the development of MCHF,[52,55] and (ii) the transition from less than seven-figure SCF results to essentially exact SCF (10 and more figures), also understood as bona-fide HF results.[50]

An application to beryllium's ground state[56] brought up an important aspect: high accuracy results demanded further CI treatment using MCHF orbitals. Subsequent MCHF codes thus introduced increasingly efficient CI algorithms.[57,58] Nonrelativistic MCHF is now an established methodology[50,59,60] supported by state-of-the-art software.[61]

Relativistic SCF (also with exchange) was first derived by Swirles[62] using a variational principle. The most complete theoretical relativistic HF and MCHF project was carried out by Grant[63–66] extending over a period of fifty years.

After the pioneer work of Desclaux on computational Dirac-Breit relativistic HF and MCHF,[67–70] Grant and coworkers produced their own reliable and efficient computer codes for relativistic HF and MCHF.[71,72] After teaming with Fischer and collaborators, substantial CI was also incorporated.[73–75]

3. MOLECULAR SELF-CONSISTENT FIELDS

Distinct from atomic self-consistent fields, where the contributions of Hartree, Slater, Fock, Fischer, Grant, Desclaux, and their respective coworkers summarize most of the relevant work, molecular self-consistent fields called for the participation of an entire community of scientists that at some point were to give rise to quantum chemistry[12] thus we will proceed straight on a path bearing on CI one way or another.

There is an important exception to the above: numerical nonrelativistic HF for diatomic molecules,[76] where accuracy comparable to numerical atomic HF has bravely been reached through the work of a few authors.[76] Introduction of electron correlation[77] is still in the future, however, it will be most interesting to see that whenever it comes about.

Molecular orbitals were pioneered by Hund[78] and Mulliken.[79] The latter publication introduced σ, π, and δ orbitals. In 1929, Lennard-Jones[80] proposed to approximate molecular orbitals as linear combinations of atomic orbitals (LCAO). Coulson[81] calculated the first molecular SCF orbitals and soon dominated the field, producing a lucid review of molecular orbitals for simple molecules.[82]

The situation changed dramatically after Roothaan published his LCAO-SCF theory for closed-shell ground state molecules,[83] initiating a new era of approximate theoretical methods.[84] In 1959, at a Conference on Molecular Quantum Mechanics in Boulder, Colorado, Clemens presented his SCF theory for open-shell molecules.[85] The star of the conference was reported to be legendary Samuel Francis Boys, whose deeds will be recognized further below.

The conference was closed with Coulson's after dinner speech.[86] In the middle of his profound and persuasive talk, Coulson had the gallantry to mention the existence of two groups: group I (*electronic computors*) and group II (*nonelectronic computors*). And he continued "I cannot help thinking that the gap between the two is so large that there is little point in bringing them together." With that, a gallery of luminaries, including Coulson and Linus Pauling, were lost to quantum chemistry yet unborn. Was that inevitable?

Perhaps so, however, a very fruitful way of thinking just vanished overnight. The flavor of those extraordinary minds can be illustrated by the chain of reasoning of an undergraduate student who predicted the correct structure of then unknown diborane,[87] bypassing extant chemical valence theory.

4. CI AND GENERAL TOOLS

It is worth bearing in mind the conceptual (theoretical) referents of a theory or method. The first conceptual referent of CI is the Hamiltonian H. In most molecular calculations the validity of H is unquestioned. The second conceptual referent of CI is the corresponding matrix-eigenvalue equation,

$$\mathbf{H}\mathbf{C}_{\mu} = E_{\mu}^{\mathrm{FCI}}\mathbf{C}_{\mu}. \tag{2}$$

$\mathbf{H}$ is the representation of H in terms of Slater determinants $D_{ijk\ldots}^{abc\ldots}$. The FCI expansion can be expressed in cluster form as[88]:

$$\Psi^{\mathrm{FCI}} = D_0c_0 + \sum_{i}\sum_{a} D_i^a c_i^a + \sum_{i<j}\sum_{a<b} D_{ij}^{ab} c_{ij}^{ab} + \sum_{i<j<k}\sum_{a<b<c} D_{ijk}^{abc} c_{ijk}^{abc} + \cdots, \tag{3}$$

where i, j, k denote occupied spinorbitals or Dirac bi-spinors, and a, b, c refer to corresponding unoccupied one-electron functions.

The third conceptual referent of CI, more important than the second one though not always recognized as such, is the orbital basis used to express the Slater determinants. Eqs. (1)–(3) together with H define the FCI method,[89] and E_{μ}^{FCI} is the FCI energy. The conceptual CI referents completely determine expectation values. In particular, the exact eigenvalues E_{μ} of Eq. (1) can be written as:

$$E_{\mu} = E_{\mu}^{\mathrm{FCI}} + \Delta E_{\mu}^{\mathrm{CBS}}, \tag{4}$$

where $\Delta E_{\mu}^{\mathrm{CBS}}$ is a complete basis set (CBS) energy error[90] that can be obtained by extrapolating well-defined patterns of convergence for the energy.[90–92] The subscript μ will now be dropped, assuming it applies to any stationary state.

The CI expansion (3) is most conveniently rewritten in terms of configuration-state-functions (CSFs) F_{gK},

$$\Psi^{\mathrm{FCI}} = \sum_{K=1}^{K_x}\sum_{g=1}^{g_{\mathrm{K}}} F_{\mathrm{gK}}\,C_{\mathrm{gK}}, \tag{5}$$

where K and g label configurations and degenerate elements, respectively. F_{gK} is a linear combination of n_{K} Slater determinants D_{iK},

$$F_{\mathrm{gK}} = O(\Gamma,\gamma)\sum_{i=1}^{g} D_{\mathrm{iK}} b_i^{g} = \sum_{i=1}^{n_{\mathrm{K}}} D_{\mathrm{iK}} c_i^{g}, \quad g = 1,\ldots,g_{\mathrm{K}}. \tag{6}$$

$O(\Gamma, \gamma)$ is a symmetric projection operator[93–95] for all appropriate symmetry operators Γ and a given N-electron irreducible representation γ.[28,96–98] Eqs.(5) and (6) provide a general and concise modular framework to handle nonrelativistic and relativistic, atomic, and molecular calculations, in a single computer program. Thus, each time our program ATMOL for wave function calculations was tested, first in 1992 with relativistic calculations, and later in 2013 for molecular calculations, correct results were obtained in a matter of days.

Other presentations of the CI expansion will be considered in Section 5 in connection with selection of configurations.

4.1 Projection Operators

Let $\mathbf{M}^2$ be an angular momentum operator $\mathbf{L}^2, \mathbf{J}^2$, or $\mathbf{S}^2$, for orbital, total, or spin angular momentum, and eigenvalues $k(k+1)$ and m, and let F be a function with angular momentum components F_k,

$$F = \sum_{k=0}^{\infty} F_k a_k. \tag{7}$$

In 1931, Gray and Wills[99] used a projection operator $\mathcal{O}(\mathbf{M}^2; k, m)$,

$$\mathcal{O}(\mathbf{M}^2; k, m) = \prod_{\ell=0}^{\ell \neq k} \frac{\mathbf{M}^2 - \ell(\ell+1)}{k(k+1) - \ell(\ell+1)}, \tag{8}$$

which, when applied to a function F, eliminates all components $\ell \neq k$[100]:

$$\mathcal{O}(\mathbf{M}^2; k, m) F = F_k a_k. \tag{9}$$

A computer program to evaluate L-S eigenfunctions by means of $\mathcal{O}(\mathbf{M}^2; k, m)$, Eq. (8), was published by Rotenberg.[101]

A considerably more efficient version of Eq. (8) was developed by Löwdin[102]:

$$\mathcal{O}(\mathbf{M}^2; k, m) = (2k+1)! \frac{(k+m)!}{(k-m)!} \sum_{\nu=0}^{k_{\max}-k} \frac{(-1)^{\nu} \mathbf{M}_{-}^{(\mathbf{k}-\mathbf{m}+\nu)} \mathbf{M}_{+}^{(\mathbf{k}-\mathbf{m}+\nu)}}{\nu!(2k+\nu+1)!}. \tag{10}$$

Eq. (10) is substantially more complicated than Eq. (8). Earlier computer implementations[96,97,103] continue to be actively improved today.[12] Although

Eq. (10) is readily parallelized and Eq. (8) is not, both are useful. First, the largest gain in parallelization is in the concurrent evaluation of different degenerate elements g of Eq. (6). Second, Eq. (10) becomes numerically unstable for high values (30–70) of the sum of all angular momentum values of the individual spinorbitals, even in quadruple precision arithmetic. For atoms up to Ne, beyond $\ell = 24$ the evaluation of Eq. (6) must be carried out by Eq. (8), which remains stable at $\ell = 30$.

In atomic L-S calculations, the projection operator $\mathcal{O}(\Gamma,\gamma)$ used in Eq. (6) is the product of orbital and spin angular momentum operators $\mathcal{O}(\mathbf{L}^2;L,M_L)$ and $\mathcal{O}(\mathbf{S}^2;S,M_S)$, where $\mathcal{O}(\mathbf{L}^2;L,M_L)$ is applied first, followed by application of its spin counterpart. Atomic jj-JM calculations (nonrelativistic or relativistic) make use of a single projection operator, $\mathcal{O}(\mathbf{J}^2;J,M_J)$. The projection operators for nonrelativistic molecular calculations are products of the projection operator for the chosen molecular point group[93,94] times a spin projection operator.

All kinds of atomic open shells, and all open shells in combination with abelian molecular groups have been verified. Program AUTOCL has several nonabelian groups tested but temporarily disconnected. Since program ATMOL uses one- and two-electron integrals from program NWchem,[31] plans for incorporation of nonabelian groups will proceed after finding an external source for the corresponding integrals.

The L-S option is the only one that still needs speed improvements. Very large configuration lists for molecular calculations run in hardly more than a few minutes. Starting with 2300 lines of Fortran code in 1966 (slightly more than one full box of cards), and more than 100,000 lines in 2010,[104] AUTOCL now consists of about 200,000 lines, producing selected symmetry-configuration lists with TEEs for atoms and molecules[11] before taking up large-scale CI.

4.2 Subclasses

A subclass[105] is formed by the set of all ordered configurations differing just in the labels of the correlation orbitals of each irrep. Within a subclass, the b_i^g and c_i^g coefficients of Eq. (6) are the same, in magnitude and sign. The set of subclasses which remains invariant upon nonsingular transformations of the correlation orbitals is called a class.

In evaluation of matrix elements, there are two outer loops running over subclasses with indices 1 and 2, denoting left and right (bra and ket) of a

Hamiltonian matrix element. If 1 is the innermost loop, it is first run just to mark those subclasses that will have nonzero matrix elements according to hole specifics and the particular irreps involved in holes and particles. Afterward, the loop 1 is run just over marked subclasses and concurrent processing of sequentially marked subclasses provides the main source of parallelism in this step.

4.3 Partition of Degenerate Spaces: Interacting Spaces

Degenerate spaces arise in open-shell configurations. In 1968, Bunge[105] had the idea to partition the degenerate spaces of singly and doubly excited configurations in two parts: one with the maximum possible dimension which has null matrix elements with the Hartree–Fock configurations, called the maximum projected noninteracting space (MAXPNI) and the other, its orthogonal complement, called the minimum projected interacting space (MINPI). MAXPNI and MINPI were first used in a calculation of atomic carbon.[105,106]

Interacting spaces of smaller size are possible.[107–109] The quotient between the dimension of MINPI and that one of the full space may exceed the value of 10 for a single reference, which is decisive in some applications: (i) to optimize basis orbitals,[11] (ii) in Multireference CI, where that quotient may considerably exceed the value of 10,[110,111] and (iii) to evaluate optimally decoupled positive-energy orbitals.[11]

For singly excited configurations, MINPI is expanded entirely in terms of singly and doubly excited determinants,[107] whereas for doubly excited configurations, MINPI is exclusively expanded by doubly excited determinants.[107]

4.4 Fast-Converging CI Expansions: NOs and Ergonals

In 1955 Löwdin introduced natural (spin)orbitals[112] χ_i and occupation numbers n_i:

$$\gamma(1,1') = \sum n_i \chi_i^*(1)\chi_i(1'), \tag{11}$$

as the eigenfunctions and eigenvalues of the reduced first-order density matrix $\gamma(1,1')$:

$$\gamma(1,1') = N\int \Psi(1,2,\ldots,N)^*\Psi(1',2,\ldots,N)d(2,3,\ldots,N). \tag{12}$$

In Eq. (11) the n_i are in nonincreasing order. Let $\mathcal{G} \equiv \{\phi_i(1), i = 1,2,\ldots,p\}$ be a given spinorbital basis $\mathcal{G}$ of size p, and let Ψ be an FCI expansion expressed in these spinorbitals:

$$\Psi = \sum D_K(\mathcal{G}) c_K. \tag{13}$$

It is not hard to see that the diagonal elements γ_{ii} of $\gamma(1,1')$ are given by:

$$\gamma_{ii} = \sum_K^{(i)} |c_K|^2, \tag{14}$$

where the summation goes over all K holding index i. The corresponding natural spinorbitals $\mathcal{S} \equiv \{\chi_i(1), i = 1,2,\ldots,p\}$ afford an alternative representation $\mathcal{S}$ of the original $\mathcal{G}$.

Being an FCI, Ψ is invariant upon nonsingular linear transformations of the spinorbital basis. Thus Ψ can also be expressed in terms of its natural spinorbitals as:

$$\Psi = \sum D_K(\mathcal{S}) a_K, \tag{15}$$

whereby,

$$n_i = \sum_K^{(i)} |a_K|^2. \tag{16}$$

After a nontrivial application of the variational theorem for finite Hermitian matrices[113,114] one arrives to:

$$\sum_{i=1}^{r} n_i \geq \sum_{i=(1)}^{(r)} \gamma_{ii}, \quad r < p, \tag{17}$$

where the sum in the right-hand side goes over **any** r indices, whence it follows:

$$\sum_{i=1}^{r} \sum_K^{(i)} |a_K|^2 \geq \sum_{i=(1)}^{(r)} \sum_K^{(i)} |c_K|^2, \quad r < p, \tag{18}$$

expressing the optimum convergence property of the natural (spin) orbitals.[89,115]

Ergonals ζ_i (or energy-optimized spinorbitals[116]) and ergonal energies ϵ_i were introduced by the author in 1973[117]:

$$h^{(p)}(1,1') = \sum \epsilon_i \zeta_i^*(1)\zeta_i(1'), \tag{19}$$

as the eigenfunctions and eigenvalues of the reduced, averaged, and projected first-order energy density matrix $h^{(p)}(1,1')$:

$$h^{(p)}(1,1') = N\int \Psi(1,2,...,N)^*\mathcal{O}^{(p)}\frac{(H(1,2,\ldots,N)+H(1',2,\ldots,N)}{2}\mathcal{O}^{(p)} \Psi(1',2,\ldots,N)d(2,3,\ldots,N), \tag{20}$$

where $\mathcal{O}^{(p)}$ projects any manifold into the space $\mathcal{G}$, and H is the usual Hamiltonian. Ψ is still invariant thanks to the use of the projection operator $\mathcal{O}^{(p)}$ in Eq. (20), therefore:

$$\Psi = \sum D_K(\mathcal{G})c_K = \sum D_K(\mathcal{S})a_K = \sum D_K(\mathcal{E})e_K. \tag{21}$$

Similarly as with NO's, one gets in Eq. (22) an inequality similar to Eq. (18) but with $\leq$ instead of $\geq$ as it should be for a smaller energy[116]:

$$\sum_{i=1}^{r}\sum_{K}^{(i)}\sum_{L} e_K e_L H_{KL}(\mathcal{E}) \leq \sum_{i=(1)}^{(r)}\sum_{K}^{(i)}\sum_{L} a_K a_L H_{KL}(\mathcal{S}), \tag{22}$$

expressing the optimum convergence property of ergonals[116] relative to NO's themselves. For atoms, orbital bases are rather good thus ergonals should be almost identical to NO's. For molecules, on the other hand, current basis sets carry rather large CBS errors thus time is ripe to verify whether ergonals afford a significant energy convergence advantage over NO's.

4.5 Multireference CI

Multireference CI (MR-CI-*n*x) calls for incorporation of all configurations of CI-*n*x out of a list of more than one reference configuration subject to elimination of repeated configurations. MR-CI in the interacting space was first advanced by Brooks and Schaefer in the late 1970s.[110,111]

The case $n = 2$ is particularly interesting when the reference list includes *m*-excited configurations, $m = 1, 2, 3, 4$ (MR-4x-CI-2x). Our code allows the choice of only those excitations within the interacting space of each reference. In such a case, selection is optional. The method will be tried on those large systems for which CI-6x becomes too time-consuming.

FCI accuracy on the symmetric dissociation of H_2O with molecular geometries expanded up to eight times from the equilibrium distance has been reported[12] together with comparisons with other MR-CI and coupled cluster calculations.

4.6 Atomic Orbital Bases With Prescribed Completeness Error

Numerical atomic calculations carried out by state-of-the-art methodology and computer programs[61,75] provide the best possible occupied orbitals together with additional correlation orbitals that can be used to obtain a fair account of electron correlation.

The accurate calculation of atomic energy levels is of interest in connection with accurate transition energies. If one sets to calculate the energy levels of the boron atom with an error inferior to 1000 μHartree (220 cm^{-1}), however, the above numerical state-of-the-art methods fail to deliver.

One reason for it is obvious: even for relatively moderate orbital basis sizes, none of these codes has been designed to deal with hundreds of millions of CSFs of an unselected CI-4x much less for a CI-6x. In order to overcome this limitation, neither selection of configurations nor CI by parts alone would be enough: both tools are needed, but they would still be insufficient.

What one really needs is much larger orbital bases than can presently be generated by numerical MCHF codes. Only these large optimized orbital bases will provide a decent CBS energy error and corresponding TEE for the FCI.

Automatic optimization of Slater-type and Gaussian-type atomic orbitals with control of approximate linear dependencies and prescribed truncation energy error recently became a reality, see section 3.1 in Ref. 11. Our procedure employs a single run and, distinct from numerical methods, is free from convergence problems.

The present version of program ATMOL introduces parallelism on integral and matrix element evaluation that speeds automatic optimization by a factor of three on a single-processor 4-core laptop. By using servers equipped with 32-core nodes more significant speedups are expected in the coming months after achieving concurrent optimization of different trials on the same problem at each stage of orbital basis build up.

5. ENERGY CONTRIBUTION OF A SINGLE CONFIGURATION

5.1 Dick Brown's Original Formula

In 1967, graduate student Brown[118] astonished student mates and his professor Harry Shull with a simple expression for the energy contribution ΔE_{gK} of CSF F_{gK} in Eq. (5). Brown showed that ΔE_{gK} is exactly given by:

$$\Delta E_{gK} = (E - H_{gK,gK})C_{gK}^2/(1 - C_{gK}^2), \tag{23}$$

provided: (i) upon removal of F_{gK} from Ψ, Eq. (5), the new variational function $\Psi(-F_{gK})$ is given by:

$$\Psi(-F_{gK}) = \mathcal{N}(\Psi - F_{gK}C_{gK}), \tag{24}$$

where $\mathcal{N}$ is a normalization factor, (ii) ΔE_{gK} is computed as:

$$\Delta E_{gK} = \langle\Psi|H|\Psi\rangle - \langle\Psi(-F_{gK})|H|\Psi(-F_{gK})\rangle, \tag{25}$$

where $E = \langle\Psi|H|\Psi\rangle$, and (iii) all variational coefficients C_{gM} remain *variational* throughout.

Eq. (23) is particularly good for small values of ΔE_{gK}, viz, when Brown's first hypothesis is expected to hold, corresponding precisely to the terms we wish to discard while recording their energy contribution. Brown's first hypothesis certainly does not hold in a nonorthogonal regime. As pointed out by Shavitt,[119] configurational energy contributions of perturbational lineage have been used by other authors. Brown's formula, instead, is probably the only formula for energy contributions of variational origin.

The application of Brown's formula requires knowledge of diagonal matrix elements $H_{gK,gK}$ which are expensive to compute, and of variational coefficients C_{gK} that become known only *after* completing a calculation. By summing over all degenerate elements g of Eq. (5) we get:

$$\Psi^{FCI} = \sum_{K=1}^{K_x} G_K B_K; \quad G_K = N_K \sum_{g=1}^{g_K} F_{gK} C_{gK}, \tag{26}$$

in terms of normalized symmetry-configurations G_K and coefficients B_K. An additional reason for this manipulation is that distinct from the C_{gK} coefficients of Eq. (5), the B_K coefficients of three- and higher-excited configurations can be predicted from B_K coefficients of lower excited ones (Section 5.2).

Selection of configurations K will now use the Brown's formula[118] for individual energy contributions ΔE_K of K without further ado:

$$\Delta E_K = (E - H_{KK}) B_K^2 / (1 - B_K^2), \tag{27}$$

where B_K is the corresponding CI coefficient, and the diagonal matrix element H_{KK} is well approximated by any diagonal element between Slater determinants D_{iK} of K:

$$H_{KK} \approx \langle D_{iK} | H | D_{iK} \rangle, \tag{28}$$

since $(E - H_{KK})$ is of the order of a few Hartree for those ΔE_K of the order of 1 μ Hartree or smaller. Thanks to this simplification, ΔE_K can be computed for any values of the degeneracy g_K and of the number of determinants n_K in Eq. (6).

Eq. (28) is applicable to disconnected Ks, viz, those n-excited Ks that can be written as products of excitations of order lower than n. All other configurations are called connected ones. The latter occur because symmetry orbitals are used.

For connected Ks, B_K coefficients cannot be well predicted a priori. For selection purposes, a rough approximation is given by[104]:

$$\Delta \mathcal{E}_K = (E - H_{KK}) \prod_{i=1}^{q} n_{K_i}, \tag{29}$$

where the n_{K_i}s are NO occupation numbers of K running up to excitation order q. On the left of (29), E has been replaced by $\mathcal{E}$ to clearly separate them from the E of Eq. (27). They have units of energy but Eq. (29) is only valid for two-electron systems, so we call $\Delta \mathcal{E}_K$ a pseudo-energy.

Connected configurations disappear after introducing a sufficient number of chosen reference configurations and applying a *separate* selection process out from each reference, which is under study and will be reported in due time. Also, all configurations within a subclass are either connected or disconnected ones.

The a priori TEE ΔE_s^{FCI} caused by discarded configurations is obtained by:

$$\Delta E_s^{FCI} = \sum_{\text{discarded K}} \Delta E_K + \Delta \mathcal{E}_K. \tag{30}$$

Selection thresholds $T(q)$ are used for Eq. (27), while $\mathcal{T}_d(q)$ and $\mathcal{T}_c(q)$ are for Eq. (29) for disconnected and connected configurations, respectively, and for each order of excitation q. The need to calculate pseudo-energies also for disconnected configurations arises because the corresponding CI coefficients always have a connected component, particularly in triple-excitations.

Since $\Delta\mathcal{E}_K$ pseudo-energies have large uncertainties, their final assessment is obtained as a result of a sensitivity analysis, viz, after carrying out a series of variational calculations with diminishing energy selection thresholds $\mathcal{T}_d(q)$ and $\mathcal{T}_c(q)$ until results become stable to within an uncertainty $\delta\mathcal{E}$. Thus Eq. (30) is finally rewritten as:

$$\Delta E_s^{\text{FCI}} = \sum_{\text{discarded K}} \Delta E_K \pm \delta\mathcal{E}. \tag{31}$$

In order to evaluate Eq. (31) we still need to find a way to calculate the coefficients B_K of Eq. (26), which will be the task of the next section.

5.2 A Priori Brown's Formulae

Sinanoğlu[120] first suggested that the $c_{ijk\ldots}^{abc\ldots}$ coefficients of n-excited determinants in Eq. (3) can be approximately obtained from the coefficients of determinants with lower excitation order. He wrote an expression for the coefficients of quadruply excited determinants in Be ground state in terms of coefficients of doubly excited determinants:

$$c_{ijkl}^{abcd} \approx c_{ij}^{ab} c_{k\ell}^{cd} + c_{ik}^{ab} c_{j\ell}^{cd} + c_{i\ell}^{ab} c_{jk}^{cd}. \tag{32}$$

The same CI coefficients are exactly connected with the coupled cluster amplitudes $t_{ijk\ldots}^{abc\ldots}$ which define the cluster operator T in the coupled cluster (CC) ansatz[121–123]:

$$\Psi^{\text{CC}} = \exp(T)D_0 = \Psi^{\text{FCI}} = (1+T)D_0. \tag{33}$$

Starting from Eq. (33), Primas[122] derived a connection between CI coefficients and CC amplitudes which was further generalized by Monkhorst and coworkers.[124,125] For example, the expression for $c_{ijk\ell}^{abcd}$ can be abbreviated as[29]:

$$\begin{aligned} \frac{1}{c_0} c_{ijk\ell}^{abcd} &= t_{ijk\ell}^{abcd} + \frac{1}{c_0^2}\left(c_{ij}^{ab} c_{k\ell}^{cd} + 17 \text{ terms}\right) - \frac{2}{c_0^3}\left(c_i^a c_j^b c_{k\ell}^{cd} + 71 \text{ terms}\right) \\ &+ \frac{1}{c_0^2}\left(c_i^a c_{jk\ell}^{bcd} + 15 \text{ terms}\right) + \frac{6}{c_0^4}\left(c_i^a c_j^b c_k^c c_\ell^d + 23 \text{ terms}\right). \end{aligned} \tag{34}$$

Apart from the presence of c_0, Eq. (34) is a special case of eq. (A4′) of Ref. 124.

In favor of notation, Eq. (26) is rewritten as another cluster CI expansion given by:

$$\begin{aligned}\Psi &= \Phi_0 B_0 + \sum_i \sum_a \Phi_i^a B_i^a + \sum_{i\le j} \sum_{a\le b} \Phi_{ij}^{ab} B_{ij}^{ab} \\ &+ \sum_{i\le j\le k} \sum_{a\le b\le c} \Phi_{ijk}^{abc} B_{ijk}^{abc} + \sum_{i\le j\le k\le \ell} \sum_{a\le b\le c\le d} \Phi_{ijk\ell}^{abcd} B_{ijk\ell}^{abcd} + \cdots\end{aligned} \tag{35}$$

where now $\le$ has taken the place of $<$ in Eq. (3), symmetry–orbitals replace spin–orbitals, and the summation over the degeneracy index g has already taken place masking through Eq. (26) the true linear variational coefficients. Formally, other than for calculation purposes, (34) is entirely equivalent to (26) as well as to (5) and (3), thus any $\Phi_{ijk\ldots}^{abc\ldots}$ is identical with some G_{K} of (26).

By neglecting $t_{ijk\ell}^{abcd}$ and the contributions involving c_i^a coefficients, and by using B instead of c, Eq. (34) translates into:

$$\begin{aligned}B_{ijk\ell}^{abcd} &= \frac{1}{B_0^2}(B_{ij}^{ab}B_{k\ell}^{cd} + B_{k\ell}^{ab}B_{ij}^{cd} - B_{ij}^{ac}B_{k\ell}^{bd} - B_{k\ell}^{ac}B_{ij}^{bd} + B_{ij}^{ad}B_{k\ell}^{bc} + B_{k\ell}^{ad}B_{ij}^{bc} \\ &- B_{ik}^{ab}B_{j\ell}^{cd} - B_{j\ell}^{ab}B_{ik}^{cd} + B_{ik}^{ac}B_{j\ell}^{bd} + B_{j\ell}^{ac}B_{ik}^{bd} - B_{ik}^{ad}B_{j\ell}^{bc} - B_{j\ell}^{ad}B_{ik}^{bc} \\ &+ B_{i\ell}^{ab}B_{jk}^{cd} + B_{jk}^{ab}B_{i\ell}^{cd} - B_{i\ell}^{ac}B_{jk}^{bd} - B_{jk}^{ac}B_{i\ell}^{bd} + B_{i\ell}^{ad}B_{jk}^{bc} + B_{jk}^{ad}B_{i\ell}^{bc}),\end{aligned} \tag{36}$$

When the degeneracy g_{K} is equal to one, and $i<j<k<\ell$, $a<b<c<d$, Eq. (36) coincides with Eq. (34) in terms of $c_{ijk\ldots}^{abc\ldots}$ coefficients of (3)[29] and, not surprisingly, ΔE_{K} values obtained by replacing (28) and (36) into (27) are fairly accurate. However, apart from $i<j<k<\ell$, there are seven additional possibilities: $i=j<k<\ell$, $i<j=k<\ell$, $i<j<k=\ell$, $i=j<k=\ell$, $i=j=k<\ell$, $i<j=k=\ell$, and $i=j=k=\ell$. These eight possibilities for occupied orbitals together with the eight choices for unoccupied orbitals give rise to 64 arrangements of these indices for four-excited coefficients.

Direct replacement of index choices in (36) yields too large ΔE_{K} values (in magnitude). However, if after replacement of index choices in (36), only terms with different indices are kept and positive signs are used throughout, each of the 64 resulting formulas together with (28) yield remarkably good ΔE_{K} values.[29]

In general, there are 16 formulas for 3x coefficients, 64 for 4x, 256 for 5x, and 1024 for 6x. All 1360 formulas together with the corresponding Fortran code were generated by another Fortran program.[29]

Abelian groups only require 9, 25, 81, and 196 formulas, respectively, however, these already occupy more than 70,000 compact lines of code so that, beyond sextuples, for both atoms and molecules, only selected MR-*m*x-CI-*n*x offers a realistic alternative.

We will use Eqs. (26), (35) just for preselecting and selecting configurations K rather than CSFs; for all other purposes we will be using the CI expansion in terms of CSFs, given by Eq. (5).

6. SUMMARY OF SCI

The central purpose of SCI is to approximate FCI energies with TEE using large basis sets for small atomic and molecular systems not exceeding 10 electrons. The treatment of larger systems will be based on the above applied to various valence CI (VCI) models implemented in highly parallelized codes.

6.1 Preselection of Configurations Without TEE

Eq. (29) gives a pseudo-energy contribution $\Delta\mathcal{E}_{\mathrm{K}}$ for connected Ks. The value of H_{KK} rarely reaches 10^6 Hartree, and it is usually below a few thousand Hartree. Thus $(E - H_{\mathrm{KK}})$ may be removed in Eq. (29), and the bare product $\prod_{i=1}^{q} n_{\mathrm{K}_i}$ may be used for preselection purposes since it does not differ from $\Delta\mathcal{E}_{\mathrm{K}}$ by more than a factor of 10^6.

Preselection of configurations is achieved by discarding all configurations with products of NO occupation numbers under preselection thresholds $\mathcal{T}_{\mathrm{ps}}(q)$:

$$\prod_{i=1}^{q} n_{\mathrm{K}_i} \leq \mathcal{T}_{\mathrm{ps}}(q). \tag{37}$$

Since products on the left side of Eq. (37) are ordered by nonincreasing occupation numbers, they form a partially ordered set, becoming an ordered set only for orbitals of the same symmetry and with all orbital occupancies not greater than one.

For sextuple excitations and fixed occupied orbitals, efficient selection requires the use of six nested loops running over the indices of unoccupied orbitals. If the two innermost loops run over the two largest occupation numbers of any given product, exit from these two loops may be enforced at the first fulfillment of Eq. (37). In the case of ordered sets, the exits are enforced at every nesting level. Notice that for excitations higher than six, in general, the number of active unoccupied orbitals is dramatically reduced.

We now proceed to evaluate a rigorous TEE $\Delta E_{\rm ps}^{\rm FCI}$ by repeated pruning of the full configuration list with successively smaller values of $\mathcal{T}_{\rm ps}(q)$ and correspondingly larger (in magnitude) values of the extant TEE, $\Delta E_{\rm sT}^{\rm FCI}$, occurring in the right side of Eq. (20), until reaching convergence to within an uncertainty $\delta E_{\rm ps}^{\rm FCI}$:

$$\Delta E_{\rm ps}^{\rm FCI} = \Delta E_{\rm s\mathcal{T}}(\lim \mathcal{T}_{\rm ps}(q) \to 0) - \Delta E_{\rm s\mathcal{T}}(\mathcal{T}_{\rm ps}(q)) \pm \delta E_{\rm ps}^{\rm FCI}. \qquad (38)$$

In Eq. (38), $\Delta E_{\rm s\mathcal{T}}(\mathcal{T}_{\rm ps}(q))$ is $\Delta E_{\rm s}^{\rm FCI}$ from Eq. (31) obtained using preselection thresholds $\mathcal{T}_{\rm ps}(q)$, which makes these quantities bona-fide energies despite that energy units do not appear in Eq. (37). According to Ref. 11, for H_2O and Ne, using large basis sets, and $\lim \mathrm{T}_{\rm ps}(q) \to 0 = 10^{-22}$, $\mathrm{T}_{\rm ps}(q) = 10^{-20}$, for $q = 3$–6, yields $\Delta E_{\rm ps}^{\rm FCI} \approx 1\mu$ Hartree.

Preselection provides significant simplification for SCI-TEE and also allows TEEs to be true and therefore more reliable.

6.2 Selection of Configurations With TEE

We always start with the FCI space $\mathcal{F}$ limited to n-excitations with $n \leq 10$.

In a first selection step, the FCI space $\mathcal{F}$ is reduced to a model space $\mathcal{M}$ after preselection of configurations with thresholds $\mathcal{T}_{\rm ps}(q)$ and selection of configurations with thresholds $\mathrm{T}^{\rm M}(q)$ and Eq. (27), and $\mathcal{T}_d^{\rm M}(q)$, $\mathcal{T}_c^{\rm M}(q)$ and Eq. (29). A given subclass is eliminated when none of its configurations are selected. The majority of subclasses are removed from the configuration list at this stage, and the corresponding TEE are accumulated.

An important simplification is that the number of symmetry orbitals for each excitation level q decreases with increasing value of q as detected "by hand" from convergence studies of $\Delta E_{\rm s}^{\rm FCI}$ for each value of q. This requires intensive use of program AUTOCL and needs to be automatized.

In a second selection step, the model space $\mathcal{M}$ is reduced to a model space $\mathcal{S}$ after selection of configurations with thresholds $\mathrm{T}^{\rm S}(q)$ and Eq. (27), and $\mathcal{T}_d^{\rm S}(q)$, $\mathcal{T}_c^{\rm M}(q)$ and Eq. (29). The majority of configurations in the model space $\mathcal{M}$ are usually removed at this stage, and the corresponding TEE are accumulated.

6.3 SCI by Parts

The need for SCI by parts arises when the number of Hamiltonian matrix elements with magnitudes above a small threshold is too large to fit in RAM, as automatically detected by program ATMOL. Its divide-and-conquer strategy and corresponding equations have been described at length.[30] Here we only remark that SCI by parts considers *all* Hamiltonian matrix elements of the eigen-problem and carries out rigorous variational calculations using Davidson's[126] eigen-solver implemented with a program[127] that was recently parallelized by Almora-Díaz.[128] In the first iteration, all CI coefficients are obtained variationally. After that, only a small and fixed subset of them are calculated variationally, the other ones remain fixed thereafter. Consequently, the final energy $E_{\mathrm{up}}^{\mathrm{S}}$ will be an upper bound to the eigenvalue of the exact eigen-problem E^{S} by an amount ΔE^{CIBP}:

$$E^{\mathrm{S}} = E_{\mathrm{up}}^{\mathrm{S}} + \Delta E^{\mathrm{CIBP}}. \tag{39}$$

6.4 Truncation Energy Error

The total TEE will be the sum of TEE quantities obtained in first going from the FCI space $\mathcal{F}$ to the model space $\mathcal{M}$, and then to the final selected space $\mathcal{S}$, plus the error due to the approximations involved in the eigen-problem. Thus we have the TEE $\Delta E_{\mathrm{s}}^{\mathrm{FCI}}$ accumulated from discarding disconnected configurations,

$$\Delta E_{\mathrm{s}}^{\mathrm{FCI}} = \Delta E_{\mathrm{s}}^{\mathrm{S}} + \Delta E_{\mathrm{s}}^{\mathrm{M}} \tag{40}$$

and the total full energy E^{FCI}

$$E^{\mathrm{FCI}} = E_{\mathrm{up}}^{\mathrm{S}} + \Delta E_{\mathrm{s}}^{\mathrm{FCI}} + \Delta E^{\mathrm{CIBP}} \pm \delta E(\mathrm{ps, s, CIBP}) \tag{41}$$

where δE(ps,s,CIBP) is a consolidated uncertainty for the steps of preselection, selection, and SCI by parts.

The energy eigenvalue E of Schrödinger's equation becomes

$$E = E^{\mathrm{FCI}} + \Delta\epsilon^{\mathrm{CBS}} \pm \delta\epsilon^{\mathrm{CBS}}, \tag{42}$$

where $\delta\epsilon^{\mathrm{CBS}}$ is by far the largest of all uncertainties, and $|\Delta\epsilon^{\mathrm{CBS}}| \gg |\Delta E_{\mathrm{s}}^{\mathrm{FCI}}|$.

6.5 Valence SCI and Related Models

In 1916, Lewis developed a very successful chemical theory of valence[129] based on a new concept: the (Lewis) electron pair. With the advent of quantum mechanics, Lewis electron pairs were interpreted as two electrons occupying the same orbital and therefore with opposite spins. Lewis pairs appear in: (i) filling core levels in an atom,[33] (ii) lone pairs[130] important in acid–base theory[130] and protein chemistry, and (iii) forming a covalent bond.[131]

Occupied orbitals are mathematically defined by either HF or Multi-reference HF, depending on the importance of keeping an independent-particle model with more than one reference configuration, like in most relativistic atomic calculations or, in general, when dealing with potential energy surfaces. Valence orbitals are those nearly degenerate occupied orbitals with highest energies; fixed-core orbitals make up the rest of occupied orbitals.

VCI is generally understood as a CI in which orbital-excitations are restricted to valence orbitals.[33,130] The success of Lewis theory for other than metallic bonding suggests that valence SCI (VSCI) and related models might perform likewise.

The key question in applying VSCI is how to limit the excitation levels for valence electrons. By analogy to the CI-*n*x notation, we will use VSCI-*n*x for VSCI limited to *n*-excitations of the valence orbitals.

We know that 10-electron FCI is feasible whenever CI-6x is essentially correct,[11] viz, when septuple to decuple excitations contribute negligibly, however, beyond 10 electrons, we would be very happy to be able to carry out VSCI-6x, and most of the time we will have to settle for VSCI-4x or MR-*m*x-VSCI-4x.

VSCI can be validated on a small number of valence electrons. Better approximations can be sought either by increasing the number of valence electrons and/or opening core shells systematically to account for core–valence and core–core significant interactions. For those (few) cases in which a convergence sensitivity analysis is possible, one would be able to eliminate the usual need to invoke cancelation of errors in energy differences.

7. MECHANICS OF THE SELECTION PROCESS

Here we explain in detail the successive tasks involved in a SCI-TEE calculation. MR-*m*x-CI-*n*x denotes multireference CI in which each reference contains up to *m*-excitations, which may reach $n = 6$, and CI may expanded to CI-*n*x.

For a given orbital basis, SCI-TEE starts with CI-2x (or MR-2x-CI-2x when more than one reference is needed) to obtain NOs and corresponding occupation numbers. When using CI-2x, the leading NOs are kept as Hartree–Fock orbitals, otherwise we use actual NOs all along.

With the NO-2x basis, CI-2x (MR-2x-CI-2x) is carried out to get CI coefficients CIC-2x and corresponding occupation numbers. These may be used either in program AUTOCL to evaluate the model space $\mathcal{M}$, or in program ATMOL to evaluate SCI-TEE for model spaces within CI-4x. The best choice is to go straight on the first choice to get NOs of CI-4x (NO-4x), since NO-4x are exceptionally superior to NO-2x. Once with an adequate $\mathcal{M}$ space, use ATMOL to obtain the best possible NO-4x, for which it might be necessary to go into CI by parts.

With NO-4x, one is ready to carry out selected CI-3x to get CIC-3x bearing in mind the corresponding NO-3x are not as good as NO-4x, so CI-3x is of no concern. Because the triples in CIC-3x are used just to predict B_K coefficients of quintuples and sextuples, the number of orbitals for triples is that normally used for quintuples and sextuples in SCI-TEE, this step should not be undertaken until the user has a clear idea about the size of orbital basis for 3x and 4x excitations. (Otherwise, full CI-3x with large orbital bases might be unmanageable.)

Having obtained CIC-3x, one has all data necessary to prune the full configuration list and to inquire the TEE of SCI up to CI-6x. Also, one may go beyond CI-6x and up to CI-10x using Eq. (29) for configuration selection.

Finally, we reached the point of actually carrying out SCI-TEE according to a scheme drawn from the results of the above inquiries. The patterns of convergence of these calculations together with those from the pruning of the full configuration list are used to assign a reasonable value for δE^{FCI}. The value of $\delta E^{\mathrm{FCI}}_{\mathrm{up}}$ is obtained by extrapolation of $E^{\mathrm{FCI}}_{\mathrm{up}}$ values obtained with increasingly larger dimensional spaces for the first stage of the eigenvalue problem in CI by parts.[30]

Once energies corrected by TEE are obtained for a set of increasingly accurate orbital bases, they are used in various CBS extrapolation formulas to get $\Delta\epsilon^{\mathrm{CBS}}$ and $\delta\epsilon^{\mathrm{CBS}}$, the latter being taken as the dispersion of extrapolated values using all possible combinations involving the energies for the largest orbital bases. In H_2O ground state at equilibrium geometry,[11,128] 93% of $\Delta\epsilon^{\mathrm{CBS}}$ comes from CI-2x, and most of the rest comes from CI-3x, suggesting that the best way to reduce the magnitude of $\delta\epsilon^{\mathrm{CBS}}$ is through the identification of reliable patterns for convergence of the energy for CI-2x and CI-3x.

Putting Eq. (41) into Eq. (42) one gets the electronic energy E sought, with an uncertainty dominated by $\delta E^{\mathrm{CBS}}_{\mathrm{patt}}$ (H_2O ground state, see Ref. 11) or by a still undetermined combination of all uncertainties involved (Ne ground state, see Ref. 11).

8. SCI ON LARGE SYSTEMS

Wholesale evaluation of matrix elements is a fascinating subject. During the last 2 years my main concern has been to produce a reliable and efficient program useful to extend our work from 10 electron systems toward a significantly larger number of electrons.

The use of symbolic formulae for efficient evaluation of matrix elements was pioneered by Hibbert,[132,133] predating Shavitt's work[134] on the graphical unitary group approach (GUGA) in molecules.

There are some advantages in the use of symbolic formulae for atomic[60,65,135] and molecular[11,109,128] work over the GUGA approach. One of them is that the formulas can be used in a priori SCI with TEE,[29,30] whereas the GUGA is oriented toward incorporation of full classes of configurations whose number scales up wildly with orbital basis size. For example, our H_2O work[11,128] uses 412 C_{2v} orbitals distributed among the four possible symmetries as 139a_1 70a_2 112b_1 91b_2. If a GUGA code is run on such a basis, it will naturally generate zillions more unneeded matrix elements than those really needed in SCI. The same may be said about using GUGA in current atomic work[60,65] with relatively moderate CI demands. My reason for bringing up this discussion in 2017 is that over and over I have heard keynote speakers aggressively declaring that there is no place in the community for those not using GUGA.

A formula is usually computed for the $g_K \cdot n_L$ ($\leq g_L \cdot n_K$) matrix elements $H_{Kg,Lh}$

$$H_{Kg,Lh} = \langle F_{Kg}|H|F_{Lh}\rangle, \tag{43}$$

where we are using the notation in Eq. (6), and Löwdin's "turn over" rule[136] is employed to drastically reduce the numbers of matrix elements between Slater determinants for each complete matrix element of Eq. (43), from $n_K \cdot n_L$ to $g_K \cdot n_L$. In general, $H_{Kg,Lh}$ involves a sum over n_I one- and two-electron symmetry integrals I_t,

$$H_{Kg,Lh} = \sum_{t=1}^{n_I} I_t C_t \tag{44}$$

where the coefficients C_t involve products of b_i^g and c_j^h expansion coefficients of Eq. (6). In the atomic case one also has to include (just for two-electron integrals I_t) additional products between coupling coefficients.

For each pair of subclasses S1 and S2 for which nonzero matrix elements might occur, all n_{nz} nonzero matrix elements are marked. Then, two inner loops over N2 and N1 configurations and skipping the unmarked pairs are used to find the number of different formulae n_F, after which in another loop of size n_F all coefficients C_t for each formula are evaluated. Next, in a loop over the n_{nz} nonzero matrix elements, the various formulae are applied using Eq. (44), with the indexing of integrals I_t having been determined generally as part of the formula.

Formulae are useful when n_{nz} is sufficiently large, which is not the case for all nonzero pairs S1,S2 in SCI. Each one of these steps has been parallelized with various outcomes. The main gain in parallelization is in the loop over S1 after marking all nonzero pairs S1,S2. In order to have the whole picture, a move from a 4-core laptop to a many-core workstation is necessary.

The handling of many more electrons than 10 required changing the determinant notation from one appropriate to carry out the projections in program AUTOCL to another one for ATMOL where occupied orbitals must come first and unoccupied ones come afterward. This change is carried at the level of subclasses, as well as the comparisons of the occupied part of determinants.

An analogous decoupling of occupied and unoccupied orbitals was present from the very beginning more than 50 years ago.[12] But there was a good reason for a program based on CSFs to postpone traumatic changes when things look satisfactory: decoupling of occupied and unoccupied spinorbitals (or Dirac-bispinors) led to the need to develop an algorithm that processes the 40 different ways to combine the comparisons between occupied spinorbitals with those between unoccupied ones which are needed in formula evaluation. In the new ATMOL, the calculation of formulae for matrix elements takes less than one thousandth of total CPU time.

Sometimes, work on programming seems never ending.

9. FINAL REMARKS

I started thinking in SCI 16 years ago after I read several announcements promising very cheap RAM. After spending 4000 euros on a laptop with 512 Mb and two disks, I first set to change programs from Fortran 77 to Fortran 90, in order to allocate/deallocate RAM to get an essentially open ended code. My scientific motivation was as clear then as it is now: very few people were then and are interested today to get fully engaged in solving the correlation problem.[89] I then felt that together with Brown's formula and some of the cards I have just discussed, I had a good opportunity to do something different and useful. New cards started arriving, above all my student César, now Dr. Almora-Díaz, and my student Herzain, soon to become Dr. Rivera-Arrieta.

Our current version of ATMOL, yet to be published, has been fully restructured to unveil opportunities for parallelism. Computer time in large calculations is still mostly spent in wholesale evaluation of matrix elements. AUTOCL and ATMOL will be made freely available online when a basic parallel version is completed and reasonably verified. The author asks the readers' indulgence on the additional time still needed to make this a reality.

At the moment we feel confident in almost any problem involving stationary states for systems with no more than 10 electrons. We are now beginning to venture to considerably larger systems. But new jobs are scarce, which causes good students to remain available. Progress in a very old field, such as the correlation problem, requires people willing to take deliberate albeit well-reasoned risks: the risks inherent in jumping into the unknown, although, just in case, we now have the opportunity to retreat to the 10 electron domain.

There are two basic questions that need to be vigorously addressed: (i) the problem of slow convergence of the CI expansion, in which I am now deeply and optimistically involved, and (ii) the need to automatize whole procedures, like the mechanics of the selection process described in Section 7. In particular, after good solutions are obtained, one should be able to automatically examine energy convergence as a function of number of orbitals within each symmetry and within families of hole excitations.

Finally, we live in a magical computer era. New affordable high speed memories will affect a revolution in many ab initio methods, particularly in SCI. Friendly software for coprocessors and GPUs are becoming increasingly available. Fortunately, as always, the most promising path to progress lies in motivating and stimulating our brains to solve useful problems.

ACKNOWLEDGMENTS

I am grateful to the Instituto de Física and to the Dirección General de Tecnologías de la Información (DGTIC) of Universidad Nacional Autónoma de México for excellent and free computer services. As always, I am deeply indebted to all my coauthors and colleagues, in Mexico and in several other countries, who shared their knowledge with me through the years.

REFERENCES

1. Schrödinger, E. Quantisierung als Eigenwertproblem (Erste Mitteilung). *Ann. Phys.* **1926**, *79*, 361.
2. Schrödinger, E. Quantisierung als Eigenwertproblem (Zweite Mitteilung). *Ann. Phys.* **1926**, *79*, 489.
3. Schrödinger, E. Quantisierung als Eigenwertproblem (Dritte Mitteilung). *Ann. Phys.* **1926**, *80*, 437.
4. Schrödinger, E. Quantisierung als Eigenwertproblem (Vierte Mitteilung). *Ann. Phys.* **1926**, *81*, 109.
5. Schrödinger, E. An Undulatory Theory of the Mechanics of Atoms and Molecules. *Phys. Rev.* **1926**, *28*, 1049.
6. Hylleraas, E. A. Reminiscences From Early Quantum Mechanics of Two-Electron Atoms. *Rev. Mod. Phys.* **1963**, *35*, 421.
7. Hylleraas, E. A. Neue Berechnung der Energie des Heliums im Grundzustande, Sowie des tiefsten Terms von Ortho-Helium. *Z. Phys.* **1929**, *54*, 347.
8. Pekeris, C. L. Ground State of Two-Electron Atoms. *Phys. Rev.* **1958**, *112*, 1649.
9. Herzberg, G. Ionization Potentials and Lamb Shift of the Ground State of ^{4}He and ^{3}He. *Proc. R. Soc. Lond. A* **1958**, *248*, 302.
10. Kandula, D. Z.; Gohle, C.; Pinkert, T. J.; Ubachs, W.; Eikema, K. S. E. Extreme Ultraviolet Frequency Comb Metrology. *Phys. Rev. Lett.* **2010**, *105*, 063001.
11. Almora-Díaz, C. X.; Rivera-Arrieta, H. I.; Bunge, C. F. Recent Progress in the Variational Orbital Approach to Atomic and Molecular Electronic Structure. *Adv. Quantum. Chem.* **2016**, *72*, 129.
12. Bunge, C. F. Specifics on the Scientific Legacy of Per-Olov Löwdin. *Adv. Quantum. Chem.* **2017**, *74*, 53.
13. Drake, G. W. F.; Cassar, M. M.; Nistor, R. A. Ground State Energies for He, H^-, and Ps^-. *Phys. Rev. A* **2002**, *65*, 054501.
14. Drake, G. High Precision Calculations for Helium. In: *Springer Handbook of Atomic, Molecular and Optical Physics*, Drake, G. W. F. (Ed.). Springer Science+Business Media, Inc.: New York, 2006; pp. 199–219.
15. Yan, Z.-C.; Drake, G. W. F. Relativistic and QED Energies in Lithium. *Phys. Rev. Lett.* **1998**, *81*, 774.
16. Yan, Z.-C.; Drake, G. W. F. Lithium Transition Energies and Isotope Shifts: QED Recoil Corrections. *Phys. Rev. A* **2002**, *66*, 042504.
17. Singer, K. The Use of the Gaussian (Exponential Quadratic) Wave Functions in Molecular Problems. I. General Formulae for the Evaluation of Integrals. *Proc. R. Soc. Lond. A* **1960**, *258*, 412.
18. Lester, W. A.; Krauss, M. Gaussian Correlation Functions: Two-Electron Systems. *J. Chem. Phys.* **1964**, *41*, 1407.
19. King, H. F. Some Theorems Concerning Symmetry, Angular Momentum, and Completeness of Atomic Geminals With Explicit r_{12} Dependence. *J. Chem. Phys.* **1967**, *46*, 705.
20. Cencek, W.; Rychlewski, J. Many-Electron Explicitly Correlated Gaussian Functions. I. General Theory and Test Results. *J. Chem. Phys.* **1993**, *98*, 1252.
21. Komasa, J.; Rychlewski, J.; Jankowski, K. Benchmark Energy Calculations on Be-Like Atoms. *Phys. Rev. A* **2002**, *65*, 042507.
22. Pachucki, K.; Komasa, J. Relativistic and QED Corrections for the Beryllium Atom. *Phys. Rev. Lett.* **2004**, *92*, 213001.
23. Stanke, M.; Kedziera, D.; Bubin, S.; Adamowicz, L. Lowest Excitation Energy of ^{9}Be. *Phys. Rev. Lett.* **2007**, *99*, 043001.

24. Sims, J. S.; Hagstrom, S. A. Hylleraas-Configuration-Interaction Nonrelativistic Energies for the ^{1}S Ground States of the Beryllium Isoelectronic Sequence. *J. Chem. Phys.* **2014**, *140*, 224312.
25. Puchalski, M.; Komasa, J.; Pachucki, J. Explicitly Correlated Wave Function for a Boron Atom. *Phys. Rev. A* **2015**, *92*, 062501.
26. Bubin, S.; Adamowicz, L. Lowest ^{2}S Electronic Excitations of the Boron Atom. *Phys. Rev. Lett.* **2017**, *118*, 043001.
27. Adamowicz, L. Personal Communication. 2015. March 20.
28. Bunge, A. V.; Bunge, C. F.; Jáuregui, R.; Cisneros, G. Symmetry Eigenfunctions for Many-Electron Atoms and Molecules: A Unified and Friendly Approach for Frontier Research and Student Training. *Comput. Chem.* **1989**, *13*, 201.
29. Bunge, C. F. Selected Configuration Interaction With Truncation Energy Error and Application to the Ne Atom. *J. Chem. Phys.* **2006**, *125*, 014107.
30. Bunge, C. F.; Carbó-Dorca, R. Select-Divide-and-Conquer Method for Large-Scale Configuration Interaction. *J. Chem. Phys.* **2006**, *125*, 014108.
31. Valiev, M.; Bylaska, E. J.; Govind, N.; Kowalski, K.; Straatsma, T. P.; van Dam, H. J. J.; Wang, D.; Nieplocha, J.; Apra, E.; Windus, T. L.; de Jong, W. A. NWChem: A Comprehensive and Scalable Open-Source Solution for Large Scale Molecular Simulations. *Comput. Phys. Commun.* **2010**, *181*, 1477.
32. Bohr, N. On the Constitution of Atoms and Molecules. *Philos. Mag.* **1913**, *26*, 1.
33. Bohr, N. Der Bau der Atome und die Physikalischen und Chemischen Eigenshaften der Elemente. *Z. Phys.* **1922**, *9*, 1.
34. Mendelejeff, D. Ueber die Besiehungen der Eigenschaften zu den Atomgewichten der Elemente. *Z. Chem.* **1869**, *12*, 405.
35. Pauli, W. Über den Zusammenhang des Abschlusses der Elektronen-gruppen im Atom mit der Komplexstruktur der Spektren. *Z. Phys.* **1925**, *31*, 765.
36. Uhlenbeck, G. E.; Goudsmit, S. Ersetzung der Hypothese vom unmechanischen Zwang durch eine Forderung bezfiglich des inneren Verhaltens jedes einzelnen Elektrons. *Naturwissenschaften* **1925**, *13*, 953.
37. Uhlenbeck, G. E.; Goudsmit, S. Spinning Electrons and the Structure of Spectra. *Nature* **1926**, *117*, 264.
38. Goudsmit, S. A. The Discovery of the Electron Spin. https://lorentz.leidenuniv.nl/history/spin/goudsmit.html. Accessed June 15, 2017.
39. Thomas, L. H. The Motion of the Spinning Electron. *Nature* **1926**, *117*, 514.
40. Jáuregui, R.; Bunge, C. F.; Ley-Koo, E. Upper Bounds to the Eigenvalues of the No-Pair Hamiltonian. *Phys. Rev. A* **1997**, *55*, 1781.
41. Hartree, D. R. The Wave Mechanics of an Atom With a Non-Coulomb Central Field. Part I. Theory and Methods. *Proc. Camb. Philos. Soc.* **1928**, *24*, 89.
42. Hartree, D. R. The Wave Mechanics of an Atom With a Non-Coulomb Central Field. Part II. Some Results and Discussion. *Proc. Camb. Philos. Soc.* **1928**, *24*, 111.
43. Dirac, P. A. M. The Quantum Theory of the Electron. *Proc. R. Soc. Lond. A* **1928**, *117*, 610.
44. Dirac, P. A. M. Quantum Mechanics of Many-Electron Systems. *Proc. R. Soc. Lond. A* **1929**, *123*, 714.
45. Slater, J. C. Interpretation of the Hydrogen and Helium Spectra. *Proc. Natl. Acad. Sci. USA* **1925**, *11*, 732.
46. Slater, J. C. The Theory of Complex Spectra. *Phys. Rev.* **1929**, *34*, 1293.
47. Slater, J. C. Note on Hartree's Method. *Phys. Rev.* **1930**, *35*, 210.
48. Fock, V. Näherungsmethode zur Lösung des Quantenmechanischen Mehrkörper-Problems. *Z. Phys.* **1930**, *61*, 126.
49. Fischer, C. *Froese Douglas Rayner Hartree: His Life in Science and Computing.* World Scientific Publishing Co. Pte. Ltd: Singapore, 2003.

50. Fischer, C. *Froese the Numerical Hartree-Fock Method.* Wiley: New York, 1977.
51. Bunge, C. F.; Ley-Koo, E.; Jáuregui, R. Justification of Relativistic Dirac-Hartree-Fock and Configuration Interaction. *Mol. Phys.* **2000**, *98*, 1067.
52. Fischer, C. Froese A Multi-Configuration Hartree-Foxk Program. *Comput. Phys. Commun.* **1970**, *1*, 151.
53. Saxena, K. M. S.; Malli, G. Accurate Numerical Hartree-Fock Self-Consistent-Field Wave Functions for Rare-Earth Ions. *Phys. Rev. A* **1971**, *3*, 1278.
54. Hartree, D. R.; Hartree, W.; Swirles, B. Self-Consistent Field, Including Exchange and Superposition of Configurations, With Some Results for Oxygen. *Phil. Trans. R. Soc. A* **1939**, *238*, 229.
55. Fischer, C. F. A Multi-Configuration Hartree-Fock Program With Improved Stability. *Comput. Phys. Commun.* **1972**, *4*, 107.
56. Fischer, C. F.; Saxena, K. M. S. Correlation Study of Be $1s^2 2s^2$ by a Separated-Pair Numerical Multiconfiguration Hartree-Fock Procedure. *Phys. Rev. A* **1974**, *9*, 1498.
57. Fischer, C. F. The MCHF Atomic-Structure Package. *Comput. Phys. Commun.* **1991**, *64*, 369.
58. Stathopoulus, A.; Fischer, C. F. A Davidson Program for Finding a Few Selected Extreme Eigenpairs of a Large, Sparse, Real, Symmetric Matrix. *Comput. Phys. Commun.* **1994**, *79*, 268.
59. Fischer, C. F.; Brage, T. *Computational Atomic Structure: An MCHF Approach.* CRC Press: Boca Raton, FL, 1997.
60. Fischer, C. F. Atomic Structure: Multiconfiguration Hartree-Fock Theories. In: *Springer Handbook of Atomic, Molecular and Optical Physics*, Drake, G. W. F. (Ed.). Springer Science+Business Media, Inc., New York, 2006; pp. 307–324.
61. Fischer, C. F.; Tachiev, G.; Gaigalas, G.; Godefroid, M. An MCHF Atomic-Structure Package for Large-Scale Calculations. *Comput. Phys. Commun.* **2007**, *176*, 559.
62. Swirles, B. The Relativistic Self-Consistent Field. *Proc. R. Soc. Lond. A* **1935**, *152*, 625.
63. Grant, I. P. Relativistic Self-Consistent Fields. *Proc. R. Soc. Lond. A* **1961**, *262*, 555.
64. Grant, I. P. Gauge Invariance and Relativistic Radiative Transitions. *J. Phys. B* **1974**, *7*, 1458.
65. Grant, I. P. Relativistic Atomic Structure. In: *Springer Handbook of Atomic, Molecular and Optical Physics*, Drake, G. W. F. (Ed.). Springer Science+Business Media, Inc., New York, 2006; pp. 325–358.
66. Grant, I. P. *Relativistic Quantum Theory of Atoms: Theory and Computation.* Springer Science & Business Media: New York, 2007.
67. Desclaux, J. P.; Mayers, D. F.; O'Brien, F. Relativistic Atomic Wave Functions. *J. Phys. B* **1971**, *4*, 631.
68. Desclaux, J. P. Relativistic Dirac-Fock Expectation Values for Atoms With $Z = 1$ to $Z = 120$. *At. Data Nucl. Data Tables* **1973**, *12*, 311.
69. Desclaux, J. P. Multiconfiguration Relativistic Dirac-Fock Program. *Comput. Phys. Commun.* **1975**, *9*, 31.
70. Desclaux, J. P. Status of Relativistic Calculations for Atoms and Molecules. *Phys. Scr.* **1980**, *21*, 436.
71. Grant, I. P.; McKenzie, B. J.; Norrington, P. H.; Mayers, D. F.; Pyper, N. C. An Atomic MCDF Package. *Comput. Phys. Commun.* **1980**, *21*, 207.
72. Dyall, K. G.; Grant, I. P.; Johnson, C. T.; Parpia, F. A.; Plummer, E. P. GRASP: A General-Purpose Relativistic Atomic Structure Program. *Comput. Phys. Commun.* **1989**, *55*, 425.
73. Parpia, F. A.; Fischer, C. F.; Grant, I. P. GRASP92: A Package for Large-Scale Relativistic Atomic Structure Calculations. *Comput. Phys. Commun.* **1996**, *94*, 249.
74. Jönsson, P.; He, X.; Fischer, C. F.; Grant, I. P. The Grasp2K Relativistic Atomic Structure Package. *Comput. Phys. Commun.* **2007**, *177*, 597.

75. Jönsson, P.; Gaigalas, G.; Bieroń, J.; Fischer, C. F.; Grant, I. P. New version: Grasp2K Relativistic Atomic Structure Package. *Comput. Phys. Commun.* **2013**, *184*, 2197.
76. Kobus, J. Overview of Finite Difference Hartree-Fock Method Algorithm, Implementation and Application. *AIP Conf. Proc.* **2012**, *1504*, 189.
77. Morrison, J. C.; Steffen, K.; Pantoja, B.; Nagaiya, A.; Kobus, J.; Ericcson, T. Numerical Methods for Solving the Hartree-Fock Equation of Diatomic Molecules II. *Commun. Comput. Phys.* **2016**, *19*, 632.
78. Hund, F. Zur Deutung einiger Erscheinungen in den Molekelspektren. *Z. Phys.* **1926**, *36*, 657.
79. Mulliken, R. S. The Assignment of Quantum Numbers for Electrons in Molecules. *Phys. Rev.* **1927**, *32*, 186.
80. Lennard-Jones, J. E. The Electronic Structure of Some Diatomic Molecules. *Trans. Faraday Soc.* **1929**, *25*, 668.
81. Coulson, C. A. Self-Consistent Field for Molecular Hydrogen. *Proc. Camb. Philos. Soc.* **1938**, *34*, 204.
82. Coulson, C. A. Representation of Simple Molecules by Molecular Orbitals. *Q. Rev. Chem. Soc. London* **1947**, *1*, 144.
83. Roothaan, C. C. J. New Developments in Molecular Orbital Theory. *Rev. Mod. Phys.* **1951**, *23*, 69.
84. Mulliken, R. S.; Roothaan, C. C. J. Broken Bottlenecks and the Future of Molecular Quantum Mechanics. *Proc. Natl. Acad. Sci. USA* **1959**, *45*, 394.
85. Roothaan, C. C. J. Self-Consistent Field Theory for Open Shells of Electronic Systems. *Rev. Mod. Phys.* **1960**, *32*, 179.
86. Coulson, C. A. Present State of Molecular Structure Calculations. *Rev. Mod. Phys.* **1960**, *32*, 170.
87. Longuet-Higgins, H. C.; Bell, R. P. The Structure of the Boron Hydrides. *J. Chem. Soc.* **1943**, 250. https://doi.org/10.1039/ JR9430000250.
88. Nesbet, R. K. Brueckner's Theory and the Method of Superposition of Configurations. *Phys. Rev.* **1958**, *109*, 1632.
89. Löwdin, P.-O. Correlation Problem in Many-Electron Quantum Mechanics. I. Review of Different Approaches and Discussion of Some Current Ideas. *Adv. Chem. Phys.* **1959**, *2*, 207. (First published as Tech. Note No. 1, Uppsala Quantum Chemistry Group, June 21, 1957).
90. Dunning, T. H., Jr.; Peterson, K. A.; Woon, D. E. Correlation Consistent Basis Sets for Molecular Calculations. In: *Encyclopedia of Computational Chemistry*; Schleyer, P. v. R. (Ed.). Wiley: New York, 1998; pp. 88–115.
91. Schwartz, C. Importance of Angular Correlations Between Atomic Electrons. *Phys. Rev.* **1962**, *126*, 1015.
92. Bunge, C. F. Electronic Wave Functions for Atoms. II. Some Aspects of the Convergence of the CI Expansion for the He Isoelectronic Series. *Theor. Chim. Acta* **1970**, *16*, 126.
93. Wigner, E. P. *Group Theory (English Translation)*. Academic Press: New York, 1959.
94. Cornwell, J. F. *Group Theory in Physics*; Vol. 1. Academic Press: New York, 1984. Vol. 1.
95. Löwdin, P.-O. Angular Momentum Wavefunctions Constructed by Projection Operators. *Rev. Mod. Phys.* **1964**, *36*, 966.
96. Jáuregui, R.; Bunge, C. F.; Bunge, A. V.; Cisneros, G. Angular Momentum Eigenfunctions for Many-Electron Calculations. *Comput. Chem.* **1989**, *13*, 223.
97. Bunge, A. V.; Bunge, C. F.; Jáuregui, R.; Cisneros, G. Spin Eigenfunctions for Many-Electron Calculations. *Comput. Chem.* **1989**, *13*, 239.
98. Cisneros, G.; Jáuregui, R.; Bunge, C. F.; Bunge, A. V. Molecular Symmetry Eigenfunctions for Many-Electron Calculations. *Comput. Chem.* **1989**, *13*, 255.

99. Gray, N. M.; Wills, L. A. Note on the Calculation of Zero Order Eigenfunctions. *Phys. Rev.* **1931**, *38*, 248.
100. Löwdin, P.-O. Quantum Theory of Many-Particle Systems. III. Extension of the Hartree-Fock Scheme to Include Degenerate Systems and Correlation Effects. *Phys. Rev.* **1955**, *97*, 1509.
101. Rotenberg, A. Calculation of Exact Eigenfunctions of Spin and Orbital Angular Momentum Using the Projection Operator Method. *J. Chem. Phys.* **1963**, *39*, 512.
102. Löwdin, P.-O. Angular Momentum Wavefunctions Constructed by Projection Operators. *Rev. Mod. Phys.* **1964**, *36*, 966. (First published as Tech. Note No. 12, Uppsala Quantum Chemistry Group, May 10, 1958).
103. Bunge, C. F.; Bunge, A. Eigenfunctions of Spin and Orbital Angular Momentum by the Projection Operator Technique. *J. Comput. Phys.* **1971**, *8*, 409.
104. Bunge, C. F. A Priori Selected Configuration Interaction With Truncation Energy Error, General Sensitivity Analysis and Application to the Ne Atom. *Mol. Phys.* **2010**, *108*, 3279.
105. Bunge, A. Electronic Wave Functions for Atoms. III. Optimal Partition of Degenerate Spaces and Ground State of Carbon. *J. Chem. Phys.* **1970**, *53*, 20.
106. Bunge, A.; Bunge, C. F. Correlation Energy of the Ground State of C. *Phys. Rev. A* **1970**, *1*, 1599.
107. Bunge, C. F.; Bunge, A. Symmetry Eigenfunctions Suitable for Many-Electron Theories and Calculations. I. Mainly Atoms. *Int. J. Quantum. Chem.* **1973**, 7, 927.
108. McLean, A. D.; Liu, B. Classification of Configurations and the Determination of Interacting and Noninteracting Spaces in Configuration Interaction. *J. Chem. Phys.* **1973**, *58*, 1066.
109. Liu, B.; Yoshimine, M. The Alchemy Configuration Interaction Method. I. The Symbolic Matrix Method for Determining Elements of Matrix Operators. *J. Chem. Phys.* **1981**, *74*, 612.
110. Brooks, B. R.; Schaefer, H. F. The Graphical Unitary Approach to the Electron Correlation Problem. Methods and Preliminary Applications. *J. Chem. Phys.* **1979**, *70*, 5092.
111. Brooks, B. R.; Laidig, W. D.; Saxe, P.; Handy, N. C.; Schaefer, H. F. The Loop-Driven Graphical Unitary Group Approach: A Powerful Method for the Variational Description of Electron Correlation. *Phys. Scr.* **1980**, *21*, 312.
112. Löwdin, P.-O. Quantum Theory of Many-Particle Systems. I. Physical Interpretations by Means of Density Matrices, Natural Spin-Orbitals, and Convergence Problems in the Method of Configurational Interaction. *Phys. Rev.* **1955**, *97*, 1474.
113. Hylleraas, E. A.; Undheim, B. Numerische Berechnung der 2S-Terme von Ortho- und Par-Helium. *Z. Phys.* **1930**, *65*, 759.
114. MacDonald, J. K. L. Successive Approximations by the Rayleigh-Ritz Variation Method. *Phys. Rev.* **1933**, *43*, 830.
115. Löwdin, P.-O. Present Situation of Quantum Chemistry. *J. Phys. Chem.* **1957**, *61*, 56.
116. Bunge, C. F. Energy Optimized Spinorbitals in Accurate Electronic Calculations. *Phys. Rev. A* **1973**, 7, 15.
117. Bunge, C. F. Ergonals: Optimum Spinorbitals in Accurate Electronic Calculations? In: *Energy Structure and Reactivity*, Smith, D. W.; McRae, W. (Eds.). Wiley: Hoboken, NJ, 1973; pp. 68–74.
118. Brown, R. E. A Configuration Interaction Study of the $^1\Sigma^+$ States of the Lithium Hydride Molecule (Ph.D. thesis); Department of Chemistry, Indiana University, 1967.
119. Shavitt, I. The Method of Configuration Interaction. In: *Modern Theoretical Chemistry*, Schaefer, H. F., III (Ed.). Vol. 3. Plenum: New York, 1977; p. 189.
120. Sinanoğlu, O. Many-Electron Theory of Atoms and Molecules. I. Shells, Electron Pairs vs. Many-Electron Correlations. *J. Chem. Phys.* **1962**, *36*, 706.

121. Coester, F. Bound States of a Many-Particle System. *Nucl. Phys.* **1958**, 7, 421.
122. Primas, H. Separability in Many-Electron Systems. In: *Modern Quantum Chemistry,* Sinanoğlu, O. (Ed.). Pt. II, Academic: New York, 1965; pp. 45–74.
123. Shavitt, I.; Bartlett, R. J. *Many Body Methods in Chemistry and Physics.* Cambridge University Press: United Kingdom, 2009.
124. Živković, T. P.; Monkhorst, H. J. Analytic Connection Between Configuration-Interaction and Coupled-Cluster Solutions. *J. Math. Phys.* **1978**, *19*, 1007.
125. Harris, F. E.; Monkhorst, H. J.; Freeman, D. L. *Algebraic and Diagrammatic Methods in Many-Fermion Theory.* Oxford University Press: New York, 1992.
126. Davidson, E. R. The Iterative Calculation of a Few of the Lowest Eigenvalues and Corresponding Eigenvectors of Large Real-Symmetric Matrices. *J. Comput. Phys.* **1975**, *17*, 87.
127. Cisneros, G.; Berrondo, M.; Bunge, C. F. DVDSON: A Subroutine to Evaluate Selected Sets of Eigenvalues and Eigenvectors of Large Symmetric Matrices. *Comput. Chem.* **1986**, *10*, 281.
128. Almora-Díaz, C. X. Highly Correlated Configuration Interaction Calculations on Water With Large Orbital Bases. *J. Chem. Phys.* **2014**, *140*, 184302.
129. Lewis, G. N. The Atom and the Molecule. *J. Am. Chem. Soc.* **1916**, *38*, 762.
130. Lewis, G. N. *Valence and the Structure of Atoms and Molecules.* The Chemical Catalog Co: New York, 1923.
131. Heitler, W.; London, F. Weschelwirkung Neutraler Atome und Homöopolare Bindung Nach der Quantenmechamik. *Z. Phys.* **1927**, *44*, 455.
132. Hibbert, A. A General Program for Calculating Angular Momentum Integrals in Atomic Structure. *Comput. Phys. Commun.* **1970**, *1*, 359.
133. Glass, R.; Hibbert, A. Relativistic Effects in Many Electron Atoms. *Comput. Phys. Commun.* **1978**, *16*, 19.
134. Shavitt, I. Matrix Element Evaluation in the Unitary Group Approach to the Electron Correlation Problem. *Int. J. Quantum. Chem.* **1978**, *14*, 5.
135. Jitrik, O.; Bunge, C. F. Atomic Configuration Interaction and Studies of He, Li, Be and Ne Ground States. *Phys. Rev. A* **1997**, *56*, 2614.
136. Löwdin, P.-O. The Normal Constants of Motion in Quantum Mechanics Treated by Projection Technique. *Rev. Mod. Phys.* **1962**, *34*, 520.

CHAPTER TWO

Recent Developments in Asymptotic Expansions From Numerical Analysis and Approximation Theory

Avram Sidi[1]
Technion – Israel Institute of Technology, Haifa, Israel
[1]Corresponding author: e-mail address: asidi@cs.technion.ac.il

Contents

1. Introduction 36
2. Trapezoidal Rule Approximations and Generalizations of the E–M Expansion 38
   2.1 Classical E–M Expansion 38
   2.2 Navot's Generalizations of Classical E–M Expansion 39
   2.3 Recent Generalizations of E–M Expansions 40
   2.4 Applications to Computation of Singular Integrals 44
   2.5 Further Remarks on E–M Expansion and Generalizations 45
3. Error Expansions for Gauss–Legendre Quadrature 46
4. Asymptotics of Legendre Polynomials and Legendre Series Coefficients 49
   4.1 Asymptotics of Legendre Polynomials and Related Integrals 50
   4.2 Asymptotics of Legendre Series Coefficients 52

References 56

Abstract

In this chapter, we discuss some recently obtained asymptotic expansions related to problems in numerical analysis and approximation theory.

- We present a generalization of the Euler–Maclaurin (E–M) expansion for the trapezoidal rule approximation of finite-range integrals $\int_a^b f(x)dx$, when $f(x)$ is allowed to have arbitrary algebraic–logarithmic endpoint singularities. We also discuss effective numerical quadrature formulas for so-called *weakly singular, singular,* and *hypersingular* integrals, which arise in different problems of applied mathematics and engineering.
- We present a full asymptotic expansion (as the number of abscissas tends to infinity) for Gauss–Legendre quadrature for finite-range integrals $\int_a^b f(x)dx$, where $f(x)$ is allowed to have arbitrary algebraic–logarithmic endpoint singularities.

Advances in Quantum Chemistry, Volume 76
ISSN 0065-3276
https://doi.org/10.1016/bs.aiq.2017.06.002

- We present full asymptotic expansions, as $n \to \infty$, (i) for Legendre polynomials $P_n(x)$, $x \in (-1, 1)$, (ii) for the integral $\int_c^d f(x)P_n(x)dx$, $-1 < c < d < 1$, and (iii) for Legendre series coefficients $e_n[f] = (n+1/2)\int_{-1}^{1} f(x)P_n(x)dx$, when $f(x)$ has arbitrary algebraic–logarithmic (interior and/or endpoint) singularities in $[-1, 1]$.

1. INTRODUCTION

In many problems of science and engineering, one is confronted with the problem of determining the asymptotic behavior of some function $f(x)$ as $x \to a$ for some fixed a; typically, $a = 0$ or $a = \infty$.

1. In some cases, the best one can do is to obtain an *upper bound* for $|f(x)|$ as $x \to a$; that is, one can have

$$f(x) = O(g(x)) \quad \text{as } x \to a, \quad g(x) \text{ a known simple function.} \tag{1}$$

Of course, this means that there exist fixed positive constants M, ϵ, and X, for which,

$$|f(x)| \le M\,|g(x)| \quad \begin{cases} \forall \;\; x \in (a-\epsilon, a+\epsilon) \;\; \text{or} \;\; \forall \;\; x \in (a, a+\epsilon) \;\; \text{if } a \text{ finite,} \\ \forall \;\; x \in (X, \infty) \;\; \text{if } a = \infty. \end{cases} \tag{2}$$

2. In other problems, one can obtain the *actual* asymptotic behavior of $f(x)$ as $x \to a$ in the form of an *asymptotic equality*, namely,

$$f(x) \sim h(x) \quad \text{as } x \to a, \quad h(x) \text{ a known simple function,} \tag{3}$$

which means that

$$\lim_{x \to a} \frac{f(x)}{h(x)} = 1. \tag{4}$$

3. Yet in other cases, one can obtain a *complete asymptotic expansion* for $f(x)$ as $x \to a$ that is in the form

$$f(x) \sim \sum_{n=0}^{\infty} c_n \phi_n(x) \quad \text{as } x \to a, \tag{5}$$

where c_n are some constants and the functions $\phi_n(x)$ are simple known functions such that the sequence $\{\phi_n(x)\}_{n=0}^{\infty}$ is an *asymptotic scale*, namely,

$$\lim_{x \to a} \frac{\phi_{n+1}(x)}{\phi_n(x)} = 0, \quad n = 0, 1, \ldots. \tag{6}$$

(For examples, the functions $\phi_n(x) = (x - a)^n$, with a finite, form an asymptotic scale as $x \to a$. Similarly, the functions $\phi_n(x) = x^{-n}$ form an asymptotic scale as $x \to \infty$).

By (5), we mean

$$f(x) - \sum_{n=0}^{N-1} c_n \phi_n(x) = O(\phi_N(x)) \quad \text{as } x \to a. \tag{7}$$

In this case, if $c_n = 0$ for $0 \leq n \leq k - 1$ and $c_k \neq 0$, we also have the asymptotic equality

$$f(x) \sim c_k \phi_k(x) \quad \text{as } x \to a. \tag{8}$$

If $c_n = 0$ for all n, then we have that $\lim_{x \to a} f(x)/\phi_n(x) = 0$ for all n. (Note that the infinite series $\sum_{n=0}^{\infty} c_n \phi_n(x)$ may be convergent or divergent and this does not present an issue we need to worry about.)

Clearly, the situation described in item 2 is more informative than that described in item 1 since it implies the latter, and the situation described in item 3 is more informative than that described in item 2 and implies the latter.

Our purpose in this work is to give a review of some recently obtained *complete* asymptotic expansions for some commonly occurring problems arising in numerical analysis and approximation theory that have also been observed to occur in other disciplines, such as applied mathematics and theoretical physics, for example.

In Section 2, we discuss the error expansion for trapezoidal rule approximations of finite-range integrals $\int_a^b f(x)dx$ (as the number of abscissas tends to infinity) when $f(x)$ is allowed to have arbitrary algebraic–logarithmic endpoint singularities in general, and present recent generalizations of the well-known Euler–Maclaurin (E–M) expansions. We also present very effective numerical quadrature formulas for some singular integrals that are derived from these expansions.

In Section 3, we present the error expansion of Gauss–Legendre quadrature formulas for finite-range integrals $\int_a^b f(x)dx$ (again, as the number of abscissas tends to infinity), where $f(x)$ is allowed to have arbitrary algebraic–logarithmic endpoint singularities.

In Section 4, we discuss the asymptotic expansions as $n \to \infty$ (i) for Legendre polynomials $P_n(x)$, $x \in (-1, 1)$, (ii) for the integral $\int_c^d f(x)P_n(x)dx$, $-1 < c$

$< d < 1$, and (iii) for the coefficients $e_n[f] = (n + 1/2)\int_{-1}^{1} f(x)P_n(x)dx$ of the Legendre series $\sum_{n=0}^{\infty} e_n[f]P_n(x)$ of a function $f(x)$, when $f(x)$ has arbitrary algebraic–logarithmic (interior and/or endpoint) singularities in $[-1, 1]$. These asymptotic expansions, in addition to being of interest by themselves, can have applications in asymptotic analyses involving Legendre expansions, such as integral equations, numerical quadrature, and in series of spherical harmonics.

Before closing, we mention that, when computing the integrals $\int_a^b f(x)\,dx$, where $f(x)$ has singularities at the endpoints $x = a$ and $x = b$, we can first use suitable variable transformations and apply the trapezoidal rule or the Gauss–Legendre quadrature to the transformed integral; by this, we can achieve very high accuracy. When $f(x)$ has asymptotic expansions of the types discussed in this work, some of the variable transformation can be tuned to enable the quadrature formulas to attain accuracies that are optimal in some asymptotic sense. We do not treat this subject here; we refer the reader to Sidi[1] and the references therein.

Finally, for simplicity of notation, in the sequel, we will write "$(h \to 0)$" or the equivalent "$(n \to \infty)$" instead of "as $h \to 0$" or the equivalent "as $n \to \infty$".

2. TRAPEZOIDAL RULE APPROXIMATIONS AND GENERALIZATIONS OF THE E–M EXPANSION

2.1 Classical E–M Expansion

We start with the well-known classical E–M expansion for finite-range integrals $I[f] = \int_a^b f(x)dx$, where $f \in C^\infty[a,b]$, which involves the trapezoidal rule approximation for $I[f]$. We summarize the subject in the following theorem.

Theorem 1. *Let*

$$I[f] = \int_a^b f(x)dx \text{ and } T_n[f] = h\left[\frac{1}{2}f(a) + \sum_{i=1}^{n-1} f(a+ih) + \frac{1}{2}f(b)\right],\ h = \frac{b-a}{n}. \tag{9}$$

Assume that $f \in C^\infty[a,b]$. Then there holds

$$T_n[f] \sim I[f] + \sum_{k=1}^{\infty} \frac{B_{2k}}{(2k)!}\left[f^{(2k-1)}(b) - f^{(2k-1)}(a)\right]h^{2k} \quad (h \to 0). \tag{10}$$

Here B_s are the Bernoulli numbers.

The following are immediate consequences of Theorem 1:

$$f'(a) \neq f'(b) \;\Rightarrow\; T_n[f] - I[f] \sim \frac{1}{12}[f'(b) - f'(a)]h^2 = O(h^2) \quad (h \to 0).$$

$$f^{(2k-1)}(a) = f^{(2k-1)}(b), \quad 1 \leq k \leq m-1 \;\Rightarrow\; T_n[f] - I[f] = O(h^{2m}) \quad (h \to 0).$$

$$f^{(2k-1)}(a) = f^{(2k-1)}(b), \quad k = 1,2,\ldots \Rightarrow\; T_n[f] - I[f] = O(h^{\mu}) \quad (h \to 0), \;\forall\; \mu > 0,$$

and this happens when $f \in C^{\infty}(-\infty, \infty)$ and is $(b - a)$-periodic, for example.

The integral $\int_a^b f(x)dx$ can be computed to high accuracy by applying the *Richardson extrapolation process* to the sequence of trapezoidal rule approximations $\{T_{2^n}[f]\}_{n=0}^{\infty}$ also taking into account the E–M expansion; this method is known as the *Romberg integration.* All this is treated in many numerical analysis books; see Atkinson,[2] Ralston and Rabinowitz,[3] and Stoer and Bulirsch,[4] for example. See also the book by Steffensen.[5] For a detailed treatment of extrapolation methods and their applications, see Sidi.[6]

2.2 Navot's Generalizations of Classical E–M Expansion

The first generalizations of the E–M expansion were given by Navot,[7,8] and these concern the cases in which $f(x)$ has endpoint singularities of algebraic and algebraic–logarithmic types, respectively. Navot's results were later rederived by Lyness and Ninham[9] using techniques involving generalized functions. We state the result pertaining to the algebraic case in the next theorem. We will deal with the algebraic–logarithmic case when discussing our recent generalizations of the E–M expansions later in this section.

Theorem 2. *Let $f(x)$ have algebraic endpoint singularities as in*

$$f(x) = (x - a)^{\alpha} g_a(x) = (b - x)^{\beta} g_b(x), \quad g_a \in C^{\infty}[a, b), \quad g_b \in C^{\infty}(a, b], \tag{11}$$

and $\Re\alpha, \Re\beta > -1$ and α, β not necessarily integers. Let also

$$I[f] = \int_a^b f(x)\, dx \quad \text{and} \quad \check{T}_n[f] = h\sum_{i=1}^{n-1} f(a + ih), \quad h = \frac{b-a}{n}. \tag{12}$$

(Note that $\check{T}_n[f]$ does not include $f(a)$ and $f(b)$; thus it is always defined.)

Then there holds

$$\check{T}_n[f] \sim I[f] + \sum_{k=0}^{\infty} \frac{\zeta(-\alpha-k)}{k!} g_a^{(k)}(a) h^{\alpha+k+1}$$
$$+ \sum_{k=0}^{\infty} (-1)^k \frac{\zeta(-\beta-k)}{k!} g_b^{(k)}(b) h^{\beta+k+1} \quad (h \to 0). \tag{13}$$

Here $\zeta(\omega)$ is the Riemann zeta function.[a]

Note that in case $\alpha = 0 = \beta$, we have $g_a(x) = f(x) = g_b(x)$, and with the help of the known results concerning the zeta function, namely,

$$\zeta(0) = -\frac{1}{2}; \ \zeta(-2m) = 0, \quad \zeta(1-2m) = -\frac{B_{2m}}{2m}, \quad m = 1, 2, \ldots, \tag{14}$$

we recover the (classical) E–M expansion of Theorem 1.

2.3 Recent Generalizations of E–M Expansions

In a recent series of papers by Sidi,[10–12] the E–M expansion has been extended to the most general case in which the integrand $f(x)$ is infinitely differentiable in the finite (open) interval (a, b) and is allowed to have arbitrary singular behavior of the algebraic–logarithmic types at the endpoints. These results, which we summarize as Theorem 3, contain all earlier ones as special cases, but are not contained in the latter. Before we state them, we mention that they are valid also when the integrals $\int_a^b f(x)\,dx$ are divergent hence do not exist in the regular sense, but they are defined in the sense of *Hadamard finite part* (HPF). In such cases, we continue to use the notation $I[f]$ to denote the HFP of the (divergent) integral $\int_a^b f(x)\,dx$. Of course, the HFP of a convergent integral is equal to the actual value of the integral.

Theorem 3. *Assume that $f \in C^{\infty}(a, b)$ and let*

$$\check{T}_n[f] = h \sum_{i=1}^{n-1} f(a + ih), \quad h = \frac{b-a}{n}. \tag{15}$$

[a] We recall that the Riemann zeta function $\zeta(\omega)$ is first defined via the convergent series $\zeta(\omega) = \sum_{k=1}^{\infty} k^{-\omega}$ for $\Re\omega > 1$, and then continued analytically to the ω-plane. It is analytic everywhere except at $\omega = 1$, where it has a simple pole with residue 1.

We consider the following two cases:

1. *Let f(x) have asymptotic expansions as $x \to a+$ and as $x \to b-$ as in*

$$\begin{aligned} f(x) &\sim K(x-a)^{-1} + \sum_{s=0}^{\infty} c_s (x-a)^{\gamma_s} \quad (x \to a+) \\ f(x) &\sim L(b-x)^{-1} + \sum_{s=0}^{\infty} d_s (b-x)^{\delta_s} \quad (x \to b-) \end{aligned} \tag{16}$$

where γ_s and δ_s are in general complex constants satisfying

$$\begin{aligned} &\Re\gamma_0 \le \Re\gamma_1 \le \Re\gamma_2 \le \cdots; \quad \gamma_s \ne -1, \quad \lim_{s\to\infty} \Re\gamma_s = \infty, \\ &\Re\delta_0 \le \Re\delta_1 \le \Re\delta_2 \le \cdots; \quad \delta_s \ne -1, \quad \lim_{s\to\infty} \Re\delta_s = \infty. \end{aligned} \tag{17}$$

Assume also that the asymptotic expansions in (16) are differentiable infinitely many times. Then

$$\begin{aligned} \check{T}_n[f] \sim I[f] + K(C - \log h) + &\sum_{\substack{s=0 \\ \gamma_s \notin \{2,4,\ldots\}}}^{\infty} c_s \zeta(-\gamma_s) h^{\gamma_s+1} \\ + L(C - \log h) + &\sum_{\substack{s=0 \\ \delta_s \notin \{2,4,\ldots\}}}^{\infty} d_s \zeta(-\delta_s) h^{\delta_s+1} \quad (h \to 0). \end{aligned} \tag{18}$$

Here C = 0.577... is Euler's constant.

2. *More generally, let f(x) have asymptotic expansions as $x \to a+$ and as $x \to b-$ as in*

$$\begin{aligned} f(x) &\sim \hat{P}(\log(x-a))(x-a)^{-1} + \sum_{s=0}^{\infty} P_s(\log(x-a))(x-a)^{\gamma_s} \quad (x \to a+) \\ f(x) &\sim \hat{Q}(\log(b-x))(b-x)^{-1} + \sum_{s=0}^{\infty} Q_s(\log(b-x))(b-x)^{\delta_s} \quad (x \to b-), \end{aligned} \tag{19}$$

where $\hat{P}(y)$, $P_s(y)$, $\hat{Q}(y)$, and $Q_s(y)$ are arbitrary polynomials in y, and γ_s and δ_s are in general complex satisfying

$$\begin{aligned} &\Re\gamma_0 \le \Re\gamma_1 \le \Re\gamma_2 \le \cdots; \quad \gamma_s \ne -1, \quad \lim_{s\to\infty} \Re\gamma_s = \infty, \\ &\Re\delta_0 \le \Re\delta_1 \le \Re\delta_2 \le \cdots; \quad \delta_s \ne -1, \quad \lim_{s\to\infty} \Re\delta_s = \infty, \end{aligned} \tag{20}$$

such that

$$\begin{aligned}\Re\gamma_{s+1} = \Re\gamma_s \quad &\Rightarrow \quad \deg P_{s+1} \le \deg P_s \quad \text{and} \\ \Re\delta_{s+1} = \Re\delta_s \quad &\Rightarrow \quad \deg Q_{s+1} \le \deg Q_s.\end{aligned} \tag{21}$$

Assume also that the asymptotic expansions in (19) are differentiable infinitely many times. Let

$$D_\omega \equiv d/d\omega; \quad \hat{P}(y) = \sum_{i=0}^{\hat{p}} \hat{c}_i y^i, \quad \hat{Q}(y) = \sum_{i=0}^{\hat{q}} \hat{d}_i y^i. \tag{22}$$

Then

$$\begin{aligned}\check{T}_n[f] \sim I[f] &+ \sum_{i=0}^{\hat{p}} \left[\sum_{r=i}^{\hat{p}} \binom{r}{i} \hat{c}_r \sigma_{r-i} \right] (\log h)^i - \sum_{i=0}^{\hat{p}} \hat{c}_i \frac{(\log h)^{i+1}}{i+1} \\ &+ \sum_{s=0}^{\infty} P_s(D_{\gamma_s}) \left[\zeta(-\gamma_s)\, h^{\gamma_s+1} \right] + \sum_{s=0}^{\infty} Q_s(D_{\delta_s}) \left[\zeta(-\delta_s)\, h^{\delta_s+1} \right] \\ &+ \sum_{i=0}^{\hat{q}} \left[\sum_{r=i}^{\hat{q}} \binom{r}{i} \hat{d}_r \sigma_{r-i} \right] (\log h)^i - \sum_{i=0}^{\hat{q}} \hat{d}_i \frac{(\log h)^{i+1}}{i+1} \quad (h \to 0).\end{aligned} \tag{23}$$

Here, σ_i are Stieltjes numbers defined via

$$\sigma_i = \lim_{n\to\infty} \left[\sum_{k=1}^{n} \frac{(\log k)^i}{k} - \frac{(\log n)^{i+1}}{i+1} \right], \quad i = 0, 1, \ldots.$$

Remark

1. By the asymptotic expansions in (16) and (19) being differentiable infinitely many times we mean that, for each integer $k = 1, 2, \ldots, f^{(k)}(x)$, the k-th derivative of $f(x)$, has asymptotic expansions as $x \to a+$ and as $x \to b-$ that are obtained by differentiating those in (16) and (19) formally term by term.
2. The function $f(x)$ in Theorem 1 is contained in part 1 of Theorem 3 as follows: Being in $C^\infty[a, b]$, $f(x)$ has Taylor series expansions at $x = a$ and $x = b$ and we have $\gamma_s = \delta_s = s$ and $c_s = f^{(s)}(a)/s!$ and $d_s = (-1)^s f^{(s)}(b)/s!$, $s = 0, 1, \ldots$. In addition, these Taylor series are differentiable infinitely many times.[b]

[b] If $f \in C^\infty[a-\epsilon, a+\epsilon]$ for some $\epsilon > 0$, then its Taylor series $\sum_{s=0}^{\infty} \frac{f^{(s)}(a)}{s!}(x-a)^s$ is differentiable infinitely many times whether it converges or diverges.

3. Similarly, the function $f(x)$ in Theorem 2 is contained in part 1 of Theorem 3 as follows: $\gamma_s = s + \alpha$, $\delta_s = s + \beta$, and $c_s = g_a^{(s)}(a)/s!$, $d_s = (-1)^s g_b^{(s)}(b)/s!$, $s = 0, 1, \ldots$.
4. The function $f(x) = g(x)(x-a)^{\alpha}\log(x-a)$ with $g \in C^{\infty}[a,b]$ (considered in Ref. 8) is contained in part 2 of Theorem 3 as follows: $\gamma_s = s+\alpha$, $\delta_s = s$ and $P_s(y) = [g^{(s)}(a)/s!]y$, $Q_s(y) = (-1)^s f^{(s)}(b)/s!$, $s = 0, 1, \ldots$.
5. The singular terms $K(x-a)^{-1}$ and $L(b-x)^{-1}$ in (16) and $\hat{P}(\log(x-a))(x-a)^{-1}$ and $\hat{Q}(\log(b-x))(b-x)^{-1}$ in (19) are not present in any of the earlier generalizations of the E–M expansion. (These were treated in Refs. 11 and 12 for the first time.)
6. $I[f]$ is the exact value of the integral $\int_a^b f(x)\,dx$ when this converges; it is the HFP of $\int_a^b f(x)\,dx$ otherwise. Here are a few examples relevant to us:

$$u(x) = (x-a)^{\omega},\quad \omega \neq -1 \;\Rightarrow\; I[u] = \frac{(b-a)^{\omega+1}}{\omega+1},$$
$$u(x) = (x-a)^{-1} \;\Rightarrow\; I[u] = \log(b-a),$$
$$u(x) = (\log(x-a))^{i}(x-a)^{\omega},\quad \omega \neq -1 \;\Rightarrow\; I[u] = \frac{d^i}{d\omega^i}\frac{(b-a)^{\omega+1}}{\omega+1},$$
$$u(x) = (\log(x-a))^{i}(x-a)^{-1} \;\Rightarrow\; I[u] = \frac{[\log(b-a)]^{i+1}}{i+1}.$$

$I[u]$ is the exact integral $\int_a^b u(x)\,dx$ if $\Re\omega > -1$, it is the HFP of $\int_a^b u(x)\,dx$ otherwise.
7. If $R(y) = \sum_{i=0}^{r} e_i y^i$ and $D_\omega \equiv d/d\omega$, then $R(D_\omega) = \sum_{i=0}^{r} e_i D_\omega^i$ is a linear differential operator of order r; therefore, for any function g that is sufficiently differentiable as a function of ω, we have $R(D_\omega)g \equiv \sum_{i=0}^{r} e_i (D_\omega^i g) = \sum_{i=0}^{r} e_i (d^i g/d\omega^i)$. Thus

$$R(D_\omega)\left[\zeta(-\omega)h^{\omega+1}\right] = h^{\omega+1}W(\log h),\quad W(y)\ \text{polynomial of degree}\ r\ \text{in}\ y.$$

As a result, the term $P_s(D_{\gamma_s})\left[\zeta(-\gamma_s)\,h^{\gamma_s+1}\right]$ in (23) is simply the product of h^{γ_s+1} and a polynomial in $\log h$ whose degree is precisely $\deg P_s$. Similarly, for $Q_s(D_{\delta_s})\left[\zeta(-\delta_s)\,h^{\delta_s+1}\right]$.
8. Note that with $f(x)$ as in part 1 of Theorem 3 and $K = L = 0$, when $\gamma_0, \delta_0 \notin \mathbb{Z}^+$, we have at worst $\check{T}_n[f] - I[f] = O(n^{-(\alpha+1)})$ as $n \to \infty$, where $\alpha = \min\{\Re\gamma_0, \Re\delta_0\}$.

We would like to emphasize that these remarks about Theorem 3 are relevant to Theorems 6 and 9 that we state in the sequel.

2.4 Applications to Computation of Singular Integrals

E–M expansions and their generalizations have been very useful in the development of numerical quadrature formulas of high accuracy for numerical computation of integrals $\int_a^b f(x)\,dx$, where $f(x)$ has a singularity at $x = t \in (a, b)$, which may be algebraic and/or logarithmic, such that $f(x)$ may or may not be integrable through $x = t$. Commonly occurring cases are those for which

$$f(x) = g(x)\log|x-t| + \tilde{g}(x), \tag{24}$$

$$f(x) = \frac{g(x)}{(x-t)^m}, \quad m = 1, 2\ldots, \tag{25}$$

$$f(x) = \frac{g(x)}{|x-t|^\beta}, \quad \beta \geq 1 \;\text{ arbitrary}, \tag{26}$$

$g(x)$ and $\tilde{g}(x)$ being well behaved in (a, b). The first case involving $\log|x-t|$ arises in so-called *weakly singular* Fredholm integral equations; the integral $\int_a^b f(x)\,dx$ exists in the regular sense in this case. The second case with $m = 1$ and $m = 2$ arises in so-called *singular* and *hypersingular* Fredholm integral equations: (i) when $m = 1$, $\int_a^b f(x)\,dx$ diverges but exists as a *Cauchy principal value* integral, and (ii) when $m = 2$, $\int_a^b f(x)\,dx$ diverges but exists as an *HPF* integral and is also called a *hypersingular* integral.

By manipulating the relevant generalized E–M expansions in suitable ways, numerical quadrature formulas for all these cases were developed and analyzed in the papers by Sidi and Israeli[13] and Sidi.[14–17] We do not go into the construction of these formulas here. We mention only the methods that were designed for functions $f(x)$ that are *periodic with period* $T = b - a$ and are infinitely differentiable for all x except $x = t \pm kT$, $k = 0, 1, 2, \ldots$. Letting $h = (b - a)/n$, we define the numerical quadrature formulas $Q_n[f]$ as follows:

- For $f(x)$ as in (24),

$$Q_n[f] = h\sum_{j=1}^{n-1} f(t+jh) + \tilde{g}(t)h + g(t)h\log\left(\frac{h}{2\pi}\right),$$

and we have the asymptotic expansion

$$Q_n[f] \sim I[f] + \sum_{k=1}^{\infty} w_k h^{2k+1} \quad (h \to 0); \quad w_k = -2\frac{\zeta'(-2k)}{(2k)!}g^{(2k)}(t).$$

Thus, $Q_n[f] - I[f] \sim w_1 h^3$ as $h \to 0$. We can now apply the Richardson extrapolation process to a sequence $\{Q_{n_i}[f]\}_{i=0}^{\infty}$ with $n_i = n_0 2^i$, $i = 0, 1, \ldots$, and some arbitrary n_0, for example, and obtain highly accurate approximations to $I[f]$.

- For $f(x)$ as in (25) with $m = 1$,

$$Q_n[f] = h\sum_{j=1}^{n} f(t + jh - h/2),$$

and we have

$$Q_n[f] - I[f] = O(h^{\mu}) \quad (h \to 0), \quad \forall \ \mu > 0,$$

which means that the error tends to zero faster than every positive power of h.

- For $f(x)$ as in (25) with $m = 2$,

$$Q_n[f] = h\sum_{j=1}^{n} f(t + jh - h/2) - \pi^2 g(t) h^{-1},$$

and wc havc

$$Q_n[f] - I[f] = O(h^{\mu}) \quad (h \to 0), \quad \forall \ \mu > 0,$$

which means that the error tends to zero faster than every positive power of h.

All three numerical quadrature formulas are used in the numerical solution of Fredholm integral equations mentioned above.

2.5 Further Remarks on E–M Expansion and Generalizations

1. The asymptotic expansions described in Theorems 1–3 concern the *trapezoidal rule* approximations $T_n[f]$ and $\check{T}_n[f]$ for $\int_a^b f(x)\,dx$. Similar but more general versions of these theorems that concern the so-called *offset trapezoidal rule* approximations that are defined as in

$$\widetilde{T}_n[f] = h\sum_{j=0}^{n-1} f(t + jh + \theta h), \quad h = \frac{b-a}{n}, \quad \theta \in [0,1],$$

have been considered in Refs. 5 (in connection with Theorem 1), 7–9 (in connection with Theorem 2), and 10–12 (in connection with Theorem 3). The cases we have mentioned in this work are precisely those corresponding to $\theta = 1$. $\theta = 1/2$ gives rise to *midpoint rule* approximations. Since the statements of the theorems for arbitrary θ are more involved, we do not discuss them here and refer the reader to the original works.

2. An important and interesting point to note concerning the asymptotic expansions given in all these sources is that, *provided f(x) is infinitely differentiable in the open interval (a, b)* [i.e., $f \in C^{\infty}(a,b)$], they are all determined completely by the asymptotic expansions of $f(x)$ as $x \to a+$ and $x \to b-$; nothing else is needed. This conclusion is valid for all our developments in the next sections too.

3. ERROR EXPANSIONS FOR GAUSS–LEGENDRE QUADRATURE

In the preceding section we were concerned with the trapezoidal rule approximations or their modifications to finite-range integrals with or without endpoint singularities of algebraic–logarithmic types. We now turn to their approximation by the Gauss–Legendre quadrature. For convenience of notation, we will consider integrals on the (standard) interval (−1, 1). We also assume that these integrals exist in the regular sense. Thus the integral $I[f]$ of $f(x)$ and the corresponding n-point Gauss–Legendre quadrature formula are

$$I[f] = \int_{-1}^{1} f(x)\,dx \quad \text{and} \quad G_n[f] = \sum_{i=1}^{n} w_{ni} f(x_{ni}), \tag{27}$$

where x_{ni} are the abscissas [i.e., the zeros of $P_n(x)$, the n-th Legendre polynomial] and w_{ni} are the corresponding weights. Recall that the x_{ni} are all in the open interval (−1, 1). This guarantees that $G_n[f]$ is well defined when $f(x)$ is continuous on (−1, 1) no matter how $f(x)$ behaves at $x = \pm 1$. The functions $f(x)$ we consider have this property.

It is known that if $f \in C^{\infty}[-1,1]$, then the error $G_n[f] - I[f]$ tends to zero as $n \to \infty$ faster than all negative powers of n, that is, $G_n[f] - I[f] = o(n^{-\mu})$ as $n \to \infty$ for every $\mu > 0$. In particular, when $f(z)$ is analytic in an open set of the z-plane that contains the interval [−1, 1] in its interior, there holds $G_n[f] - I[f] = O(e^{-\sigma n})$ as $n \to \infty$ for some $\sigma > 0$; see Davis and Rabinowitz [Ref. 18, p. 312].

When $f(x)$ has integrable endpoint singularities, however, the error tends to zero slowly, its rate of decay depending on the strength of the singularities. For example, when $f(x) = (1-x)^{\alpha} g(x)$, with $\Re\alpha > -1$ but $\alpha \neq 0, 1, \ldots,$ and

$g \in C^\infty[-1,1]$, it is known that the error is $O(n^{-2\alpha-2})$ as $n \to \infty$; see Davis and Rabinowitz [Ref. 18, p. 313]. A complete asymptotic expansion for $G_n[f]$ for this case was first derived by Verlinden in Ref. 19, theorem 1, by imposing on $f(z)$ some analyticity conditions. This expansion is reproduced in Theorem 4 that follows.

Theorem 4. *Let $f(x) = (1-x)^\alpha g(x)$, with $\Re\alpha > -1$ but $\alpha \neq 0, 1, \ldots$, and $g(z)$ analytic in an open set that contains the interval $[-1, 1]$ in its interior. Then $G_n[f]$ has the asymptotic expansion*

$$G_n[f] \sim I[f] + \sum_{k=1}^{\infty} a_k h^{\alpha+k} \quad (n \to \infty); \quad h = (n+1/2)^{-2}. \tag{28}$$

Here, a_k are some constants independent of n.

Verlinden also considers $G_n[f]$ for a more general case involving an algebraic–logarithmic endpoint singularity, namely, $f(x) = \log(1-x)(1-x)^\alpha g(x)$, and shows that the asymptotic expansion of $G_n[f]$ in this case is obtained by differentiating that of Theorem 4 with respect to α term by term.

We now apply Verlinden's theorem (Theorem 4) with $g(x) \equiv 1$.

Theorem 5. *Let $f_\omega^\pm(x) = (1 \pm x)^\omega$, $\Re\omega > -1$, but ω not an integer. Then $I[f_\omega^+] = I[f_\omega^-] = 2^{\omega+1}/(\omega + 1)$ and $G_n[f_\omega^+] = G_n[f_\omega^-]$ and*

$$G_n[f_\omega^\pm] \sim I[f_\omega^\pm] + \sum_{k=1}^{\infty} u_k(\omega) h^{\omega+k} \quad (n \to \infty), \tag{29}$$

for some functions $u_k(\omega)$ that are analytic in ω for $\Re\omega > -1$.[c] *The $u_k(\omega)$ are the same for both $f_\omega^+(x)$ and $f_\omega^-(x)$. If ω is a nonnegative integer, then $G_n[f_\omega^\pm] = I[f_\omega^\pm]$ for $n \geq (\omega + 1)/2$ and $u_k(\omega) = 0$ for all $k \geq 1$.*[d]

In Theorem 6 that we state next we consider functions $f(x)$ that have arbitrary algebraic–logarithmic endpoint singularities at one or both endpoints ± 1 that are basically as those in Theorem 3, without any analyticity assumption being made. Thus the class of functions we consider is much more general than that considered in Ref. 19 and contains the latter as a subclass. This theorem is due to Sidi,[20] and its proof makes use of Theorem 5.

[c] So far, no simple expression for the functions $u_k(\omega)$ is known.

[d] This follows from the fact that $G_n[f] = I[f]$ for all functions $f(x)$ that are polynomials of degree at most $2n - 1$.

Theorem 6. *Assume that* $f \in C^{\infty}(-1,1)$ *and let* $h = (n+1/2)^{-2}$ *as before, and define* $\mathbb{Z}^{+} = \{0,1,2,\ldots\}$. *Let the functions* $u_k(\omega)$ *be precisely those that appear in Theorem 5. We consider the following two cases:*

1. *Let* $f(x)$ *have asymptotic expansions as* $x \to -1$ *(from the right) and as* $x \to 1$ *(from the left) as in*

$$f(x) \sim \sum_{s=0}^{\infty} c_s(1+x)^{\gamma_s} \quad (x \to -1+),$$
$$f(x) \sim \sum_{s=0}^{\infty} d_s(1-x)^{\delta_s} \quad (x \to 1-), \tag{30}$$

where γ_s *and* δ_s *are in general complex and satisfy*

$$-1 < \Re\gamma_0 \le \Re\gamma_1 \le \Re\gamma_2 \le \cdots; \quad \lim_{s\to\infty} \Re\gamma_s = +\infty,$$
$$-1 < \Re\delta_0 \le \Re\delta_1 \le \Re\delta_2 \le \cdots; \quad \lim_{s\to\infty} \Re\delta_s = +\infty. \tag{31}$$

Assume also that the asymptotic expansions in (30) are differentiable infinitely many times. Then

$$G_n[f] \sim I[f] + \sum_{\substack{s=0 \\ \gamma_s \notin \mathbb{Z}^+}}^{\infty} c_s \sum_{k=1}^{\infty} u_k(\gamma_s) h^{\gamma_s+k}$$
$$+ \sum_{\substack{s=0 \\ \delta_s \notin \mathbb{Z}^+}}^{\infty} d_s \sum_{k=1}^{\infty} u_k(\delta_s) h^{\delta_s+k} \quad (n \to \infty). \tag{32}$$

2. *More generally, let* $f(x)$ *have asymptotic expansions as* $x \to -1$ *(from the left) and as* $x \to 1$ *(from the right) as in*

$$f(x) \sim \sum_{s=0}^{\infty} P_s(\log(1+x))(1+x)^{\gamma_s} \quad (x \to -1+),$$
$$f(x) \sim \sum_{s=0}^{\infty} Q_s(\log(1-x))(1-x)^{\delta_s} \quad (x \to 1-), \tag{33}$$

where $P_s(y)$ *and* $Q_s(y)$ *are arbitrary polynomials in* y, *and* γ_s *and* δ_s *are in general complex and satisfy*

$$-1 < \Re\gamma_0 \le \Re\gamma_1 \le \Re\gamma_2 \le \cdots; \quad \lim_{s\to\infty} \Re\gamma_s = +\infty,$$
$$-1 < \Re\delta_0 \le \Re\delta_1 \le \Re\delta_2 \le \cdots; \quad \lim_{s\to\infty} \Re\delta_s = +\infty, \tag{34}$$

such that

$$\begin{aligned} \Re\gamma_{s+1} = \Re\gamma_s \quad &\Rightarrow \quad \deg P_{s+1} \leq \deg P_s \ \ \textit{and} \\ \Re\delta_{s+1} = \Re\delta_s \quad &\Rightarrow \quad \deg Q_{s+1} \leq \deg Q_s. \end{aligned} \tag{35}$$

Assume also that the asymptotic expansions in (33) are differentiable infinitely many times. Then

$$\begin{aligned} G_n[f] \sim I[f] \; &+ \sum_{s=0}^{\infty} \sum_{k=1}^{\infty} P_s(D_{\gamma_s})\left[u_k(\gamma_s) h^{\gamma_s + k}\right] \\ &+ \sum_{s=0}^{\infty} \sum_{k=1}^{\infty} Q_s(D_{\delta_s})\left[u_k(\delta_s) h^{\delta_s + k}\right] \quad (n \to \infty). \end{aligned} \tag{36}$$

For the precise meaning of the terms $P_s(D_{\gamma_s})[u_k(\gamma_s)h^{\gamma_s+k}]$ and $Q_s(D_{\delta_s})[u_k(\delta_s)h^{\delta_s+k}]$, we refer the reader to Remark 7 following the statement of Theorem 3.

Note that, with $f(x)$ as in part 1 of Theorem 6, when $\gamma_0, \delta_0 \notin \mathbb{Z}^+$, we have at worst $G_n[f] - I[f] = O(n^{-2(\alpha+1)})$ as $n \to \infty$, where $\alpha = \min\{\Re\gamma_0, \Re\delta_0\}$.

When $f(x)$ is infinitely differentiable in the (closed) interval $[-1, 1]$, then $f(x)$ is precisely of the form described in (30) with the γ_s and δ_s in $\mathbb{Z}^+$, which implies that $u_k(\gamma_s) = 0$ and $u_k(\delta_s) = 0$ for all $k = 1, 2, \ldots$. That is, the asymptotic expansion in (32) is empty; therefore, $G_n[f] - I[f] = O(n^{-\mu})$ as $n \to \infty$ for every $\mu > 0$. This means that the error in $G_n[f]$ tends to zero faster than every negative power of n, which is a known result mentioned also above.

As is the case concerning the E–M expansion and all of its generalizations, an important and interesting point to note concerning the asymptotic expansions given in all these sources is that, *provided $f(x)$ is infinitely differentiable in the open interval* $(-1, 1)$ [i.e., $f \in C^\infty(-1,1)$], they are all determined completely by the asymptotic expansions of $f(x)$ as $x \to -1+$ and $x \to 1-$; nothing else is needed.

4. ASYMPTOTICS OF LEGENDRE POLYNOMIALS AND LEGENDRE SERIES COEFFICIENTS

Orthogonal polynomial expansions of functions appear in most branches of science and engineering and their partial sums serve as approximations to the functions involved. Of course, the quality of these approximations depends on the behavior of the coefficients in these expansions, and the behavior of the coefficients is determined by the smoothness properties of the functions being approximated. Here we are concerned with Legendre polynomial expansions on the interval $(-1,\ 1)$ specifically. As usual, $P_n(x)$ is the n-th Legendre polynomial standardized such that $P_n(1) = 1$, so that

$$\int_{-1}^{1} P_m(x)P_n(x)dx = \frac{1}{n+1/2}\delta_{m,n}, \quad m, n = 0, 1, \ldots. \tag{37}$$

4.1 Asymptotics of Legendre Polynomials and Related Integrals

We begin with the asymptotic analysis of Legendre polynomials in the open interval $(-1,\ 1)$. It is known that, for large n and for $x \in (-1,\ 1)$, the Legendre polynomial $P_n(x)$ behaves like a trigonometric function, as in

$$P_n(\cos\theta) = \left(\frac{2}{\pi n \sin\theta}\right)^{1/2} \sin\left(n\theta + \frac{\theta}{2} + \frac{\pi}{4}\right) \\ + O\left(n^{-3/2}\right) \quad (n \to \infty), \quad x = \cos\theta. \tag{38}$$

The following complete asymptotic expansion is given in Sidi [Ref. 21, theorem 3.2]:

Theorem 7. *There exist analytic functions $\phi_k(z)$ that are regular for $|z| = 1$, $z \neq \pm 1$, such that, with $\hat{n} = n + 1/2$ and arbitrary fixed $\epsilon \in (0,\ \pi/2)$,*

$$P_n(\cos\theta) \sim \Re\left\{ e^{i\hat{n}\theta} \sum_{k=0}^{\infty} \frac{\phi_k(e^{i\theta})}{\hat{n}^{k+1/2}} \right\} \quad (n \to \infty), \quad \text{uniformly for } \epsilon \le \theta \le \pi - \epsilon. \tag{39}$$

That is, for each $p = 0,\ 1,\ \ldots,$ and for $0 < \theta < \pi$, there holds

$$P_n(\cos\theta) = \Re\left\{ e^{i\hat{n}\theta} \left[\sum_{k=0}^{p-1} \frac{\phi_k(e^{i\theta})}{\hat{n}^{k+1/2}} + R_{p,n}(\theta) \right] \right\}, \tag{40}$$

where

$$R_{p,n}(\theta) = O\left(\hat{n}^{-p-1/2}\right) \quad (n \to \infty), \quad \text{uniformly for } \epsilon \le \theta \le \pi - \epsilon. \tag{41}$$

Actually, with $D_\theta = d/d\theta$, we have

$$\phi_k(e^{i\theta}) = (-1)^k \frac{2}{\pi^{1/2}} \binom{-1/2}{k} e^{i(\theta-\pi)/2} \sum_{s=0}^{k} \binom{k}{s} B_{k-s}^{(1/2)} \cdot \left\{ \left(\frac{i}{2} D_\theta\right)^s \left[\left(1 - e^{i2\theta}\right)^{-1/2} \right] \right\}$$
$$= (-1)^k \frac{2}{\pi^{1/2}} \binom{-1/2}{k} e^{i(\theta-\pi)/2} B_k^{(1/2)} \left(\frac{i}{2} D_\theta\right) \left[\left(1 - e^{i2\theta}\right)^{-1/2} \right]. \tag{42}$$

Here $B_s^{(\sigma)}$ and $B_s^{(\sigma)}(u)$ are generalized Bernoulli numbers and polynomials.[e]

It is easy to see that (38) is obtained from the first ($k = 0$) term of the summation in (40).

An expansion of the form similar to that in (40) is given in Szegő [Ref. 23, p. 196, theorem 8.21.9], and this expansion involves the powers $n^{-k-1/2}$ and does not provide the coefficients $\phi_k(z)$ explicitly. This should be compared with our expansion in (40) that involves the powers $(n+1/2)^{-k-1/2}$, for which the $\phi_k(z)$ are given explicitly.

The next theorem concerns the asymptotics of integrals of the form $\int_c^d f(x) P_n(x)\, dx$ with $-1 < c < d < 1$ and it is given in Ref. 21, theorem 4.2.

Theorem 8. *Let $-1 < c < d < 1$, and assume that $f \in C^\infty(c,d)$. Let $\alpha = \cos^{-1} d$ and $\beta = \cos^{-1} c$; clearly $0 < \alpha < \beta < \pi$. Assume that, as $x \to c+$ and as $x \to d-$, f(x) is such that $F(\theta) = \sin\theta \cdot f(\cos\theta)$ has the asymptotic expansions*

$$\begin{aligned} F(\theta) &\sim \sum_{s=0}^{\infty} U_s (\theta - \alpha)^{\rho_s} \quad (\theta \to \alpha+); \quad \alpha = \cos^{-1} d > 0, \\ F(\theta) &\sim \sum_{s=0}^{\infty} V_s (\beta - \theta)^{\sigma_s} \quad (\theta \to \beta-); \quad \beta = \cos^{-1} c < \pi, \end{aligned} \tag{43}$$

where

$$\begin{aligned} -1 < \Re\rho_0 \le \Re\rho_1 \le \Re\rho_2 \le \cdots; \quad \lim_{s\to\infty} \Re\rho_s = \infty, \\ -1 < \Re\sigma_0 \le \Re\sigma_1 \le \Re\sigma_2 \le \cdots; \quad \lim_{s\to\infty} \Re\sigma_s = \infty, \end{aligned} \tag{44}$$

---

[e] The generalized Bernoulli polynomials $B_s^{(\sigma)}(u)$ are defined via (see, for example, Andrews et al. [Ref. 22, p. 615])

$$\left(\frac{t}{e^t - 1}\right)^\sigma e^{ut} = \sum_{s=0}^{\infty} B_s^{(\sigma)}(u) \frac{t^s}{s!}, \quad |t| < 2\pi.$$

They satisfy $B_s^{(\sigma)}(\sigma - u) = (-1)^s B_s^{(\sigma)}(u)$; hence $B_s^{(\sigma)}(\sigma/2) = 0$ for $s = 1, 3, 5, \ldots$. $B_s^{(\sigma)}(0)$ are called the generalized Bernoulli numbers and are denoted by $B_s^{(\sigma)}$. Note that $B_0^{(\sigma)} = 1$ for all σ. In addition, $B_k^{(\sigma)}(u) = \sum_{s=0}^{k} \binom{k}{s} B_{k-s}^{(\sigma)} u^s$ for all k.

and U_s and V_s are nonzero constants. Assume also that these asymptotic expansions can be differentiated termwise an infinite number of times. With the functions $\phi_k(z)$ as in Theorem 7, for arbitrary $\theta \in [\alpha, \beta]$, let

$$\phi_{kj}(\theta) = \frac{1}{j!}\frac{d^j}{d\theta^j}\phi_k(e^{i\theta}), \quad j, k = 0, 1, \ldots, \tag{45}$$

and

$$\begin{aligned}
G_\mu^{(+)}(\theta;\omega) &= \frac{1}{2}\sum_{\substack{j,k\geq 0\\ j+k=\mu}} \mathrm{i}^{\omega+j+1}\,\phi_{kj}(\theta)\,\Gamma(\omega+j+1),\\
\hat{G}_\mu^{(+)}(\theta;\omega) &= \frac{1}{2}\sum_{\substack{j,k\geq 0\\ j+k=\mu}} (-\mathrm{i})^{\omega+j+1}\,\overline{\phi_{kj}(\theta)}\,\Gamma(\omega+j+1),\\
G_\mu^{(-)}(\theta;\omega) &= \frac{1}{2}\sum_{\substack{j,k\geq 0\\ j+k=\mu}} (-1)^j\,(-\mathrm{i})^{\omega+j+1}\,\phi_{kj}(\theta)\,\Gamma(\omega+j+1),\\
\hat{G}_\mu^{(-)}(\theta;\omega) &= \frac{1}{2}\sum_{\substack{j,k\geq 0\\ j+k=\mu}} (-1)^j\,\mathrm{i}^{\omega+j+1}\,\overline{\phi_{kj}(\theta)}\,\Gamma(\omega+j+1).
\end{aligned} \tag{46}$$

Then, with $\hat{n} = n + 1/2$, as $n \to \infty$,

$$\begin{aligned}
\int_c^d f(x)P_n(x)\,dx \sim\; & e^{i\hat{n}\alpha}\sum_{s=0}^{\infty} U_s \sum_{\mu=0}^{\infty}\frac{G_\mu^{(+)}(\alpha;\rho_s)}{\hat{n}^{\rho_s+\mu+3/2}} + e^{-i\hat{n}\alpha}\sum_{s=0}^{\infty} U_s \sum_{\mu=0}^{\infty}\frac{\hat{G}_\mu^{(+)}(\alpha;\rho_s)}{\hat{n}^{\rho_s+\mu+3/2}}\\
& + e^{i\hat{n}\beta}\sum_{s=0}^{\infty} V_s \sum_{\mu=0}^{\infty}\frac{G_\mu^{(-)}(\beta;\sigma_s)}{\hat{n}^{\sigma_s+\mu+3/2}} + e^{-i\hat{n}\beta}\sum_{s=0}^{\infty} V_s \sum_{\mu=0}^{\infty}\frac{\hat{G}_\mu^{(-)}(\beta;\sigma_s)}{\hat{n}^{\sigma_s+\mu+3/2}}.
\end{aligned} \tag{47}$$

Note again that the asymptotic expansion in (47) is determined completely by the asymptotic expansions in (43), nothing else being needed. In addition, it is the sum of four asymptotic expansions that are multiplied by $\cos n\alpha$, $\sin n\alpha$, $\cos n\beta$, and $\sin n\beta$.

4.2 Asymptotics of Legendre Series Coefficients

Let $\sum_{n=0}^{\infty} e_n[f]P_n(x)$ be the Legendre series of the function $f(x)$ on $(-1, 1)$, where

$$e_n[f]=(n+1/2)\int_{-1}^{1}P_n(x)f(x)\,dx,\quad n=0,1,\ldots. \tag{48}$$

It is known that when $f(x)$ and $|f(x)|^2$ are integrable on $(-1, 1)$, we have (see, for example, Szegő[23] or Freud[24]),

$$e_n[f]=o(\sqrt{n})\quad (n\to\infty). \tag{49}$$

When $f\in C^r[-1, 1]$ for some integer $r\geq 0$, then (see Sidi[25], Introduction)

$$e_n[f]=O\left(n^{-r+1/2}\omega_{f^{(r)}}(2/n)\right)\quad (n\to\infty), \tag{50}$$

where $\omega_g(\delta)$ stands for the modulus of continuity of $g(x)$ on $[-1, 1]$. (For moduli of continuity, see, for example, Davis [Ref. 26, pp. 7–8], or Lorentz [Ref. 27, pp. 43–46].) Clearly, when $f(x)$ is continuously differentiable only r times on $[-1, 1]$, the best we can say about $e_n[f]$ is (50), and that the smaller r is, the slower the convergence of the series to $f(x)$ becomes. Of course, neither (49) nor (50) give us the best possible information about the behavior of $e_n[f]$ as $n\to\infty$.

From (50), it is easy to see that when $f\in C^\infty[-1,1]$, $e_n[f]$ satisfies (51) for every $r>0$, and this implies that $e_n[f]$ tends to zero as $n\to\infty$ faster than every negative power of n, that is,

$$e_n[f]=O(n^{-\mu})\quad (n\to\infty),\quad \forall\ \mu>0. \tag{51}$$

In particular, when $f(z)$ is analytic in an open set of the z-plane that contains the interval $[-1, 1]$ in its interior, there holds

$$e_n[f]=O(e^{-\sigma n})\quad (n\to\infty),\quad \text{for some } \sigma>0. \tag{52}$$

4.2.1 Asymptotics of $e_n[f]$ in Presence of Endpoint Singularities

We now turn to functions $f(x)$ that are infinitely differentiable in the (open) interval $(-1, 1)$ but can have regular or general singular behavior at one or both of the endpoints $x=\pm 1$ as described in Theorem 6. In Theorem 10, we present a complete asymptotic expansion as $n\to\infty$ for $e_n[f]$ that is derived in Sidi.[25]

We begin with the following result that is analogous to Theorem 5. See Sidi [Ref. 25, theorem 2.1].

Theorem 9. *Let* $f_\omega^\pm(x)=(1\pm x)^\omega$, $\Re\omega>-1$, *but* ω *not an integer. Then*

$$(-1)^n e_n[f_\omega^+]=e_n[f_\omega^-]=2^\omega(2n+1)(-\omega)_n/(\omega+1)_{n+1},\quad n=0,1,\ldots, \tag{53}$$

and

$$(-1)^n e_n[f_\omega^+] = e_n[f_\omega^-] \sim \sum_{k=1}^{\infty} \nu_k(\omega) h^{\omega+k+1/2} \quad (n \to \infty); \quad h = (n+1/2)^{-2}, \tag{54}$$

for some functions $\nu_k(\omega)$ that are analytic in ω for $\Re\omega > -1$ that are given as in

$$\nu_k(\omega) = 2^{\omega+1} \frac{\Gamma(1+\omega)}{\Gamma(-\omega)} \frac{B_{2k}^{(\sigma)}(\sigma/2)}{(2k)!} \frac{\Gamma(2k+2\omega+2)}{\Gamma(2\omega+2)},$$
$$k = 0, 1, \ldots, \quad \sigma = -2\omega - 1. \tag{55}$$

Here $B_s^{(\sigma)}(u)$ is the s-th generalized Bernoulli polynomial. When $\omega \in \mathbb{Z}^+$, there holds $\nu_k(\omega) = 0$ for each $k = 0, 1, \ldots$; in this case, we also have $e_n[f_\omega^\pm] = 0$ for all $n > \omega$.

In Theorem 10, we use the notation of Theorem 6.

Theorem 10. *Assume that $f \in C^\infty(-1,1)$ and $h = (n+1/2)^{-2}$, and define $\mathbb{Z}^+ = \{0,1,2,\ldots\}$. Let the functions $\nu_k(\omega)$ be precisely those that appear in Theorem 9. We consider the following two cases:*

1. *Let f(x) have asymptotic expansions as $x \to -1$ and as $x \to 1$ precisely as in (30) and (31) of Theorem 6, and assume that these asymptotic expansions are infinitely differentiable. Then*

$$e_n[f] \sim \sum_{\substack{s=0 \\ \gamma_s \notin \mathbb{Z}^+}}^{\infty} c_s \sum_{k=1}^{\infty} \nu_k(\gamma_s) h^{\gamma_s+k+1/2} + (-1)^n \sum_{\substack{s=0 \\ \delta_s \notin \mathbb{Z}^+}}^{\infty} d_s \sum_{k=1}^{\infty} \nu_k(\delta_s) h^{\delta_s+k+1/2} \quad (n \to \infty). \tag{56}$$

2. *More generally, let f(x) have asymptotic expansions as $x \to -1$ and as $x \to 1$ precisely as in (33)–(35) of Theorem 6, and assume that these asymptotic expansions are infinitely differentiable. Then*

$$e_n[f] \sim \sum_{s=0}^{\infty} \sum_{k=1}^{\infty} P_s(D_{\gamma_s})[\nu_k(\gamma_s) h^{\gamma_s+k+1/2}] + (-1)^n \sum_{s=0}^{\infty} \sum_{k=1}^{\infty} Q_s(D_{\delta_s})[\nu_k(\delta_s) h^{\delta_s+k+1/2}] \quad (n \to \infty). \tag{57}$$

For the precise meaning of the terms $P_s(D_{\gamma_s})[\nu_k(\gamma_s)h^{\gamma_s+k}]$ and $Q_s(D_{\delta_s})[\nu_k(\delta_s)h^{\delta_s+k}]$, we refer the reader to Remark 7 following the statement of Theorem 3.

Note that, with $f(x)$ as in part 1 of Theorem 10, when $\gamma_0, \delta_0 \notin \mathbb{Z}^+$, we have at worst $e_n[f] = O(n^{-2(\alpha+1)})$ as $n \to \infty$, where $\alpha = \min\{\Re\gamma_0, \Re\delta_0\}$.

4.2.2 Asymptotics of $e_n[f]$ in Presence of Interior and Endpoint Singularities

Theorem 10 can be extended to situations in which $f(x)$ has a number of singularities in the (open) interval $(-1, 1)$. This problem has been treated in detail in Sidi[21] by assuming algebraic singularities. Since this treatment is complicated, we will be content with a brief description of the assumptions and a sketch of the results.

Let us assume first that $f(x)$ has asymptotic expansions at $x = \pm 1$ as given in (30). Let us assume also that $f(x)$ has algebraic singularities in $(-1, 1)$ at the points $x_1 > x_2 > \cdots > x_m$ and that $f \in C^\infty(-1,1)$ everywhere else in $(-1, 1)$. Let us also assume that $f(x)$ has asymptotic expansions as $x \to x_r\pm$, that is, as $x \to x_r$ from the right and from the left, of the forms

$$f(x) \sim \sum_{s=0}^{\infty} W_{rs}^{(\pm)} |x - x_r|^{\delta_{rs}^{(\pm)}} \quad (x \to x_r \pm), \tag{58}$$

and let us allow these expansions to be different. Making the change of variable $x = \cos\theta$ in (48), we can express $e_n[f]$ in the form

$$e_n[f] = (n + 1/2) \int_0^\pi F(\theta) P_n(\cos\theta)\, d\theta; \;\; F(\theta) = \sin\theta \cdot f(\cos\theta). \tag{59}$$

Then $F \in C^\infty(0, \pi)$ except at $\theta_1 < \theta_2 < \cdots < \theta_m$, where $\theta_r = \cos^{-1} x_r$, $r = 1, \ldots, m$, and it has asymptotic expansions as $\theta \to \theta_r\pm$, of the form

$$F(\theta) \sim \sum_{s=0}^{\infty} T_{rs}^{(\pm)} |\theta - \theta_r|^{\gamma_{rs}^{(\pm)}} \quad (\theta \to \theta_r \pm), \tag{60}$$

where the $\gamma_{rs}^{(\pm)}$ satisfy

$$-1 < \Re\gamma_{r0}^{(\pm)} \le \Re\gamma_{r1}^{(\pm)} \le \Re\gamma_{r2}^{(\pm)} \cdots; \;\; \lim_{s\to\infty} \Re\gamma_{rs}^{(\pm)} = 0. \tag{61}$$

Note that $\gamma_{rs}^{(\pm)}$ are determined by the $\delta_{rs}^{(\pm)}$. The asymptotic expansion of $e_n[f]$ can be obtained by using Theorems 8 and 10 and so-called neutralizing functions. We do not go into any detail here and refer the reader to Sidi [Ref. 21, theorem 4.2]. We only mention that each singular point $\theta = \theta_r$ contributes two asymptotic expansions in powers of $\hat{n} = n + 1/2$, one multiplied by $e^{i\hat{n}\theta_r}$ and the other multiplied by $e^{-i\hat{n}\theta_r}$. The contribution from the endpoints $x = \pm 1$ is precisely that given in Theorem 9. See Ref. 21 for details.

REFERENCES

1. Sidi, A. Variable Transformations and Gauss-Legendre Quadrature for Integrals With Endpoint Singularities. *Math. Comput.* **2009**, *78*, 1593–1612.
2. Atkinson, K. E. *An Introduction to Numerical Analysis*, 2nd ed.; John Wiley & Sons Inc.: New York, 1989.
3. Ralston, A.; Rabinowitz, P. *A First Course in Numerical Analysis*, 2nd ed.; McGraw-Hill: New York, 1978.
4. Stoer, J.; Bulirsch, R. *Introduction to Numerical Analysis*, 3rd ed.; Springer-Verlag: New York, 2002.
5. Steffensen, J. F. *Interpolation*. Chelsea: New York, 1950.
6. Sidi, A. Practical Extrapolation Methods: Theory and Applications. no. 10. *Cambridge Monographs on Applied and Computational Mathematics;* Cambridge University Press: Cambridge, 2003.
7. Navot, I. An Extension of the Euler-Maclaurin Summation Formula to Functions With a Branch Singularity. *J. Math. Phys.* **1961**, *40*, 271–276.
8. Navot, I. A Further Extension of the Euler-Maclaurin Summation Formula. *J. Math. Phys.* **1962**, *41*, 155–163.
9. Lyness, J. N.; Ninham, B. W. Numerical Quadrature and Asymptotic Expansions. *Math. Comput.* **1967**, *21*, 162–178.
10. Sidi, A. Euler-Maclaurin Expansions for Integrals With Endpoint Singularities: A New Perspective. *Numer. Math.* **2004**, *98*, 371–387.
11. Sidi, A. Euler-Maclaurin Expansions for Integrals With Arbitrary Algebraic Endpoint Singularities. *Math. Comput.* **2012**, *81*, 2159–2173.
12. Sidi, A. Euler-Maclaurin Expansions for Integrals With Arbitrary Algebraic-Logarithmic Endpoint Singularities. *Constr. Approx.* **2012**, *36*, 331–352.
13. Sidi, A.; Israeli, M. Quadrature Methods for Periodic Singular and Weakly Singular Fredholm Integral Equations. *J. Sci. Comput.* **1988**, *3*, 201–231. Originally appeared as Technical Report No. 384, Computer Science Dept., Technion-Israel Institute of Technology, (1985), and also as ICASE Report No. 86-50 (1986).
14. Sidi, A. Comparison of Some Numerical Quadrature Formulas for Weakly Singular Periodic Fredholm Integral Equations. *Computing* **1989**, *43*, 159–170.
15. Sidi, A. Compact Numerical Quadrature Formulas for Hypersingular Integrals and Integral Equations. *J. Sci. Comput.* **2013**, *54*, 145–176.
16. Sidi, A. Analysis of Errors in Some Recent Numerical Quadrature Formulas for Periodic Singular and Hypersingular Integrals Via Regularization. *Appl. Numer. Math.* **2014**, *81*, 30–39.
17. Sidi, A. Richardson Extrapolation on Some Recent Numerical Quadrature Formulas for Singular and Hypersingular Integrals and Its Study of Stability. *J. Sci. Comput.* **2014**, *60*, 141–159.
18. Davis, P. J.; Rabinowitz, P. *Methods of Numerical Integration*, 2nd ed.; Academic Press: New York, 1984.

19. Verlinden, P. Acceleration of Gauss-Legendre Quadrature for an Integrand With an Endpoint Singularity. *J. Comput. Appl. Math.* **1997**, 77, 277–287.
20. Sidi, A. Asymptotic Expansions of Gauss-Legendre Quadrature Rules for Integrals With Endpoint Singularities. *Math. Comput.* **2009**, *78*, 241–253.
21. Sidi, A. Asymptotic Expansions of Legendre Series Coefficients for Functions With Interior and Endpoint Singularities. *Math. Comput.* **2011**, *80*, 1663–1684.
22. Andrews, G. E.; Askey, R.; Roy, R. *Special Functions*; Cambridge University Press: Cambridge, 1999.
23. Szegő, G. Orthogonal Polynomials, 4th ed.; *American Mathematical Society Colloquium Publications*, Vol. 23; American Mathematical Society: Providence, Rhode Island, 1975.
24. Freud, G. *Orthogonal Polynomials*; Pergamon Press: New York, 1971.
25. Sidi, A. Asymptotic Expansions of Legendre Series Coefficients for Functions With Endpoint Singularities. *Asymptot. Anal.* **2009**, *65*, 175–190.
26. Davis, P. J. *Interpolation and Approximation*; Dover: New York, 1975.
27. Lorentz, G. G. *Approximation of Functions*; Chelsea Publishing Company: New York, 1986.

CHAPTER THREE

The Kinetic Energy Pauli Enhancement Factor and Its Role in Determining the Shell Structure of Atoms and Molecules

Eduardo V. Ludeña[*,†,1], **Darío Arroyo**[*], **Edison X. Salazar**[‡], **Jorge Vallejo**[*]

[*]Center of Nanotechnology Research and Development, CIDNA, Escuela Superior Politécnica del Litoral, ESPOL, Guayaquil, Ecuador

[†]Centro de Química, Instituto Venezolano de Investigaciones Científicas, IVIC, Caracas, Venezuela

[‡]Theoretical Chemistry, Zernike Institute for Advanced Materials, University of Groningen, Groningen, The Netherlands

[1]Corresponding author: e-mail address: popluabe@yahoo.es

Contents

1. Introduction 60
2. The Concepts of Shell Structure and the Pauli Potential 61
   2.1 Shell Structure 61
   2.2 The Pauli Potential 63
3. The Pauli Enhancement Factor and Shell Structure 64
   3.1 Representation in Terms of One-Particle Orbitals 64
   3.2 Representation by Means of Local-Scaling Transformations 67
   3.3 The Explicit Functional Representation for T_p in Terms of the Liu–Parr Expansion 71
4. Discussion 74
Acknowledgments 75
References 75

Abstract

We deal with different representations of the noninteracting kinetic energy functional for the purpose of examining their effect upon the generation of shell structure in atoms. We decompose the noninteracting functional into a Weizsacker term plus a Pauli term where the latter is written as a product of the Thomas–Fermi $\rho^{5/3}(r)$ times the Pauli enhancement factor $F_p[\rho]$. We examine the behavior of $F_p[\rho]$ when it is given in terms of a Hartree–Fock orbital representation, of density-dependent orbitals generated through local-scaling transformations, and of the Liu–Parr power series

ISSN 0065-3276
https://doi.org/10.1016/bs.aiq.2017.05.002

expansion. In the latter, we compare the cases when the expansion coefficients have been expanded in an all-shell vs a shell-by-shell procedure. We apply these approximations to the aluminum atom. In particular, for this case, we examine in these different approximations, the role of the Pauli enhancement factor for the production of shell structure.

1. INTRODUCTION

In Kohn–Sham theory the introduction of the noninteracting kinetic energy functional T_s into the energy expression leads, upon functional differentiation, to a collection of one-particle equation whose solutions are the Kohn–Sham orbitals. This arises from the fact that $T_s[\{\phi_i\}_{i=1}^N]$ is naturally a functional of the occupied one-particle orbitals. However, one of the aims of density functional theory is to express the energy solely as a functional of the one particle density. Of course, the stumbling block is that as yet no satisfactory expression for $T_s[\rho]$ has been found, where $T_s[\rho]$ is a sufficiently accurate functional of only the one-electron density. Attaining this functional is essential for the development of, e.g., orbital-free molecular dynamics.

The shell structure of the electron distribution of atoms and molecules is an important ingredient in determining many chemical and physical properties of these many-electron systems. For example, the periodicity observed in the properties of the elements filling up the periodic table is linked to the formation of filled inner atomic shells and to the presence of valence electrons characterizing the groups or families. Although the existence of shell structure had been successfully hypothesized before the advent of quantum mechanics, its proper understanding and justification was only attained within its context. The shell structure arises from the Pauli exclusion principle and it is implicit already in the Schrödinger equation. It appears naturally in one-electron densities extracted from approximate theoretical calculations based on orbitals, whether they belong to the traditional quantum chemistry methods or to the DFT ones. Of course, its existence has also been confirmed by experimental work.

The aim of this chapter is to assess how well different representations of $T_s[\rho]$ meet with the requirement of reproducing shell structure. $T_s[\rho]$ can be exactly decomposed into the Weizsäcker plus the Pauli terms

$$T_s[\rho] = T_W[\rho] + T_P[\rho] \tag{1}$$

where

$$T_W[\rho] = (1/8)\int d^3\mathbf{r}|\nabla_{\mathbf{r}}\rho(\mathbf{r})|^2/\rho(\mathbf{r}) \tag{2}$$

$$T_P[\rho] = \int d^3\mathbf{r}\rho^{5/3}(\mathbf{r})F_P[\rho;\mathbf{r}] \tag{3}$$

Bearing in mind that $T_W[\rho]$ is related to the Fisher information entropy,[1,2] which is a local term,[3,4] shell structure may be ascribed to the behavior of the Pauli enhancement factor $F_P[\rho;\mathbf{r}]$, a nonlocal term.[5] For this reason, we examine the aptness of various representations of the Pauli enhancement factor to describe shell structure of atoms. We consider an orbital approximation, based on the Hartree–Fock orbitals of Clementi–Roetti. We also examine the exact implicit functional for $F_P[\rho;\mathbf{r}]$ generated by means of local-scaling transformations. Finally, we examine the Liu–Parr power series approximation.

In Section 2 we discuss the development of shell structure and Pauli potential concepts. In Section 3 we present results showing the relationship between the Pauli enhancement term and shell structure in atoms. Finally, in Section 4 we present some conclusions.

2. THE CONCEPTS OF SHELL STRUCTURE AND THE PAULI POTENTIAL

2.1 Shell Structure

In what follows we briefly outline, somewhat chronologically, the development of some of the ideas and experimental facts relating charge density to shell structure.

In 1927, Pauling[6] proposed a view of many-electron atoms in terms of electrons which filled atomic shells according to the Pauli exclusion principle.[7,8] In these early years in the development of quantum mechanics only the hydrogen atom had been solved by Schrödinger[9] and thus Pauling's proposal was somewhat revolutionary. However, assuming electron shells in atoms existed had previously been adopted by Unsöld for the interpretation of X-ray spectra.[10] The concentration of electronic charge in shells effectively shielded the positive charge in the nucleus and allowed treatment of an individual electron placed outside these shells as an electron in a hydrogen-like atom having a screened nuclear potential. This idea was further developed by Slater[11] with the introduction of atomic shielding constants.

For a spherically symmetrical atom $D(r)dr = 4\pi r^2\rho(r)dr$ is the average electron charge in a spherical shell of thickness dr. This quantity shows charge concentration and depletion. Experimental evidence of the existence of shell structure was obtained by reproducing the radial distribution $D(r)$ of atoms from electron diffraction data.[12] On the other hand, the density $\rho(r)$ decreases monotonically with r and does not present shell structure traits.[13,14] A model based on piece-wise exponentially decreasing densities was applied to atoms by Wang and Parr.[15] The asymptotic behavior of $\rho(r)$ at the nucleus and at infinity has been determined by Hoffmann-Ostenhof et al.[16,17] The electron one-particle density has also been extracted from single-crystal X-ray diffraction data using multipole expansions around the ionic centers.[18] Sagar et al. in 1988 using Hartree–Fock orbitals for atoms showed that the maxima in $D(r)$ correlate well with the shell radii from the Bohr–Schrödinger theory of an atom.[19] They also were able to relate the potential $\nabla^2\rho^{1/2}/(2\rho^{1/2})$ to the shell structure in atoms when $\rho(r)$ is obtained from the Clementi–Roetti HF orbitals.[20]

The convexity of the atomic charge distribution was shown to hold for some but not all atoms with $Z \leq 54$.[21] This result was extended by Esquivel et al. to show also the nonconvexity of ρ from Xe to U.[22] Moreover, the concept of pseudo-convexity has been used to define the general structural property of atomic densities as this concept embodies all the structural features empirically attributed to ρ.[23]

Kohout et al. were able to show that the quantity $-|\nabla\rho(r)|/\rho(r)$ is sufficient to determine the whole shell structure of atoms.[24] Schmider et al. have found that there exists a correlation between the extremal points and the roots of the Laplacian of the structure factor, namely, $\nabla^2F(k)$, with Z.[25] The behavior of ideal shells (namely, those designed to hold the exact number of electrons required by the Aufbau principle) has been examined both in position and momentum space by Schmider et al.[26]

Parr and Zhou[27] propose "absolute hardness," as unifying concept for identifying shells and subshells in nuclei, atoms, molecules, and metallic clusters.

Another interesting line of work related to the properties of the one-particle density is based on Shannon information entropy of quantum-mechanical systems in central potentials.[28] Rigorous lower and upper bounds to the position and momentum Shannon entropies of many-electron systems have been determined.[29–31] The connection between Shannon information entropies and classical orthogonal polynomials has been investigated.[32,33] The shell-filling properties of information measures have also been analyzed.[34]

Shell structure has also been found experimentally in quantum dots. In effect, by probing with far-infrared and capacitance spectroscopy it has been possible to distinguish, by their characteristic excitation frequencies, quantum dot Helium from quantum dot Lithium.[35] Shell structure has also been observed in sodium nanowires.[36]

Shell structure may also be related to the localization properties of electrons as described by the ELF function. Shell separation is related to the ELF's minima and the presence of shells to its maxima.[37] Another indicator of shell structure is provided by the average local electrostatic potential function, $V(r)/\rho(r)$; shell boundaries are found to be related to the successively increasing maxima of this indicator.[38] More recently, a localized electron detector, LED, has been advanced. Its aim is to show the three-dimensional structure of the different bonds identified by topological analysis of the density.[39–41] Other chemical indicators such as the curvature of the electron position uncertainty, described in terms of functions of the density have also recently been examined.

In a recent work, Finzel[42–47] has tackled the problem of incorporating the Pauli exclusion principle in the design of kinetic energy functionals for orbital-free treatments, through the concept of ideal atomic shells. This very simple idea captures, somehow, important characteristics that one should include in energy functionals expressed in terms of the one-particle density.

2.2 The Pauli Potential

The origin of the concept of the Pauli potential may be traced back to the work of March and Murray in 1966[48] who noticed that the potentials obtained as functional derivatives of the von Weizsäcker plus the Thomas–Fermi terms were insufficient to fully explain the behavior of electrons in finite metals. More directly, it arises from the Schrödinger equation for the square root of the density[49–51] where it takes the place of the local potential coming from the corrections to the Thomas–Fermi functional. The Pauli potential also emerges as an implicit functional of the one-particle density in the derivation carried out by Kryachko and Ludeña in 1992, in the context of local-scaling transformations, of the equations for the square root of the density.[52] It was shown in 1994 that the Pauli potential must satisfy the inequality $v_p(\rho, \mathbf{r}) \geq 0$.[53] The effect of harmonic confinement on the Pauli potential and its relation to the differential virial theorem has been examined by March.[54]

This same author has reviewed the concept of the Pauli potential emphasizing its importance in relation to density functional theory, DFT.[55]

The recent importance of the Pauli potential is related to the formulation of orbital-free molecular dynamics, OF-MD, where, in addition to the Kohn–Sham potential of the usual DFT, there arises also the Pauli potential from the noninteracting kinetic energy functional. Clearly, an OF-MD treatment of many-electron systems has the advantage over the traditional Kohn–Sham method that it does not rely on orbitals. The equations one has to deal with in OF-MD are related to the density or to its square root. However, these equations contain the Pauli potential as an additional term. For this reason, the proper representation of the Pauli potential becomes a crucial problem in this formalism. Recently, research has focussed on the conditions, such as the cusp condition, that this potential must satisfy.[56] The Pauli potential has also been investigated in the context of Natural Orbital Functional Theory.[57] More recently, the relation between the Pauli potential and shell structure has been investigated by Finzel.[44] In a very interesting recent work, differential equations for the Pauli potential have been advanced. Clearly, such an approach presents an alternative to the traditional Kohn–Sham orbital theory.[58]

3. THE PAULI ENHANCEMENT FACTOR AND SHELL STRUCTURE

3.1 Representation in Terms of One-Particle Orbitals

Clearly, there is no problem in writing down an exact expression for T_s in terms of orbitals:

$$T_s[\{\phi_i(\mathbf{r})\}] = \frac{1}{2}\sum_{i=1}^{N} \int d^3\mathbf{r} \;\; \nabla_{\mathbf{r}}\phi_i^*(\mathbf{r})\nabla_{\mathbf{r}}\phi_i(\mathbf{r}) \tag{4}$$

Assuming that for atoms the orbitals are products of radial parts times spherical harmonics, namely,

$$\phi_i(\mathbf{r}) = R_i(r)\,Y_{l_i,m_i}(\theta,\varphi) \tag{5}$$

upon angular integration, Eq. (4) adopts the form:

$$T_s[\{R_i(r)\}] = \int drr^2\frac{1}{2}\left\{\sum_{i=1}^{N}\left(\frac{dR_i(r)}{dr}\right)^2 + \sum_{i=1}^{N}\frac{l_i(l_i+1)}{r^2}R_i^2(r)\right\} \tag{6}$$

Further, assuming that the one-particle density is averaged over the angular part such that $\rho(\mathbf{r}) \rightarrow \rho(r) = \sum_{i=1}^{N} (R_i(r))^2$ and $\nabla_{\mathbf{r}}\rho(\mathbf{r}) = \hat{1}_r \frac{d}{dr} \sum_{i=1}^{N} (R_i(r))^2$, we can rewrite the von Weizsäcker term as:

$$T_W[\{R_i(r)\}] = \frac{1}{2} \int drr^2 \left\{ \sum_{i=1}^{N} \sum_{j=1}^{N} R_i(r) \frac{dR_i(r)}{dr} R_j(r) \frac{dR_j(r)}{dr} / \sum_{i=1}^{N} (R_i(r))^2 \right\} \tag{7}$$

From Eqs. (6) and (7), after some algebra we can rewrite $T_P = T_s - T_W$ as follows:

$$\begin{aligned} T_P[\{R_i(r)\}] = & \frac{1}{4} \int drr^2 \frac{1}{\rho(r)} \sum_{i=1}^{N} \sum_{j=1}^{N} \left(R_i(r) \frac{dR_j(r)}{dr} - R_j(r) \frac{dR_i(r)}{dr} \right)^2 \\ & + \frac{1}{2} \int drr^2 \sum_{i=1}^{N} \frac{l_i(l_i+1)}{r^2} R_i^2(r) \end{aligned} \tag{8}$$

In order to keep the notation similar to that used for local-scaling transformations (LS-DFT, see below)[59] we express the Pauli enhancement factor $F_p[\rho(r);r]$ of Eq. (3), which in this case is a functional of the radial orbitals, as:

$$F_P[\{R_i(r)\}] = (\tau_N[\{R_i(r)\}] + \kappa_N[\{R_i(r)\}]) \tag{9}$$

where

$$\tau_N[\{R_i(r)\}] = \frac{1}{2\rho^{8/3}(r)} \sum_{i=1}^{N-1} \sum_{j=i+1}^{N} W_{ij}[\{R_i(r)\}] \tag{10}$$

$$\kappa_N[\{R_i(r)\}] = \frac{1}{2\rho^{5/3}(r)} \sum_{i=1}^{N} \frac{l_i(l_i+1)}{r^2} R_i^2(r) \tag{11}$$

In Eq. (10) W_{ij} is defined as:

$$W_{ij}[\{R_i(r)\}] = \left(R_i(r) \frac{dR_j(r)}{dr} - R_j(r) \frac{dR_i(r)}{dr} \right)^2 \tag{12}$$

3.1.1 Application to the Al Atom

Aluminum, $Z = 13$ has the following electronic configuration for its ground state: $[1s^2 2s^2 2p^6 3s^2 3p^1]$.

The explicit expression for the modulating factor $\tau_N[\{R_i(r)\}]$ of aluminum is given by

$$\tau_N[\{R_i(r)\}] = \{\tau_{KL} + \tau_{KM} + \tau_{LL} + \tau_{LM} + \tau_{MM}\} \tag{13}$$

where we have collected the terms corresponding to each atomic shell and inter-shell. These terms are defined by

$$\begin{aligned} \tau_{IJ} &= \frac{1}{2\rho_g^{8/3}(f)}\omega_{IJ} \\ \omega_{KL} &= 4W_{1s2s} + 12W_{1s2p} \\ \omega_{KM} &= 4W_{1s3s} + 2W_{1s3p} \\ \omega_{LL} &= 12W_{2s2p} \\ \omega_{LM} &= 4W_{2s3s} + 2W_{2s3p} + 4W_{2p3s} + 6W_{2p3p} \\ \omega_{MM} &= 2W_{3s3p} \end{aligned} \tag{14}$$

In what follows we examine in Fig. 1 the behavior of the modulating factor $\tau_N[\{R_i(r)\}]$ for the Al atom by comparing it with the radial distribution $4\pi\, r^2\rho_{HF}(r)$ (τ in the present case is up-scaled by a factor of 20). We observe that the first and second maxima of $\tau[\{R_i(r)\}]$ are placed, respectively, at the first and second minima of the radial distribution. It is clearly seen that these points correspond to the inter-shell boundaries. In the outer region we see that τ grows strongly and approaches infinity thus creating the

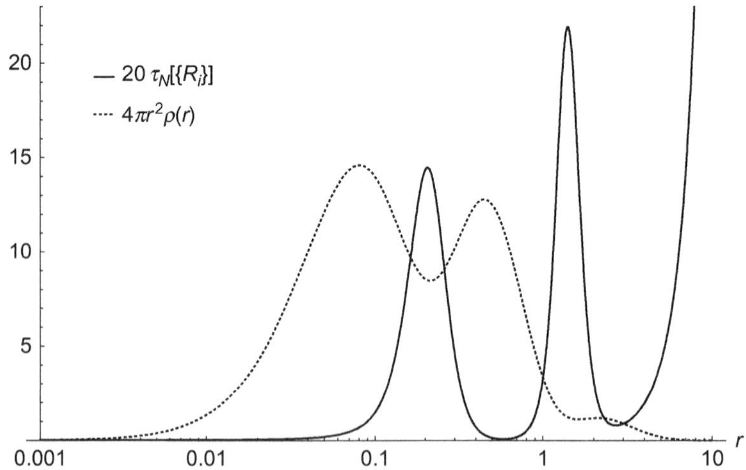

Fig. 1 The full Pauli kinetic energy modulating factor $\tau_N[\{R_i\}]$ for Al (*full line*) calculated by means of Eqs. (13) and (14). The Hartree–Fock radial density of Al is also depicted (*dashed line*).

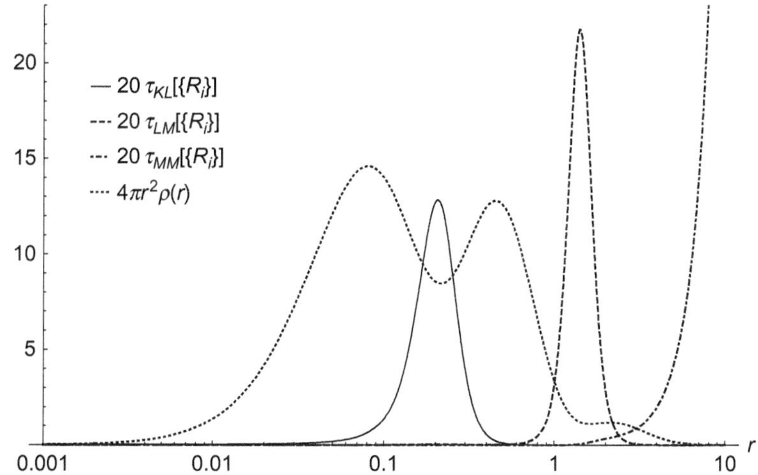

Fig. 2 Graphs of the Pauli kinetic energy modulating factor contributions for Al calculated by means of Eq. (14) for the *KL* and *LM* shell boundaries. The Hartree–Fock radial density of Al is also depicted.

basin that confines electrons to the outer shell. We may infer from Fig. 1 that the Pauli kinetic energy modulating factor bears an important role in the formation of the atomic shells. As seen from this figure, this factor is an essential ingredient for the generation of shell structure in the Al atom. In Fig. 2 we examine the behavior of the different inter-shell components of $\tau[\{R_i(r)\}]$, namely, τ_{KL}, τ_{KL}, and τ_{MM}. It is seen that the first maximum is almost entirely described by τ_{KL}. Similarly, the second one is accounted for by τ_{LM}. The outer border is represented entirely by τ_{MM}. In Fig. 3 we present the separate contribution of τ_{KM} and τ_{LL}. It is clearly seen in Fig. 3 that their contributions are negligible.

3.2 Representation by Means of Local-Scaling Transformations

Although no exact representation of T_s as an explicit functional of the one-particle density has been found so far, López–Boada et al.[59–61] have shown that it is possible to derive an exact albeit implicit form of this functional by resorting to general arguments involving local-scaling transformations. In this derivation, the functional $T_s[\rho]$ takes the form

$$T_s[\rho] = T_W[\rho] + \frac{1}{2}\int d^3\mathbf{r}\rho^{5/3}(\mathbf{r})A_N[\rho(\mathbf{r});\mathbf{r}] \quad (15)$$

where $A_N[\rho(\mathbf{r});\mathbf{r}]$ is an implicit functional of the one-particle density (see later). Note that the von Weizsacker term $T_W[\rho]$, which is an explicit

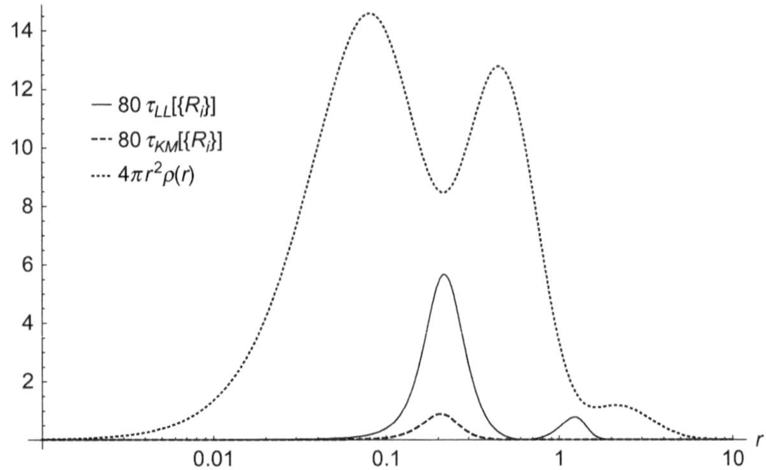

Fig. 3 Graph of the Pauli kinetic energy enhancement factor contribution for Al calculated by means of Eq. (14) for the terms τ_{LL} and τ_{KM} The Hartree–Fock radial density of Al is also depicted.

functional of the one-particle density emerges in a natural fashion in this derivation. The term $A_N[\rho(\mathbf{r}); \mathbf{r}]$ which, except for a numerical factor is equivalent to the Pauli kinetic energy enhancement factor, is however, nonlocal and is closely related to the shell structure of the system. In the case of atoms, for example, $A_N[\rho(\mathbf{r}); \mathbf{r}]$ describes the hills and basins which lead to potentials that localize electrons in shells.

Note that the local-scaling transformation version of density functional theory (LS-DFT)[62–64] is not based on the Hohenberg–Kohn theorems, and thus it is quite different from the Hohenberg–Kohn–Sham version of DFT. LS-DFT is a constructive approach, which allows us to generate energy density functionals that satisfy the variational principle. The ensuing functionals, however, have an implicit dependence on the one-particle density. Local-scaling transformations are coordinate transformations that can be expressed as functions of the one-particle density.[63,65]

A local-scaling transformation carries the vector $\mathbf{r}$ into the vector $\lambda(\mathbf{r})\mathbf{r} \equiv \mathbf{f}$ which has the same direction as $\mathbf{r}$. It can be shown that this transformed vector is $\mathbf{f} \equiv \mathbf{f}([\rho], \mathbf{r})$, namely, it is a function of the one-particle density. As a result, any function which depends on $\mathbf{r}$ may be transformed into another function which depends on $\mathbf{f}([\rho], \mathbf{r})$.

Applying this transformation to the arbitrary "generating" atomic orbital set $\{\phi_{g,i}(\mathbf{r}) = R_{g,i}(r) Y_{l_i,m_i}(\theta,\phi)\}_{i=1}^{N}$ one obtains the following set of transformed orbitals: $\{\phi_{\rho,i}(\mathbf{r}) = R_{\rho,i}(r) Y_{l_i,m_i}(\theta, \phi)\}_{i=1}^{N}$ where the locally scaled (density-dependent) radial functions are given by

$$R_{\rho,i}(r) = \sqrt{\frac{\rho(r)}{\rho_g(\lambda(r)r)}} R_{g,i}(\lambda(r)r), \tag{16}$$

In Eq. (16) only the radial (or angular-averaged densities) $\rho_g(r) = \sum_{i=1}^{N} |R_{g,i}(r)|^2$ and $\rho(r) = \sum_{i=1}^{N} |R_{\rho,i}(r)|^2$ corresponding to the initial and transformed orbitals, respectively, appear in the expression due to the fact that the vectors maintain the same direction.

A local-scaling transformation relates the densities $\rho(r)$ and $\rho_g(r)$ by means of the following first-order differential equation:

$$\frac{\rho(r)}{\rho_g(\lambda(r)r)} = \lambda(r)^3 \{1 + \mathbf{r} \cdot \nabla_{\mathbf{r}} \ln \lambda(r)\} \tag{17}$$

The single Slater determinant $\mathbf{\Phi}_\rho$ formed from these density-dependent orbitals is

$$\mathbf{\Phi}_\rho(\mathbf{r}_1, s_1, \ldots, \mathbf{r}_N, s_N) = \frac{\det}{N!^{1/2}} [\phi_{\rho,1}(\mathbf{r}_1)\sigma_1(s_1) \ldots \phi_{\rho,N}(\mathbf{r}_N)\sigma(s_N)] \tag{18}$$

Let us consider now the noninteracting kinetic energy functional given as the expectation value of the kinetic energy operator with respect to the single Slater $\mathbf{\Phi}_\rho$:

$$T_s[\mathbf{\Phi}_\rho] = \frac{1}{2} \sum_{i=1}^{N} \int_0^\infty dr r^2 \left[\left(\frac{dR_{\rho,i}(r)}{dr} \right)^2 + \frac{l_i(l_i+1)}{r^2} (R_{\rho,i}(r))^2 \right]. \tag{19}$$

The noninteracting kinetic energy expression given by Eq. (19) is an implicit functional of the one-particle density ρ, namely, $T_s[\mathbf{\Phi}_\rho] \equiv T_s[\rho]$ which after some manipulation becomes

$$\begin{aligned} T_s[\rho] = & \frac{1}{8} \int_0^\infty dr\, r^2 \frac{(\nabla \rho(r))^2}{\rho(r)} \\ & + \int_0^\infty dr\, r^2 \rho^{5/3}(r) \left(1 + \vec{r} \cdot \nabla \ln \lambda(r)\right)^{4/3} \tau_N[\rho] \\ & + \int_0^\infty dr\, r^2 \rho^{5/3}(r) \left(1 + \vec{r} \cdot \nabla \ln \lambda(r)\right)^{-2/3} \kappa_N[\rho]. \end{aligned} \tag{20}$$

where the modulating factors $\tau_N[\rho]$ and $\kappa_N[\rho]$ (with $f = \lambda(r)r$) are

$$\tau_N[\rho] = \frac{1}{\rho_g^{8/3}(f)} \frac{1}{2} \sum_{i<j} W_{ij}[\rho] \tag{21}$$

and

$$\kappa_N[\rho] = \frac{1}{\rho_g^{5/3}(f)} \sum_{i=1}^{N} \frac{l_i(l_i+1)}{2} \left(\frac{R_{g,i}(f)}{f} \right)^2, \tag{22}$$

where

$$W_{ij}[\rho] = \left(R_{g,n_i,l_i}(f) \frac{d\, R_{g,n_j,l_j}(f)}{df} - R_{g,n_j,l_j}(f) \frac{d\, R_{g,n_i,l_i}(f)}{df} \right)^2. \tag{23}$$

The Pauli enhancement factor $F_P([\rho]; \mathbf{r})$ is given by

$$\begin{aligned} F_P([\rho]; \mathbf{r}) &= \left(1 + \vec{r} \cdot \nabla \ln \lambda(r)\right)^{4/3} \tau_N \\ &+ \left(1 + \vec{r} \cdot \nabla \ln \lambda(r)\right)^{-2/3} \kappa_N \end{aligned} \tag{24}$$

These results have a striking similarity with those given by Eqs. (9), (10), and (11). The difference arises from the presence of density-dependent factors appearing in Eq. (24).

The explicit expression for the modulating factor $\tau[\rho]$ of aluminum is similar to those of Eqs. (13) and (14). It must be kept in mind, however, that in Eq. (24) $\tau[\rho]$ is a functional of the one-particle density and not of the orbitals. The generating orbitals, or broadly speaking, the generating polynomials are functions which depend implicitly on the one-particle density and which change when an approximate ρ evolves toward the exact one. The symmetry of the generating orbitals is the important feature introduced into the local-scaling expression.

In a previous work,[65] the construction of these local-scaling terms was carried out using generalized Slater-type orbitals as the generating one-particle functions. These orbitals are defined by:

$$R_{g,n_j,l_j}(r) \equiv N_{g,n_j,l_j}(f) f_{n_j,l_j}(r) \exp\left(-\alpha_{n_j,l_j} r^{\beta}\right) \tag{25}$$

where $f_{n_j,l_j}(r)$ are orthogonal polynomials.

We refer the reader to Fig. 4 (taken from Ref. 65), where the behavior of the modulating factor $\tau_N[\rho]$ for the Al atom is presented along with the radial distribution $4\pi\, r^2 \rho_{HF}(r)$. Note that $\tau[\rho]$ has been down-scaled by a factor of 200 in view of the fact that $\tau[\{R_i\}]$ and $\tau[\rho]$ differ by a factor (see Eq. 13). We observe, however, that as in the case for $\tau[\{R_i\}]$ the first and second maxima

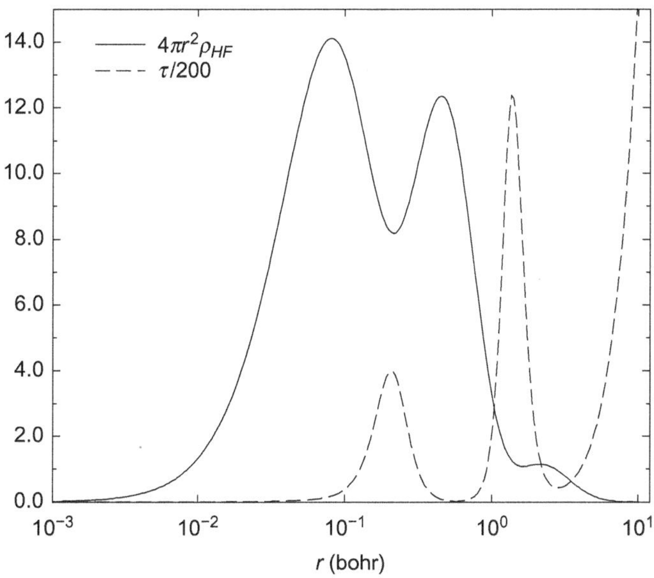

Fig. 4 The full Pauli kinetic energy modulating factor $\tau_N[\rho]$ for Al (*dashed line*) calculated by local-scaling transformations. The Hartree–Fock radial density of Al is also depicted (*full line*). *From Ludeña, E. V.; Karasiev, V.; López-Boada, R.; Valderrama, E.; Maldonado, J. Local-Scaling Transformation Version of Density Functional Theory: Application to Atoms and Diatomic Molecules.* J. Comput. Chem. ***1999,*** 20 *(1), 155–183.*

are placed, respectively, at the first and second minima of the radial distribution. Moreover, the tail behavior of this graph resembles that of the previous case in that it goes to infinity for large r. In Fig. 5, a detailed graph depicting the behavior of τ_{KL} and τ_{LM} is presented. Finally, in Fig. 6, the individual behavior of the terms τ_{KM}, τ_{LL}, and τ_{LM} is plotted.

3.3 The Explicit Functional Representation for T_p in Terms of the Liu–Parr Expansion

In recent work[66] we have applied the Liu and Parr[67] power series expansion originally used for the noninteracting kinetic energy functional T_s to represent, in the present case, the Pauli kinetic energy functional:

$$T_p[\rho] = \sum_{j=1}^{n} C_j \left[\int d^3\mathbf{r}\rho^{[1+2/3j]}(\mathbf{r}) \right]^j \tag{26}$$

It follows from this expression that the Pauli enhancement factor is given by

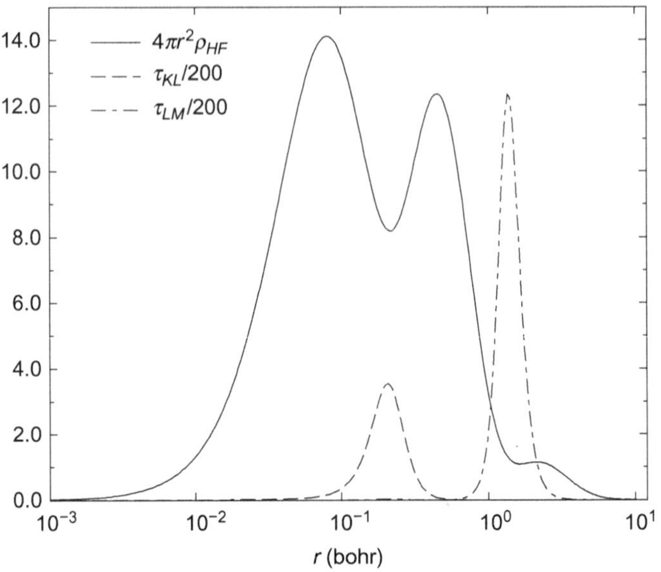

Fig. 5 Graphs of the Pauli kinetic energy modulating factor contributions for Al (*dashed line*) calculated by local-scaling transformations corresponding to the *KL* and *LM* shell boundaries. The Hartree–Fock radial density of Al is also depicted (*full line*). *From Ludeña, E. V.; Karasiev, V.; López-Boada, R.; Valderrama, E.; Maldonado, J. Local-Scaling Transformation Version of Density Functional Theory: Application to Atoms and Diatomic Molecules.* J. Comput. Chem. ***1999,*** 20 *(1), 155–183.*

$$F_p[\rho(\mathbf{r});\mathbf{r}] = \sum_{j=1}^{n} C_j \rho^{(2/3)(1/j-1)}(\mathbf{r}) \left[\int d^3\mathbf{r}\rho^{[1+2/3j]}(\mathbf{r}) \right]^{j-1} \tag{27}$$

The justification for applying the Liu–Parr expansion to the kinetic energy functional T_p is that upon simple coordinate scaling T_s given by Eq. (4) scales as $T_s^\lambda = \lambda^2 T_s$. Similarly, the von Weizsäcker term also scales as $T_W^\lambda = \lambda^2 T_W$. It follows that $T_p^\lambda = \lambda^2 T_p$. Hence, the application of the Liu–Parr expansion to T_P is justified.

In the original Liu–Parr expansion for T_s the coefficients were fixed by an optimal least-square fit to the "exact," Hartree–Fock values for atoms from H to Kr. The aptness of this expansion to represent adequately the shell structure of atoms was examined recently.[68] By and large, the three-term expansion of Liu and Parr reproduces the shape of the enhancement factor of the atoms considered quite closely, provided that a local contribution $\nu\nabla^2\rho(\mathbf{f})$ (where ν is a parameter) is added. Notice that the addition of this term does not contribute to the energy. The three-term Liu–Parr expansion, however, does not lead to satisfactory values of the

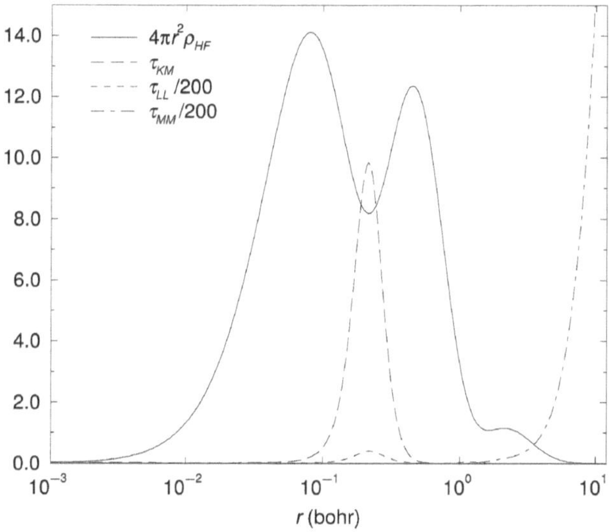

Fig. 6 Graph of the Pauli kinetic energy enhancement factor contribution for Al (*dashed line*) calculated by local-scaling transformations corresponding to the *KM*, *LL*, and *MM* terms. The Hartree–Fock radial density of Al is also depicted (*full line*). *From Ludeña, E. V.; Karasiev, V.; López-Boada, R.; Valderrama, E.; Maldonado, J. Local-Scaling Transformation Version of Density Functional Theory: Application to Atoms and Diatomic Molecules.* J. Comput. Chem. ***1999,*** 20 *(1), 155–183.*

noninteracting kinetic energy T_s for the thirty six atoms examined going from H to Kr.[68]

In Fig. 7, we present the enhancement factor $F_p[\rho]$ calculated for the Liu–Parr expansion with $n = 7$. We also present graphs for the radial distribution and for the Hartree–Fock approximation to the enhancement term $\tau[\{R_i\}]$. It is seen from Fig. 7 that the form of the enhancement factor for the aluminum atom, when it is calculated via the shell-by-shell optimization, is broadly located in the region where $\tau_{LM}[\{R_i\}]$ has its maximum. The fit is even looser when the factor uses the all-shell optimized coefficients.

In Table 1 we present the results of the Liu–Parr expansion for the noninteracting kinetic energies $T_s[\rho]$ and the Pauli kinetic energy $T_p[\rho]$. It is seen from this table that the approximations to the exact Hartree–Fock values of these quantities is very close such that the errors are $\Delta T_s = 0.0064$ Hartrees and $\Delta T_p = 0.0130$ Hartrees, respectively. These values correspond to the best approximation obtained to date. The fact that the enhancement factor does not perfectly fit the exact form given by the Hartree–Fock functional indicates that in the optimization process carried shell-by-shell the function adapted itself as much as possible to reproduce the form of $\tau_{LM}[\{R_i\}]$.

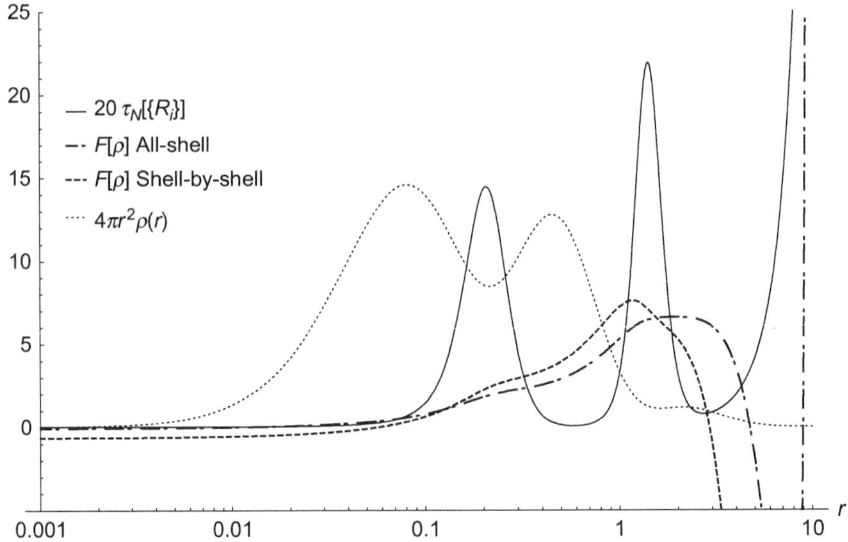

Fig. 7 The Pauli kinetic energy modulating factor $F_p[\rho]$ for Al corresponding the Liu–Parr power series expansion for n =7 evaluated with all-shell and shell-by-shell optimized coefficients. The Hartree–Fock radial density $\tau[\{R_i\}]$ of Al is also depicted.

Table 1 Noninteracting Kinetic Energy Values T_s and Pauli Kinetic Energy Values T_p (in Hartrees) for the Liu–Parr Power Series Approximations With All-Shells and Shell-by-Shell Coefficient Optimization for the Aluminum Atom

n	T_s^{HF}	T_s^{appa}	T_p^{appb}	T_p^{HF}	T_p^{appa}	T_p^{appb}
3	241.8720	245.9413	241.7640	85.1129	89.1755	84.9982
4	241.8720	241.7826	241.8630	85.1129	85.0169	85.0976
5	241.8720	241.6821	241.8240	85.1129	84.9163	85.0578
6	241.8720	241.4830	241.6767	85.1129	84.7166	85.1010
7	241.8720	241.8686	241.8656	85.1129	85.1029	85.0999

[a]All-shells.
[b]Shell-by-shell.
For comparison the respective Hartree–Fock values are included.

4. DISCUSSION

We have examined in the present work the various approximations to the Pauli enhancement factors which may be obtained by means of orbital representations, local-scaling representations, and Liu–Parr power series expansions and have related them to the appearance of shell structure in atoms. We have taken as a working example the aluminum atom.

The relation of an orbital representation with shell structure can be established in a direct fashion. Orbitals carry in them the symmetry which allows for an exact description of shell structure. This is the reason why calculation based on orbitals such as Hartree–Fock of Kohn–Sham DFT lead to the correct inclusion of shell structure. On the other hand, when we want to do away with orbitals and construct kinetic energy functionals which depend on the one-particle density, it is seen that the result is not at all obvious. It depends on the way the functional is constructed. For example, when we use generating orbitals having the correct orthogonality conditions, such as in the LS-DFT application discussed in this work, the resulting functionals (which are implicit functional of the one-particle density) maintain the shell structure characteristics as the original orbital ones. Thus, in this case, shell structure is incorporated in the functionals through the symmetry properties of the generating orbitals.

The problem at hand is, however, how to construct explicit functionals of the one-particle density which nevertheless keep the shell structure traits of the exact functionals. It is seen that through the use of the Liu–Parr power series expansions, this aim is only partially fulfilled. The shell-by-shell optimization leads to excellent values of the $T_s[\rho]$ and $T_p[\rho]$ functionals. However, the sharp curves denoting shell structure are not totally defined. The present results show that there is still much work to be done in the endeavor of obtaining both accurate and shell-structure-reproducing functionals.

ACKNOWLEDGMENTS

E.V.L. would like to gratefully acknowledge financial support to CIDNA from ESPOL authorities. E.X.S. is grateful to EACEA for allowing him to be part of the Erasmus Mundus Joint Master Degree Theoretical Chemistry and Computational Modelling (TCCM).

REFERENCES

1. Romera, E.; Dehesa, J. S.; Yánez, R. J. The Weizsäcker Functional: Some Rigorous Results. *Int. J. Quantum Chem.* **1995**, *56*(5), 627–632.
2. Romera, E.; Sánchez-Moreno, P.; Dehesa, J. S. The Fisher Information of Single-Particle Systems With a Central Potential. *Chem. Phys. Lett.* **2005**, *414*(4), 468–472.
3. Bohórquez, H. J.; Boyd, R. J. On the Local Representation of the Electronic Momentum Operator in Atomic Systems. *J. Chem. Phys.* **2008**, *129*(2), 024110.
4. Anderson, J. S.; Ayers, P. W. Hernandez, J. I. R.. How Ambiguous is the Local Kinetic Energy?*J. Phys. Chem. A* **2010**, *114*(33), 8884–8895.
5. Ludeña, E. V. On the Nature of the Correction to the Weizsäcker Term. *J. Chem. Phys.* **1982**, *76*(6), 3157–3160.

6. Pauling, L. The Theoretical Prediction of the Physical Properties of Many-Electron Atoms and Ions. Mole Refraction, Diamagnetic Susceptibility and Extension in Space. *Proc. R. Soc. A* **1927**, *114*, 181–211.
7. Pauli, W. Uber den Zusammenhang des Abschlusses der Elektronengruppen im Atom mit der Komplexstruktur der Spektren. *Z. Phys.* **1925**, *31*, 765.
8. Pauli, W. Exclusion Principle and Quantum Mechanics. 1946. Nobel Lecture.
9. Schrödinger, E. Quantisierung als Eigenwertproblem. *Ann. Phys.* **1926**, *385*(13), 437–490.
10. Ünsold, A. Beiträge zur Quantenmechanik der Atome. *Ann. Phys.* **1927**, *387*, 355–393.
11. Slater, J. C. Atomic Shielding Constants. *Phys. Rev.* **1930**, *36*, 57–64.
12. Bartell, L. S.; Brockway, L. O. Electron Diffraction Study of the Structure of Trimethylphosphine. *J. Chem. Phys.* **1960**, *32*(2), 512–515.
13. Weinstein, H.; Politzer, P.; Srebrenik, S. A Misconception Concerning the Electronic Density Distribution of an Atom. *Theoret. Chim. Acta* **1975**, *38*(2), 159–163.
14. Angulo, J. C.; Dehesa, J. S. Atomic Systems With a Completely Monotonic Electron Density. *Phys. Rev. A* **1991**, *44*(3), 1516.
15. Wang, W. P.; Parr, R. G. Statistical Atomic Models With Piecewise Exponentially Decaying Electron Densities. *Phys. Rev. A* **1977**, *16*, 891–902.
16. Hoffmann-Ostenhof, T.; Hoffmann-Ostenhof, M.; Ahlrichs, R. Schrödinger Inequalities and Asymptotic Behavior of Many-Electron Densities. *Phys. Rev. A* **1978**, *18*(2), 328.
17. Hoffmann-Ostenhof, M.; Thirring, W.; Hoffmann-Ostenhof, T. Simple Bounds to the Atomic One-Electron density at the Nucleus and to Expectation Values of One-Electron Operators. *J. Phys. B: At. Mol. Phys.* **1978**, *11*(19), L571.
18. Vidal, J. P.; Vidal-Valat, G.; Galtier, M.; Kurki-Suonio, K. X-Ray Study of the Charge Distribution in MgF2. *Acta Crystallogr. Sect. A: Cryst. Phys. Diffr. Theor. Gen. Crystallogr.* **1981**, *37*(6), 826–837.
19. Simas, A. M.; Sagar, R. P.; Ku, A. C. Smith, V. H., Jr. The Radial Charge Distribution and the Shell Structure of Atoms and Ions. *Can. J. Chem.* **1988**, *66*(8), 1923–1930.
20. Sagar, R. P.; Ku, A. C.; Smith, V. H., Jr. An Examination of the Shell Structure of Atoms and Ions as Revealed by the One-Electron Potential. *Can. J. Chem.* **1988**, *66*(4), 1005–1012.
21. Angulo, J. C.; Dehesa, J. S.; Gálvez, F. J. Atomic-Charge Convexity and the Electron Density at the Nucleus. *Phys. Rev. A* **1990**, *42*(1), 641.
22. Esquivel, R. O.; Sagar, R. P.; Smith, V. H., Jr.. Chen, J.; Stott, M. J. Pseudoconvexity of the Atomic Electron Density: A Numerical Study. *Phys. Rev. A* **1993**, *47*(6), 4735.
23. Esquivel, R. O.; Chen, J.; Stott, M. J.; Sagar, R. P. Smith, V. H., Jr.. Pseudoconvexity of the Atomic Electron Density: Lower and Upper Bounds. *Phys. Rev. A* **1993**, *47*(2), 936.
24. Kohout, M.; Savin, A.; Preuss, H. Contribution to the Electron Distribution Analysis. I. Shell Structure of Atoms. *J. Chem. Phys.* **1991**, *95*(3), 1928–1942.
25. Schmider, H.; Sagar, R. P. Smith, V. H., Jr.. Does the Atomic Structure Factor Show Atomic Structure?*J. Chem. Phys.* **1991**, *94*(6), 4346–4351.
26. Schmider, H.; Sagar, R. P. Smith, V. H., Jr.. An Investigation of Atomic Structure in Position and Momentum Space by Means of "ideal shells"*Can. J. Chem.* **1992**, *70*(2), 506–512.
27. Parr, R. G.; Zhou, Z. Absolute Hardness: Unifying Concept for Identifying Shells and Subshells in Nuclei, Atoms, Molecules and Metallic Clusters. *Acc. Chem. Res.* **1993**, *26*, 256–258.
28. Aptekarev, A. I.; Dehesa, J. S.; Yáñez, R. J. Spatial Entropy of Central Potentials and Strong Asymptotics of Orthogonal Polynomials. *J. Math. Phys.* **1994**, *35*(9), 4423–4428.
29. Yáñez, R. J.; Angulo, J. C.; Dehesa, J. S. Information Entropies of Many-Electron Systems. *Int. J. Quantum Chem.* **1995**, *56*(5), 489–498.

30. Angulo, J. C.; Antolín, J.; Sen, K. D. Fisher-Shannon Plane and Statistical Complexity of Atoms. *Phys. Lett. A* **2008**, *372*(5), 670–674.
31. Guerrero, A.; Sánchez-Moreno, P.; Dehesa, J. S. Upper Bounds on Quantum Uncertainty Products and Complexity Measures. *Phys. Rev. A* **2011**, *84*(4), 042105.
32. Jesus, S. D.; Martinez-Finkelshtein, A.; Sánchez-Ruiz, J. Quantum Information Entropies and Orthogonal Polynomials. *J. Comput. Appl. Math.* **2001**, *133*, 23–46.
33. Dehesa, J. S.; Martínez-Finkelshtein, A.; Sorokin, V. N. Information-Theoretic Measures for Morse and PöschlTeller Potentials. *Mol. Phys.* **2006**, *104*(4), 613–622.
34. López-Rosa, S.; Angulo, J. C.; Dehesa, J. S. Spreading Measures of Information-Extremizer Distributions: Applications to Atomic Electron Densities in Position and Momentum Spaces. *Eur. Phys. J. D* **2009**, *51*(3), 321–329.
35. Fricke, M.; Lorke, A.; Kotthaus, J. P.; Medeiros-Ribeiro, G.; Petroff, P. M. Shell Structure and Electron-Electron Interaction in Self-Assembled InAs Quantum Dots. *Europhys. Lett.* **1996**, *36*(3), 197.
36. Yanson, A. I.; Yanson, I. K.; van Ruitenbeek, J. M. Observation of Shell Structure in Sodium Nanowires. *Nature* **1999**, *400*(6740), 144–146.
37. Kohout, M.; Savin, A. Atomic Shell Structure and Electron Numbers. *Int. J. Quantum Chem.* **1996**, *60*(4), 875–882.
38. Sen, K. D.; Gayatri, T. V.; Toufar, H. Shell Structure of Atoms. *J. Mol. Struct. (THEOCHEM)* **1996**, *361*(1), 1–13.
39. Bohórquez, H. J.; Boyd, R. J. A Localized Electrons Detector for Atomic and Molecular Systems. *Theor. Chem. Acc.* **2010**, *127*(4), 393–400.
40. Bohórquez, H. J.; Matta, C. F.; Boyd, R. J. The Localized Electrons Detector as an Ab Initio Representation of Molecular Structures. *Int. J. Quantum Chem.* **2010**, *110*(13), 2418–2425.
41. Bohórquez, H. J.; Boyd, R. J.; Matta, C. F. Molecular Model With Quantum Mechanical Bonding Information. *J. Phys. Chem. A* **2011**, *115*(45), 12991–12997.
42. Finzel, K.; Kohout, M. How Does the Ambiguity of the Electronic Stress Tensor Influence Its Ability to Reveal the Atomic Shell Structure. *Theor. Chem. Acc.* **2013**, *132*(11), 1–13.
43. Finzel, K. Shell-Structure-Based Functionals for the Kinetic Energy. *Theor. Chem. Acc.* **2015**, *134*(9), 106.
44. Finzel, K. A Simple Approximation for the Pauli Potential Yielding Self-Consistent Electron Densities Exhibiting Proper Atomic Shell Structure. *Int. J. Quantum Chem.* **2015**, *115*(23), 1629–1634.
45. Finzel, K. Reinvestigation of the Ideal Atomic Shell Structure and Its Application in Orbital-Free Density Functional Theory. *Theor. Chem. Acc.* **2016**, *135*(4), 1–6.
46. Finzel, K. Local Conditions for the Pauli Potential in Order to yield Self-Consistent Electron Densities Exhibiting Proper Atomic Shell Structure. *J. Chem. Phys.* **2016**, *144*(3), 034108.
47. Finzel, K. Approximating the Pauli Potential in Bound Coulomb Systems. *Int. J. Quantum Chem.* **2016**, *116*(16), 1261–1266.
48. March, N. H.; Murray, A. M. Electronic Wave Functions Round a Vacancy in a Metal. *Proc. R. Soc. Lond. A Math. Phys. Sci* **1960**, *256*(1286), 400–415.
49. Levy, M.; Perdew, J. P.; Sahni, V. Exact Differential Equation for the Density and Ionization Energy of a Many-Particle System. *Phys. Rev. A* **1984**, *30*(5), 2745.
50. March, N. H. Differential Equation for the Ground-State Density in Finite and Extended Inhomogeneous Electron Gases. *Phys. Lett. A* **1985**, *113*(2), 66–68.
51. March, N. H. The Local Potential Determining the Square Root of the Ground-State Electron Density of Atoms and Molecules From the Schrödinger Equation. *Phys. Lett. A* **1986**, *113*(9), 476–478.

52. Kryachko, E. S.; Ludeña, E. V. Formulation of N-and ν-Representable Density functional Theory V. Exchange Only Self-Consistent Field. *Int. J. Quantum chem.* **1992**, *43*(6), 769–782.
53. Levy, M.; Görling, A. Recent Constrained-Search Advances for Approximating Density Functionals. *Philos. Mag. B* **1994**, *69*(5), 763–769.
54. March, N. H. The Pauli Potential in Relation to the Differential Virial Theorem With Application to Experiments on Ultracold Atomic Gases of Fermions. *Phys. Lett. A* **2008**, *372*(44), 6667–6669.
55. March, N. H. Concept of the Pauli Potential in Density Functional Theory. *J. Mol. Struct.: THEOCHEM* **2010**, *943*(1), 77–82.
56. Levämäki, H.; Nagy, A.; Kokko, K.; Vitos, L. Cusp Relation for the Pauli Potential. *Phys. Rev. A* **2014**, *90*(6), 062515.
57. Piris, M.; March, N. H. Is the Hartree-Fock Prediction that the Chemical Potential μ of Non-Relativistic Neutral Atoms is Equal to Minus the Ionisation Potential I Sensitive to Electron Correlation? *Phys. Chem. Liq.* **2015**, *53*(6), 696–705.
58. Levämäki, H.; Nagy, A.; Kokko, K.; Vitos, L. Alternative to the Kohn-Sham Equations: The Pauli Potential Differential Equation. *Phys. Rev. A* **2015**, *92*(6), 062502.
59. López-Boada, R.; Ludeña, E. V.; Karasiev, V.; Colle, R. Generation of Explicit Electron Correlation Functional by Means of Local Scaling Transformations. *Int. J. Quantum Chem.* **1998**, *69*(4), 439–450.
60. López-Boada, R.; Ludeña, E. V. Hartree-Fock Calculations in the Context of the Local-Scaling Transformation Version of Density Functional Theory. Applications to the Lithium and Beryllium Atoms. *Int. J. Quantum Chem.* **1998**, *96*, 485.
61. López-Boada, R.; Karasiev, V.; Ludeña, E. V.; Colle, R. Atomic Kinetic and Exchange Energy Functionals by Means of Local-Scaling Transformations. *Int. J. Quantum Chem.* **1998**, *69*(4), 503–512.
62. Kryachko, E. S.; Ludeña, E. V. L. *Energy Density Functional Theory of Many-Electron Systems*. Kluwer Academic Publishers: Dordrecht, 1990.
63. Kryachko, E. S.; Ludeña, E. V. Formulation of N-and ν-Representable Density-Functional Theory. I. Ground States. *Phys. Rev. A* **1991**, *43*(5), 2179.
64. Kryachko, E. S.; Ludeña, E. V. Density Functional Theory: Foundations Reviewed. *Phys. Rep.* **2014**, *544*(2), 123–239.
65. Ludeña, E. V.; Karasiev, V.; López-Boada, R.; Valderrama, E.; Maldonado, J. Local-Scaling Transformation Version of Density Functional Theory: Application to Atoms and Diatomic Molecules. *J. Comput. Chem.* **1999**, *20*(1), 155–183.
66. Ludeña, E. V.; Salazar, E. X.; Cornejo, M. H.; Arroyo, D. E.; Karasiev, V. V. Representation of the Pauli Kinetic Energy Functional in Terms of the Liu-Parr Power Series Expansion: Application to Atoms. *Phys. Rev. A* **2017**, (submitted for publication).
67. Liu, S.; Parr, R. G. Expansions of Density Functionals in Terms of Homogeneous Functionals: Justification and Nonlocal Representation of the Kinetic Energy, Exchange Energy, and Classical Coulomb Repulsion Energy for Atoms. *Phys. Rev. A* **1997**, *55*, 1792.
68. Salazar, E. X.; Guarderas, P. F.; Ludeña, E. V.; Cornejo, M. H.; Karasiev, V. V. Study of Some Simple Approximations to the Noninteracting Kinetic Energy Functional. *Int. J. Quantum Chem.* **2016**, *116*(17), 1313–1321.

CHAPTER FOUR

On the Laguerre Representation of Coulomb Functions and the Relation to Orthogonal Polynomials

Jessica A. Del Punta[*,†,‡], **Gustavo Gasaneo**[*], **Lorenzo U. Ancarani**[†,1]
[*]Universidad Nacional del Sur—IFISUR, Bahía Blanca, Argentina
[†]Théorie, Modélisation, Simulation, SRSMC, UMR CNRS 7565, Université de Lorraine, Metz, France
[‡]Universidad Nacional del Sur, Bahía Blanca, Argentina
[1]Corresponding author: e-mail address: ugo.ancarani@univ-lorraine.fr

Contents

1. Introduction 80
2. Definitions and Notations of Coulomb Wave Functions 82
3. Series Representation of Coulomb Wave Functions in Terms of Laguerre-Type Functions 84
   3.1 The Regular Coulomb Function 85
   3.2 The Irregular Coulomb Functions 86
4. Connection With Orthogonal Polynomials 91
   4.1 Generalized Pollaczek Polynomials 91
   4.2 Meixner–Pollaczek Polynomials 92
5. Summary 96
Appendix: Integral Involving Whittaker Function $W_{\chi,\frac{\mu}{2}}(z)$ 96
References 100

Abstract

We investigate the two-body Coulomb radial problem, providing extensions of known results and establishing a novel connection to orthogonal polynomials. The expansion in Laguerre-type functions of positive energy Coulomb solutions allows one to separate out the radial coordinate from the physical parameters. For the regular Coulomb wave function analytical coefficients are known to be directly connected to Pollaczek polynomials. It turns out that, simultaneously for the attractive and repulsive case, they can also be related to Meixner–Pollaczek polynomials. This allows us to provide a novel interpretation of these coefficients; considering the charge as a variable, we are able to establish orthogonality and completeness properties for these charge functions. We also investigate analytically Laguerre-type expansions of the irregular, incoming and outgoing Coulomb solutions; through a careful limit process we provide the expansion coefficients in closed form.

Advances in Quantum Chemistry, Volume 76
ISSN 0065-3276
https://doi.org/10.1016/bs.aiq.2017.06.005

1. INTRODUCTION

Atomic physics can be globally separated into two different branches. One is dedicated to the study of bound state systems. Many different tools have been developed to deal with one electron and many electron atoms. The second branch is associated with all those processes related to collision problems. Here again there is a number of available tools developed to successfully deal with the difficult mathematical issues related to the space unbound nature of the continuum states involved. In both branches, the solutions of the hydrogenic atom played and still plays a fundamental role. Indeed, many of the strategies make use, directly or indirectly, of the fact that the hydrogen wave functions are known analytically for both bound and continuum states. For bound states, in particular, polynomials associated with the Coulomb problem play a crucial role. This somehow extends to other atoms simply because for confined in space wave functions, the behavior assumption of an exponential times a polynomial looks physically and mathematically sound. Thus many properties of bound states can be related to the polynomials used to represent them. For positive energy states, on the other hand, atomic wave functions must satisfy asymptotically well-established conditions such as standing, outgoing, or incoming wave behaviors. When trying to use the solutions of the Coulomb problem to represent such behaviors, one encounters a number of mathematical difficulties. Among them, a formal difficulty arises from the fact that there is only one Coulomb solution being regular at the origin. In addition when trying to solve many electron problems, dealing with Coulomb potentials is mathematically cumbersome in many ways, not forgetting numerical convergence issues. One other obstacle may come from the evaluation of the necessary matrix elements that leads to nontrivial multivariable hypergeometric functions. One solution to avoid such integration problems, and actually to altogether avoid dealing with continuum type wave functions, is to consider expanding them on discrete basis (L^2 representations). The way the density of positive energies states is generated depends on the chosen discretization. Efficient strategies to represent the continuum such as the J-matrix (see, e.g., Ref. 1–5) or the convergent close coupling (see, e.g., Ref. 6 and references therein) use Laguerre-type basis: each element is given by the product of an exponential times a Laguerre polynomial whose order defines the basis index. By construction, the whole set of functions is regular at the origin and, at large distances, the exponential enforces a decreasing behavior which is regulated by a parameter. When used to expand, for example, Coulomb solutions, the required number of basis functions will depend on the largest radial distance up to which one needs to go. As such a distance grows the number of expansion coefficients increases too, simply because the never vanishing nature of a continuum function is represented by asymptotically vanishing basis functions. The generalized Fourier expansions in Laguerre-type functions has the following interesting property: it allows one to separate out the radial coordinate dependence from the physical parameters which enter only the expansion coefficients. Hence, the study of the physics may be done by inspection of the coefficients, and this separation makes life easier in applications such as collision studies.

In this contribution we extend some of the results corresponding to the J-matrix method. In particular, we are interested in generalized Fourier expansions of Coulomb wave functions in Laguerre-type functions, and focus on those solutions which are not regular at the origin. For the regular Coulomb solution, the expansion coefficients s_n are well established and have been studied in details.[4] Here we complete the investigation by providing in closed form the coefficients also for the irregular solutions possessing incoming or outgoing behavior at large distances. It is known that the presence of Whittaker functions in the Coulomb wave functions requires special attention because of the positive integer nature of the angular momentum. This issue has been tackled by Buchholz[7] through a subtle limit process. When looking at the generalized Fourier expansions we are faced with a similar problem, and we also tackle the difficult mathematical situation.[8]

This paper also provides a novel connection between generalized Fourier expansion coefficients and orthogonal polynomials. A relation between the coefficients s_n and generalized Pollaczek polynomials has been established, initially only for the repulsive case[4] and then extended to include also the attractive case.[9–12] Here, we extend the connection by identifying the expansion coefficients to Meixner–Pollaczek polynomials in a unified manner for both the attractive and repulsive cases. We find out that the coefficients can be also interpreted as "charge" functions, and two interesting properties (completeness and orthogonality) can be derived. As a consequence, two completeness relations can be established for the regular Coulomb function.

The rest of the chapter is arranged as follows. In Section 2, we provide the definitions and notations of the radial Coulomb problem. We proceed, in Section 3, with the study of expansions in Laguerre-type functions of Coulomb solutions. After recalling known results for the sine-like regular case, we provide novel expressions for those solutions which are irregular at the origin. Section 4 is dedicated to the connection of the expansion coefficients s_n to Pollaczek and Meixner–Pollaczek polynomials. A brief summary is provided in Section 5.

2. DEFINITIONS AND NOTATIONS OF COULOMB WAVE FUNCTIONS

We start by labeling

$$\mathbf{H}_{\mathbf{r}}^{\mathbf{C}} = -\frac{1}{2\mu}\frac{d^2}{dr^2} + \frac{\ell(\ell+1)}{2\mu\, r^2} + \frac{Z}{r} \tag{1}$$

the radial Coulomb Hamiltonian operator (ℓ is a given angular momentum). The solutions of the Coulomb Schrödinger equation

$$\left[\mathbf{H}_{\mathbf{r}}^{\mathbf{C}} - E\right]\Psi_C(\ell,k;r) = 0 \tag{2}$$

with $E > 0$, are known as Coulomb wave functions[13–15]. Defining the momentum k through $k^2 = 2\mu E$, we set

$$\eta(Z) = \frac{\mu Z}{k}, \tag{3a}$$

$$\sigma_C(\ell,Z) = \mathrm{Arg}[\Gamma(\ell+1+i\eta(Z))], \tag{3b}$$

$$N_C(\ell) = \frac{(2k)^{\ell+1}}{2}\,\frac{|\Gamma(\ell+1+i\eta(Z))|}{\Gamma(2\ell+2)}\, e^{-\frac{\pi}{2}\eta(Z)}, \tag{3c}$$

$$\tilde{N}_C^{(\pm)}(\ell) = e^{\frac{\pi}{2}\eta(Z)}\, e^{\pm i\left[\sigma_C(\ell,Z) - \ell\frac{\pi}{2}\right]} \tag{3d}$$

$$\varphi_C(\ell,k;r) = kr - \eta(Z)\ln(2kr) - \frac{\pi}{2}\ell + \sigma_C(\ell,Z), \tag{3e}$$

η standing for the so-called Sommerfeld parameter (it measures the strength of the Coulomb interaction for a given energy) and σ_C the Coulomb phase shift.

As the continuum functions, solution of Eq. (2) describe the scattering of a particle by a Coulomb potential, it is usual to consider a pair of independent solutions $F^{(s)}$, $G^{(c)}$ with sine-like (s) and cosine-like (c) asymptotic behavior, or equivalently two solutions $H^{(\pm)}$ having incoming $(-)$/outgoing $(+)$ behavior at large distances r. The function $F^{(s)}$ is the only one regular at the origin $r=0$. Analytical expressions for Coulomb wave functions can be given in terms of confluent hypergeometric functions of the first ${}_1F_1$ and second kind U[14] or, equivalently, in terms of Whittaker functions $M_{\chi,\ \mu/2}$, $W_{\chi,\ \mu/2}$.[7,14]

The solution

$$F^{(s)}(\ell,k;r)=\frac{N_C(\ell)}{(2ik)^{\ell+1}}\,M_{i\,\eta(Z),\ell+\frac{1}{2}}(2ikr)$$
$$=N_C(\ell)\,r^{\ell+1}e^{-ikr}\,{}_1F_1(\ell+1-i\eta(Z),2\ell+2;2ikr), \tag{4}$$

is a real function, is regular at the origin ($F^{(s)}(\ell,\ k;\ 0)=0$), and has a sine-like asymptotic behavior, $F^{(s)}(\ell,k;r)\overset{r\to\infty}{\sim}\sin(\varphi_C)$. The solution

$$G^{(c)}(\ell,k;r)=i\,\frac{N_C(\ell)}{(2ik)^{\ell+1}}\,M_{i\,\eta(Z),\ell+\frac{1}{2}}(2ikr)+\tilde{N}_C^{(-)}(\ell)\,W_{i\,\eta(Z),\ell+\frac{1}{2}}(2ikr) \tag{5}$$

is also a real function, is irregular at the origin, and has a cosine-like asymptotic behavior, $G^{(c)}(\ell,k;r)\overset{r\to\infty}{\sim}\cos(\varphi_C)$.

Combining these two functions it is possible to give another pair of independent solutions

$$H^{(\pm)}(\ell,k;r)=G^{(c)}(\ell,k;r)\pm i\,F^{(s)}(\ell,k;r). \tag{6}$$

In terms of Whittaker functions they are given by

$$H^{(+)}(\ell,k;r)\ =\ \tilde{N}_C^{(+)}(\ell)\,W_{-i\,\eta(Z),\ell+\frac{1}{2}}(-2ikr), \tag{7a}$$

$$H^{(-)}(\ell,k;r)\ =\ \tilde{N}_C^{(-)}(\ell)\,W_{i\,\eta(Z),\ell+\frac{1}{2}}(2ikr)=\left[H^{(+)}(\ell,k;r)\right]^*, \tag{7b}$$

they are complex functions, both irregular at $r=0$, and for $r \longrightarrow \infty$ behave as $H^{(\pm)}(\ell,k;r) \overset{r\to\infty}{\sim} e^{\pm i\varphi_C}$. The following relations between the Coulomb wave functions is easily deduced from Eq. (6):

$$F^{(s)}(\ell,k;r) = \frac{H^{(+)}(\ell,k;r) - H^{(-)}(\ell,k;r)}{2i}, \tag{8a}$$

$$G^{(c)}(\ell,k;r) = \frac{H^{(+)}(\ell,k;r) + H^{(-)}(\ell,k;r)}{2}. \tag{8b}$$

3. SERIES REPRESENTATION OF COULOMB WAVE FUNCTIONS IN TERMS OF LAGUERRE-TYPE FUNCTIONS

We are interested in generalized Fourier expansions[16,17]

$$\Psi_C(\ell,k;r) = \sum_{n=0}^{\infty} a_n\, \phi_n^L(\ell,\beta;r) \tag{9}$$

in terms of Laguerre-type functions

$$\phi_n^L(\ell,\beta;r) = N_{n,\ell}\, e^{-\beta r}\, (2\beta r)^{\ell+1}\, L_n^{2\ell+1}(2\beta r) \tag{10a}$$

$$= \frac{1}{N_{n,\ell}\ \Gamma(2\ell+2)}\, e^{-\beta r}\, (2\beta r)^{\ell+1}\, {}_1F_1(-n, 2\ell+2; 2\beta r). \tag{10b}$$

where L_n^{α} are associated Laguerre polynomials[14,18], β is a real positive parameter, and the normalization factor is given by $N_{n,\ell} = [n!/\Gamma(2\ell+2+n)]^{1/2}$. Since Laguerre-type functions are solutions of a Sturm–Liouville problem[16], they form an orthogonal and complete basis set,

$$\int_0^{\infty} \phi_p^L(\ell,\beta;r)\, \frac{1}{r}\, \phi_n^L(\ell,\beta;r)\, dr = \delta_{p,n}, \tag{11a}$$

$$\sum_{n=0}^{\infty} \phi_n^L(\ell,\beta;r_1)\, \frac{1}{r_1}\, \phi_n^L(\ell,\beta;r_2) = \delta(r_1 - r_2). \tag{11b}$$

As a consequence, the coefficients a_n are given by

$$a_n = \int_0^{\infty} \phi_n^L(\ell,\beta;r)\, \frac{1}{r}\, \Psi_C(\ell,k;r)\, dr. \tag{12}$$

Except for the angular momentum ℓ, the parameters Z, μ, k appear *only* in the coefficients a_n of the series expansion. The representation (9) of Ψ_C

provides therefore a clear separation of the dependence on the variable r from that on these physical parameters. In other words, the inputs of Schrödinger equation appear only in the coefficients a_n. This separation makes easier any physical analysis and a number of further calculations, including those of matrix elements that appear, for example, in collision studies.

We first present some known results for the case of the sine-like Coulomb wave function and then extend the study providing closed form expressions for the coefficients (12) corresponding to the series representation of the other solutions, namely $G^{(c)}$ and $H^{(\pm)}$.

3.1 The Regular Coulomb Function

For the sine-like Coulomb wave function $F^{(s)}$ the coefficients, indicated by s_n, have been presented and studied by Yamani and Fishman.[4] According to our notation, their explicit form is

$$\begin{aligned} s_n = (-1)^n \frac{N_C(\ell)}{N_{n,\ell}} \left(\frac{2\beta}{\beta^2 + k^2} \right)^{\ell+1} \\ \times \omega^{-n-i\eta(Z)} {}_2F_1\left(-n, \ell+1-i\eta(Z), 2\ell+2; 1-\omega^2\right) \end{aligned} \tag{13}$$

where we have introduced the parameter

$$\omega = \frac{\beta + ik}{\beta - ik}, \qquad \zeta = \text{Arg}(\omega), \tag{14}$$

and ${}_2F_1$ is the Gauss hypergeometric function.[14,19]

Yamani and Fishman[4] showed that the matrix representation of $[\mathbf{H}_\mathbf{r}^\mathbf{C} - E]$ in terms of not-normalized Laguerre-type functions $\phi_n^L / N_{n,\ell}$ is a tridiagonal matrix $\mathbf{J}$, called J-matrix. In our notation, the elements read

$$\begin{aligned} J_{p,n} &= \int_0^\infty \phi_p^L(\ell, \beta; r) \, [\mathbf{H}_\mathbf{r}^\mathbf{C} - E] \, \phi_n^L(\ell, \beta; r) \, dr \\ &= A_{n+1} \delta_{p,n+1} + B_n(Z) \, \delta_{p,n} + A_n \, \delta_{p,n-1}, \end{aligned} \tag{15}$$

where

$$A_n = \begin{cases} 1, & \text{if } n = 0, \\ \dfrac{N_{n,\ell}}{N_{n-1,\ell}} \dfrac{(\beta^2 + k^2)(2\ell + 1 + n)}{4\mu\beta}, & \text{if } n > 0, \end{cases} \tag{16a}$$

$$B_n(Z) = \frac{2\mu Z\beta + (\beta^2 - k^2)(\ell + 1 + n)}{2\mu\beta}. \tag{16b}$$

They also showed that the J-matrix is equivalent to a three-terms recurrence relation for the coefficients s_n. To obtain it, we replace the series representation

$$F^{(s)}(\ell,k;r)=\sum_{n=0}^{\infty} s_n\,\phi_n^L(\ell,\beta;r) \tag{17}$$

into the Coulomb differential equation (2), multiply both sides from the left by $\phi_m^L(\ell,\beta;r)$, and integrate over r, to find

$$\begin{aligned}\int_0^{\infty}\sum_{n=0}^{\infty} s_n\,\phi_m^L(\ell,\beta;r)\left[\mathbf{H}_{\mathbf{r}}^{\mathbf{C}}-E\right]\phi_n^L(\ell,\beta;r)\,dr \\ =\sum_{n=0}^{\infty} s_n\int_0^{\infty}\phi_m^L(\ell,\beta;r)\left[\mathbf{H}_{\mathbf{r}}^{\mathbf{C}}-E\right]\phi_n^L(\ell,\beta;r)\,dr=0.\end{aligned} \tag{18}$$

Note that above we made explicit the interchange of the series and the integral because, for the irregular Coulomb functions, this interchange will no longer be valid. Using result (15), we obtain the recurrence relation

$$A_{n+1}\,s_{n+1}+B_n(Z)\,s_n+A_n\,s_{n-1}=0, \tag{19}$$

setting, for $n=0$, $s_{-1}=0$. Thus, once the first element of the sequence (i.e., s_0) is known, one can find all the subsequent coefficients s_n. For this reason we give its explicit form

$$s_0=\frac{N_{0,\ell}}{2}\,|\Gamma(\ell+1+i\eta(Z))|\,e^{-\frac{\pi}{2}\eta(Z)}\,\omega^{-i\eta(Z)}\left(\frac{4\beta k}{\beta^2+k^2}\right)^{\ell+1}. \tag{20}$$

3.2 The Irregular Coulomb Functions

We now turn to the series representation (9) of functions $H^{(\pm)}$ and $G^{(c)}$ which, we recall, are irregular at the origin. Since Laguerre-type functions are regular at the origin such expansions are, on the other hand, convergent to zero at that point, so that some care will be needed. We start by performing the integral

$$h_n^{(\pm)}(\ell)=\int_0^{\infty}\phi_n^L(\ell,\beta;r)\,\frac{1}{r}\,H^{(\pm)}(\ell,k;r)\,dr \tag{21}$$

to obtain a Laguerre expansion of the incoming/outgoing wave $H^{(\pm)}$. The resulting series

$$\widetilde{H}^{(\pm)}(\ell,k;r)=\sum_{n=0}^{\infty}h_n^{(\pm)}(\ell)\,\phi_n^L(\ell,\beta;r) \tag{22}$$

converges pointwise to $H^{(\pm)}$ for $r>0$, and as a consequence of the regularity of the Laguerre-type functions at $r=0$, we have $\widetilde{H}^{(\pm)}(r=0)=0$.

In the series (22) we made explicit the ℓ parameter dependence in the coefficients $h_n^{(\pm)}(\ell)$ because particular attention must be paid for nonnegative integer values of $2\ell+1$. Indeed, in this case, a limit process must be performed for the Whittaker functions $W_{\pm i\eta(Z),\frac{2\ell+1}{2}}(\pm 2ikr)$ appearing in the definition (7) of $H^{(\pm)}$. For $2\ell+1$ not an integer, one may use the relation (equation (18a) in section 2 of Ref. 7)

$$W_{\pm i\eta(Z),\frac{2\ell+1}{2}}(\pm 2ikr)$$
$$=\frac{\pi}{\sin[\pi(2\ell+1)]}\left\{-\frac{\mathcal{M}_{\pm i\eta(Z),\frac{2\ell+1}{2}}(\pm 2ikr)}{\Gamma(-\ell\mp i\eta(Z))}+\frac{\mathcal{M}_{\pm i\eta(Z),-\frac{2\ell+1}{2}}(\pm 2ikr)}{\Gamma(\ell+1\mp i\eta(Z))}\right\} \tag{23}$$

where

$$\mathcal{M}_{\chi,\frac{\mu}{2}}(z)=\frac{1}{\Gamma(1+\mu)}M_{\chi,\frac{\mu}{2}}(z).$$

For any physical value of the angular momentum we have $2\ell+1\in\mathbb{N}\cup\{0\}$, and the above expressions for $\mathcal{M}_{\pm i\eta(Z),\pm\frac{2\ell+1}{2}}$ and $W_{\pm i\eta(Z),\frac{2\ell+1}{2}}$ are not correct. Nevertheless the limit for $2\ell+1$ approaching a nonnegative integer number exists; one must calculate

$$W_{\pm i\eta(Z),\frac{2\ell+1}{2}}(\pm 2ikr):=\lim_{\epsilon\to 0}W_{\pm i\eta(Z),\frac{2(\ell+\epsilon)+1}{2}}(\pm 2ikr), \tag{24}$$

a limit that Buchholz[7] solved by applying l'Hôpital's rule.

In order to find the coefficients $h_n^{(\pm)}$ of the series expansion (22) we need to perform

$$I_W^{(\pm)}(\ell)=\int_0^{\infty}\phi_n(\ell,\beta;r)\,\frac{1}{r}\,W_{\mp i\eta(Z),\ell+\frac{1}{2}}(\mp 2ikr)\,dr. \tag{25}$$

It suffices to provide $h_n^{(+)}(\ell) = \tilde{N}_C^{(+)}(\ell)\, I_W^{(+)}(\ell)$, since the coefficients $h_n^{(-)}(\ell)$ corresponding to the expansion of $H^{(-)}$ are easily deduced by conjugation, i.e., $h_n^{(-)}(\ell) = \left[h_n^{(+)}(\ell)\right]^*$. The Appendix is dedicated to the evaluation of $I_W^{(+)}$ for both cases $2\ell+1 \notin \mathbb{N} \cup \{0\}$ and $2\ell+1 \in \mathbb{N} \cup \{0\}$. For $2\ell+1 \notin \mathbb{N} \cup \{0\}$, one finds

$$
\begin{aligned}
& h_n^{(+)}(\ell) \\
& = \frac{\tilde{N}_C^{(+)}(\ell)\,\pi}{\sin[\pi(2\ell+1)]} \left\{ \frac{2i\,\tilde{N}_C^{(+)}(\ell)}{\Gamma(-\ell+i\eta(Z))} s_n + \frac{2\beta}{N_{n,\ell}\,\Gamma(2\ell+2)\Gamma(-2\ell)} \left(-\frac{\beta}{ik}\right)^{\ell} \right. \\
& \left. \times \frac{1}{\beta-ik} F_2\left(1, -n, -\ell+i\eta(Z), 2\ell+2, -2\ell; \frac{2\beta}{\beta-ik}, -\frac{2ik}{\beta-ik}\right) \right\},
\end{aligned} \tag{26}
$$

where F_2 is one of the Appell hypergeometric functions[20]. For $2\ell+1 \in \mathbb{N} \cup \{0\}$, we calculate a limit process,

$$
h_n^{(+)}(\ell) = \lim_{\epsilon \to 0} h_n^{(+)}(\ell+\epsilon),
$$

and end up with a much more cumbersome expression

$$
\begin{aligned}
h_n^{(+)}(\ell) = & \frac{\tilde{N}_C^{(+)}(\ell)}{N_{n,\ell}\,\Gamma(2\ell+2)} \left(\frac{\beta}{k} i\right)^{\ell} X \\
& \times \left[\frac{\Gamma(2\ell+1)}{\Gamma(\ell+1+i\eta(Z))} \sum_{q=0}^{2\ell} \frac{(-\ell+i\eta(Z))_q}{(-2\ell)_q} Y^q \,{}_2F_1(-n, q+1, 2\ell+2; X) \right. \\
& + \frac{(-1)^{2\ell+1} Y^{2\ell+1}}{\Gamma(-\ell+i\eta(Z))} \sum_{q=0}^{\infty} (\ell+1+i\eta(Z))_q \frac{Y^q}{q!} \left\{ {}_2F_1(-n, 2\ell+2+q, 2\ell+2; X) \right. \\
& \times [\psi(q+1) - \mathrm{Log}(Y) - \psi(\ell+1+q+i\eta(Z))] \\
& + \frac{nX}{4(\ell+1)^2} \left[-q(1-X)^{n-1} {}_2\Theta_1^{(1)} \left(\begin{matrix} 1,1 \mid 2\ell+2, -n+1, -q+1 \\ 2\ell+3 \mid 2, 2\ell+3 \end{matrix} \middle| ; X^*, X^* \right) \right. \\
& \left.\left.\left. + (2\ell+2+q)\, {}_2\Theta_1^{(1)} \left(\begin{matrix} 1,1 \mid 2\ell+2, -n+1, 2\ell+3+q \\ 2\ell+3 \mid 2, 2\ell+3 \end{matrix} \middle| ; X, X \right) \right] \right\} \right]
\end{aligned} \tag{27}
$$

where

$$X = \frac{2\beta}{\beta - ik}, \quad Y = -\frac{2ik}{\beta - ik}.$$

Above $\psi(z) = \Gamma'(z)/\Gamma(z)$ is the digamma function,[14,19] ${}_2\Theta_1^{(1)}$ is a generalized Kampé de Fériet hypergeometric function,[21] and $(a)_n = \Gamma(a+n)/\Gamma(a)$ is the Pochhammer symbol defined in terms of the gamma function.[19]

Making use of the expression (26) found for the first case, we present in Fig. 1 the real and imaginary parts of an approximation

$$H_N^{(+)}(\ell, k; r) = \sum_{n=0}^{N} h_n^{(+)} \phi_n^L(\ell, \beta; r) \tag{28}$$

of the outgoing Coulomb wave function $H^{(+)}$. We compare the results obtained taking $N = 25$ and $N = 35$ with the function $H^{(+)}$ given by formula (7). As expected, the more Laguerre functions we use, the more accurate the approximation of $H^{(+)}$ is, and the range of accuracy increases. Also as a consequence of the behavior $H^{(+)} \xrightarrow{r \to 0} r^{-\ell}$, we find that as ℓ increases, more terms are needed in (28) to reproduce the function near the origin.

Using the series representation of Appell function F_2

$$F_2(a, b_1, b_2, c_1, c_2; z_1, z_2) = \sum_{k=0}^{\infty} \frac{(a)_k (b_1)_k}{(c_1)_k} \, {}_2F_1(a+k, b_2, c_2; z_2) \frac{z_1^k}{k!}. \tag{29}$$

and the contiguous relations of Gauss hypergeometric functions,[14,19] one can establish a relation between three consecutive coefficients $h_n^{(+)}$. Contrary to the one satisfied by s_n, such relation for the $h_n^{(+)}$ is not homogeneous; moreover, the resulting extra term is not simple. The fact that the recurrence relation for s_n is no longer valid for $h_n^{(+)}$ is connected with the remark made after formula (18): the series and the integral interchange performed for the coefficients s_n cannot be done with the expansion of $H^{(+)}$.

From relation (8) between the Coulomb wave functions, we obtain

$$s_n = \frac{h_n^{(+)}(\ell) - h_n^{(-)}(\ell)}{2i}, \tag{30a}$$

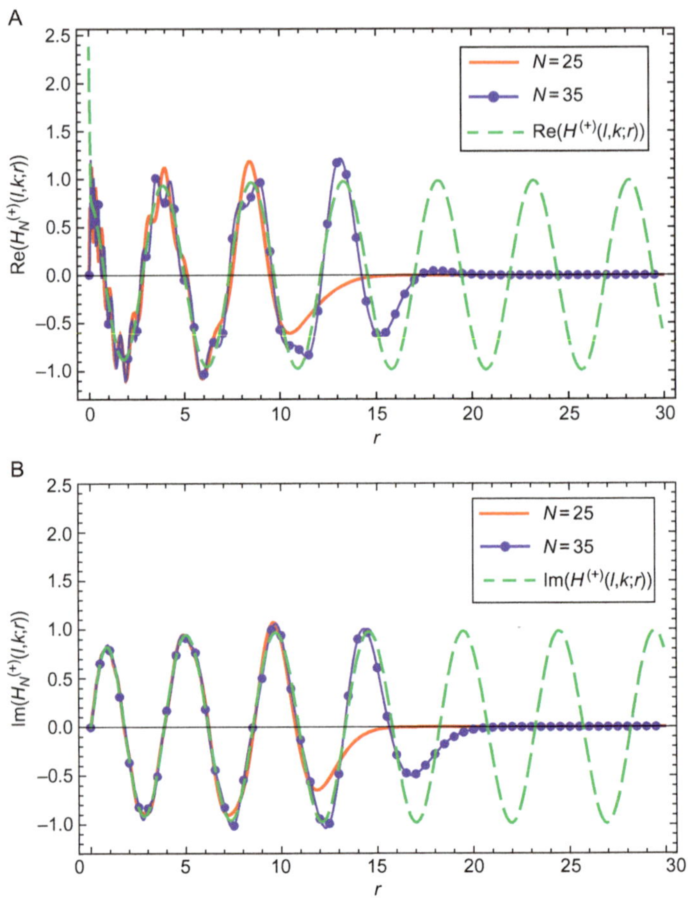

Fig. 1 Real (*top panel*) and imaginary (*bottom panel*) parts of an approximation $H_N^{(+)}$ of the Coulomb wave function $H^{(+)}$ (*dashed line*), taking $N = 25$ (*full line*) and $N = 35$ (*line with dots*). We have taken the following parameters: $Z = -1$, $\mu = 1$, $k = 1.23$, $\ell = 0.3$, and $\beta = 4$.

$$c_n(\ell) = \frac{h_n^{(+)}(\ell) + h_n^{(-)}(\ell)}{2}, \tag{30b}$$

where $c_n(\ell)$ define the expansion coefficients of $G^{(c)}$. Relation (30b) allows us to deduce directly an expression for the $c_n(\ell)$ elements without performing the corresponding integrals (12). As in the case of the coefficients $h_n^{(\pm)}(\ell)$ a distinction must be made depending on the value of $2\ell + 1$. The expression for the case $2\ell + 1 \in \mathbb{N} \cup \{0\}$ is again cumbersome, while for $2\ell + 1 \notin \mathbb{N} \cup \{0\}$ the coefficients take the form

$$c_n(\ell) = \frac{2}{\sin[\pi(2\ell+1)]} \Bigg\{ - \left[e^{\pi\eta(Z)} \sinh[\pi\eta(Z)] + \sin^2(\pi\ell) \right] s_n$$

$$+ \frac{\pi \ \beta \ e^{\frac{\pi}{2}\eta(Z)}}{N_{n,\ell}|\Gamma(\ell+1+i\eta(Z))|} \frac{1}{\Gamma(2\ell+2)\Gamma(-2\ell)} \left(\frac{\beta}{k}\right)^{\ell} \frac{1}{\beta - ik}$$

$$\times F_2\left(1, -n, -\ell + i\eta(Z), 2\ell+2, -2\ell; \frac{2\beta}{\beta - ik}, -\frac{2ik}{\beta - ik}\right) \Bigg\}.$$

4. CONNECTION WITH ORTHOGONAL POLYNOMIALS

The three-terms recurrence relation (19) satisfied by coefficients s_n can be related to generalized Pollaczek[18,19,22] or Meixner–Pollaczek[23,24] polynomials, depending on which element we consider as the variable.

4.1 Generalized Pollaczek Polynomials

The connection with Pollaczek polynomials $P_n^{\lambda}(x; a, b)$ was established and deeply investigated by different authors.[4,9,11,12] For μ, $Z, \beta > 0, \ell > -2$, and $k \in (0, +\infty)$ they found

$$s_n = \frac{N_{n,\ell}}{N_{0,\ell}} s_0 \ P_n^{\ell+1}\left(-\frac{\beta^2 - k^2}{\beta^2 + k^2}; \frac{\mu Z}{\beta}, -\frac{\mu Z}{\beta}\right). \tag{31}$$

From the orthogonality relation satisfied by Pollaczek polynomials we can provide one for the coefficients s_n. We start from Ref. 22

$$\int_{-1}^{1} P_m^{\lambda}(x; a, b) \ P_n^{\lambda}(x; a, b) \ w_P(x) \ dx = \frac{\Gamma(2\lambda + n)}{n!(\lambda + n + a)} \delta_{m,n}, \tag{32a}$$

with weight function

$$w_P(x) = \frac{1}{\pi} e^{(2\theta - \pi)t} (2\sin\theta)^{2\lambda - 1} |\Gamma(\lambda + it)|^2 \tag{32b}$$

where

$$x = \cos\theta, \qquad t = \frac{ax + b}{\sqrt{1 - x^2}}.$$

With the particular choice of the variable and the parameters

$$x = -\frac{\beta^2 - k^2}{\beta^2 + k^2}, \qquad \lambda = 2\ell + 1, \qquad a = \frac{\mu Z}{\beta}, \qquad b = -\frac{\mu Z}{\beta},$$

the weight function can be expressed in terms of s_0,

$$w_P(k) = \frac{\Gamma(2\ell+2)}{\pi} \frac{\beta^2 + k^2}{\beta k} s_0^2 \tag{33}$$

where the momentum k dependence is made explicit. From (31) we deduce an orthogonality relation for the energy dependent (recall $k^2 = 2\mu E$) coefficients $s_n = s_n(k)$

$$\int_0^{+\infty} s_n(k)\, s_m(k) \frac{1}{\beta^2 + k^2} dk = \frac{\pi}{4\beta(\ell + 1 + n) + 4\mu Z} \delta_{m,n}.$$

For the case of an attractive Coulomb potential ($Z < 0$), the coefficients s_n are no longer related to Pollaczek polynomials. Yamani and Reinhardt[9] proposed a new set of polynomials that they named "attractive Coulomb-Pollaczek" polynomials. Ten years later, Bank and Ismail[12] presented a complete study of these polynomials and their properties. In both attractive and repulsive cases, the charge Z is considered as a parameter and the momentum k (or the energy E) is the implicit polynomial variable.

4.2 Meixner–Pollaczek Polynomials

We can also make a connection with Meixner–Pollaczek polynomials[23,24] which are orthogonal polynomials defined by

$$P_n^{\lambda}(x;\theta) = \frac{(2\lambda)_n}{n!} e^{in\theta} {}_2F_1\left(-n, \lambda + ix, 2\lambda; 1 - e^{-2i\theta}\right) \tag{34a}$$

for

$$x \in \mathbb{R}, \quad \theta \in (0, \pi), \quad \lambda > 0. \tag{34b}$$

Let us remark that the variable x is only present in the second parameter of the hypergeometric function. The recurrence relation characterizing these polynomials is

$$P_{-1}^{\lambda}(x;\theta) = 0, \tag{35a}$$

$$P_0^{\lambda}(x;\theta) = 1, \tag{35b}$$

$$(n+1)P_{n+1}^{\lambda}(x;\theta) = 2[x\sin\theta + (n+\lambda)\cos\theta]P_n^{\lambda}(x;\theta) \\ -(n+2\lambda-1)P_{n-1}^{\lambda}(x;\theta). \tag{35c}$$

The orthogonality property[24]

$$\int_{-\infty}^{+\infty} P_m^{\lambda}(x;\theta)P_n^{\lambda}(x;\theta)w_M(x)\,dx = 2\pi\frac{\Gamma(2\lambda+n)}{(2\sin\theta)^{2\lambda}\,n!}\delta_{m,n} \tag{36}$$

requires a weight function

$$w_M(x) = e^{(2\theta-\pi)x}|\Gamma(\lambda+ix)|^2. \tag{37}$$

Consider again the three terms recurrence relation (19) and multiply it by $4\mu\beta/(N_{n,\ell}\,(\beta^2+k^2))$ to obtain the equivalent relation

$$(n+1)\frac{s_{n+1}}{N_{n+1,\ell}} = 2\left[-\frac{2\mu\beta Z}{\beta^2+k^2} + \left(-\frac{\beta^2-k^2}{\beta^2+k^2}\right)(\ell+1+n)\right]\frac{s_n}{N_{n,\ell}} \\ -(2\ell+1+n)\frac{s_{n-1}}{N_{n-1,\ell}}. \tag{38}$$

Next, we define a new coefficient

$$b_n = \frac{s_n/N_{n,\ell}}{s_0/N_{0,\ell}} = \frac{N_{0,\ell}}{N_{n,\ell}}\frac{s_n}{s_0}, \tag{39}$$

for which we have

$$b_{-1} = 0, \tag{40a}$$

$$b_0 = 1, \tag{40b}$$

$$(n+1)b_{n+1} = 2\left[-\frac{2\mu\beta Z}{\beta^2+k^2} + \left(-\frac{\beta^2-k^2}{\beta^2+k^2}\right)(\ell+1+n)\right]b_n - (2\ell+1+n)\,b_{n-1}. \tag{40c}$$

Comparing (40) with (35), and setting

$$x = -\eta(Z) \overset{(3a)}{=} -\frac{\mu Z}{k}, \qquad \lambda = \ell+1$$

and ζ as in (14), we find

$$\cos(\pi - \zeta) = -\frac{\beta^2 - k^2}{\beta^2 + k^2}, \qquad \sin(\pi - \zeta) = \frac{2\beta k}{\beta^2 + k^2}.$$

We conclude that, taking $\theta = \pi - \zeta$, relation (40) for the coefficients b_n becomes the one characterizing Meixner–Pollaczek polynomials. From (34b) we have the restrictions $\ell > -1, \theta \in (0, \pi), \eta(Z) \in \mathbb{R}$ so that

$$\sin\theta = \sin(\pi - \zeta) = \frac{2\beta k}{\beta^2 + k^2} > 0.$$

Thus for $\ell > -1$, μ, β, $k > 0$, and $Z \in \mathbb{R}$ we find

$$b_n = P_n^\lambda(-\eta(Z); \pi - \zeta)$$

or equivalently, using (39),

$$s_n = \frac{N_{n,\ell}}{N_{0,\ell}} s_0 P_n^{\ell+1}(-\eta(Z); \pi - \zeta) \tag{41}$$

This constitutes an alternative interpretation for the s_n coefficients as functions of the charge $Z \in \mathbb{R}$, including both attractive and repulsive Coulomb potentials in the same family.

These "charge" functions satisfy two interesting properties. The first one is a consequence of the orthogonality property known for the Meixner–Pollaczek polynomials. Rewritten in terms of the Coulomb problem parameters, the weight function (37) becomes

$$w_M(Z) = 4\,\Gamma(2\ell + 2) \left(\frac{\beta^2 + k^2}{4\beta k} \right)^{2\ell + 2} s_0^2. \tag{42}$$

Then from relation (41) and the orthogonality property (36), we deduce an orthogonality relation for the coefficients $s_n = s_n(Z)$,

$$\int_{-\infty}^{+\infty} s_n(Z)\, s_m(Z)\, dZ = \frac{\pi k}{2\mu}\, \delta_{m,n}. \tag{43a}$$

The second property is a consequence of relation (41): the coefficients $s_n(Z)$ form a complete basis set, i.e.,

$$\sum_{n=0}^{\infty} s_n(Z_1) s_n(Z_2) = \frac{\pi k}{2\mu}\, \delta(Z_1 - Z_2). \tag{43b}$$

In Fig. 2, we plot as a function of the charge Z, three coefficients s_n: $n = 2$, $n = 3$, and $n = 4$. The figure illustrates the polynomial behavior

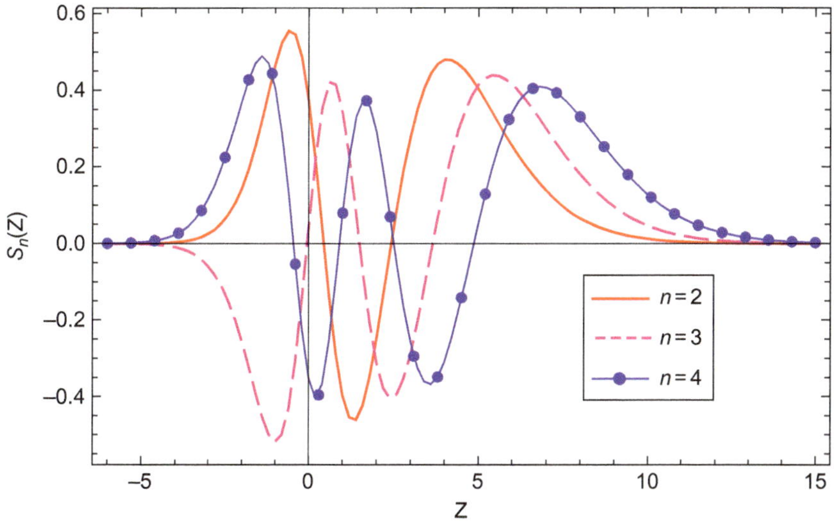

Fig. 2 Plot of three consecutive coefficients s_n as a function of the charge Z. We take $n = 2$ (*full line*), $n = 3$ (*dashed line*), and $n = 4$ (*line with dots*). The values of the parameters are $\mu = 1.7$, $k = 1.25$, $\ell = 1.3$, and $\beta = 2.4$.

of these coefficients in an " inner" region $|Z| < Z_n$, for an appropriate Z_n. Notice also how in the "asymptotic" region $|Z| > Z_n$ the $s_n(Z)$ vanish.

This novel interpretation of the coefficients $s_n(Z)$ as orthogonal functions with the charge as the variable invites us to explore the sine-like Coulomb wave function in terms of the charge. Indeed, the closure relations (11b) and (43b) combined with the series expansion (17) and the orthogonality relations (11a) and (43a), allow us to perform the following integrals over Z for the regular Coulomb wave function. Integrating over Z, we obtain a completeness relation with respect to the radial variable

$$\int_{-\infty}^{\infty} F^{(s)}(Z,\ell,k;r_1)\,\frac{1}{r_1}\,F^{(s)}(Z,\ell,k;r_2)\,dZ = \frac{\pi k}{2\mu}\;\delta(r_1 - r_2), \tag{44a}$$

while integrating over r

$$\int_{0}^{\infty} F^{(s)}(Z_1,\ell,k;r)\,\frac{1}{r}\,F^{(s)}(Z_2,\ell,k;r)\,dr = \frac{\pi k}{2\mu}\;\delta(Z_1 - Z_2) \tag{44b}$$

provides completeness considering the charge as the variable. We should mention that these integrals were deduced in Ref. 25 in a different context, where the authors proposed to consider a set $\{S_{\gamma,\ell}(x)\}_{\gamma\in\mathbb{R}}$ of charge

Coulomb Sturmian functions that happen to be regular Coulomb wave functions with the charge as the index.

5. SUMMARY

In this contribution we investigated some properties of the Laguerre representation for two-body radial Coulomb wave functions. There exists only one of them possessing standing wave behavior and being regular at the origin. There exists another function possessing standing wave behavior at large distances but it is irregular at the origin; finally, there are two other functions possessing outgoing or incoming behavior, both of them being irregular at the origin. Among these four Coulomb solutions, only for the regular one, the Laguerre representation has been presented and described in the literature.[11] Here we provide the expansions for the other three.

As a second contribution, we established the connection between Meixner–Pollaczek polynomials and the regular Coulomb solution for any value, positive or negative, of the charge. The Coulomb wave function can be transformed into charge Sturmian functions, and orthogonality and closure relations are then derived. When expanding those regular functions in Laguerre basis, the transformed functions turn to be Meixner–Pollaczek polynomials. This means that all the physical and analytical properties are condensed into these polynomials for any value of the energy, and positive and negative charges.

APPENDIX: INTEGRAL INVOLVING WHITTAKER FUNCTION $W_{\chi,\frac{\mu}{2}}(z)$

In this appendix we calculate

$$I_W^+(\ell) = \int_0^\infty \phi_n(\ell,\beta;r)\,\frac{1}{r}\,W_{-i\eta(Z),\ell+\frac{1}{2}}(-2ikr)\,dr. \tag{A.1}$$

and consider two cases.

A.1 Case $2\ell+1 \notin \mathbb{N}\cup\{0\}$

For the case $2\ell+1\notin\mathbb{N}\cup\{0\}$, the definitions of the Laguerre-type function $\phi_n(\ell, \beta;r)$ (10) and of the Whittaker function (23), give

$$
\begin{aligned}
I_W^{(+)}(\ell) &= \frac{(2\beta)^{\ell+1}}{N_{n,\ell}\,\Gamma(2\ell+2)} \frac{\pi}{\sin[\pi(2\ell+1)]} \\
&\quad \times \left\{ -\frac{(-2ik)^{\ell+1}}{\Gamma(-\ell+i\eta(Z))\,\Gamma(2\ell+2)} \right. \\
&\quad \times \int_0^\infty e^{-(\beta-ik)r} r^{2\ell+1}\, {}_1F_1(-n, 2\ell+2; 2\beta r) \\
&\quad \times {}_1F_1(\ell+1+i\eta(Z), 2\ell+2; -2ikr)\, dr \\
&\quad + \frac{(-2ik)^{-\ell}}{\Gamma(\ell+1+i\eta(Z))\,\Gamma(-2\ell)} \\
&\quad \left. \times \int_0^\infty e^{-(\beta-ik)r}\, {}_1F_1(-n, 2\ell+2; 2\beta r)\, {}_1F_1(-\ell+i\eta(Z), -2\ell; -2ikr)\, dr \right\} \\
&= \frac{(2\beta)^{\ell+1}}{N_{n,\ell}\,\Gamma(2\ell+2)} \frac{\pi}{\sin[\pi(2\ell+1)]} \\
&\quad \times \left\{ -\frac{(-2ik)^{\ell+1}}{\Gamma(-\ell+i\eta(Z))\,(\beta-ik)^{2\ell+2}} \right. \\
&\quad \times F_2(2\ell+2, -n, \ell+1+i\eta(Z), 2\ell+2, 2\ell+2; X, Y) \\
&\quad + \frac{(-2ik)^{-\ell}}{\Gamma(\ell+1+i\eta(Z))\,\Gamma(-2\ell)\,(\beta-ik)} \\
&\quad \left. \times F_2(1, -n, -\ell+i\eta(Z), 2\ell+2, -2\ell; X, Y) \right\}
\end{aligned}
\tag{A.2}
$$

where we have introduced $X = \frac{2\beta}{\beta - ik}$ and $Y = -\frac{2ik}{\beta - ik}$. To obtain the second equality in (A.2), use was made of the result given by Saad and Hall[26] for integrals of a product of confluent hypergeometric functions.

Alternative expressions can be obtained if we apply known identities[19,26] to the first Appell F_2 function or use the series representation (29) in terms of Gauss hypergeometric functions.[19,20] One representation, useful below, is

$$
\begin{aligned}
&I_W^{(+)}(\ell)\\
&=\frac{(2\beta)^{\ell+1}}{N_{n,\ell}\,\Gamma(2\ell+2)}\,\frac{\pi}{\sin[\pi(2\ell+1)]}\,\frac{(-2ik)^{-\ell}}{(\beta-ik)\,\Gamma(\ell+1+i\eta(Z))\,\Gamma(-\ell+i\eta(Z))}\\
&\quad\times\left\{-(1-X)^n Y^{2\ell+1}\sum_{q=0}^{\infty}\Gamma(\ell+1+i\eta(Z)+q)\frac{Y^q}{q!}\right.\\
&\quad\times{}_2F_1\left(-n,-q,2\ell+2;\frac{X}{X-1}\right)\\
&\quad\left.+\sum_{q=0}^{\infty}(1)_q\,\frac{\Gamma(-\ell+i\eta(Z)+q)}{\Gamma(-2\ell+q)}\frac{Y^q}{q!}\;{}_2F_1(-n,1+q,2\ell+2;X)\right\}.
\end{aligned}
\tag{A.3}
$$

Also we find a formulation involving the coefficients s_n defined by (13)

$$
\begin{aligned}
I_W^{(+)}(\ell)=&\ \frac{1}{\Gamma(\ell+1+i\eta(Z))}\,\frac{\pi}{\sin[\pi(2\ell+1)]}\\
&\times\left\{\frac{2i\,\widetilde{N}_C^{(+)}(\ell)}{\Gamma(-\ell+i\eta(Z))}\,s_n+\frac{2\beta}{N_{n,\ell}\,\Gamma(2\ell+2)\,\Gamma(-2\ell)(\beta-ik)}\right.\\
&\left.\times\left(-\frac{\beta}{ik}\right)^{\ell}F_2(1,-n,-\ell+i\eta(Z),2\ell+2,-2\ell;X,Y)\right\}.
\end{aligned}
\tag{A.4}
$$

A.2 Case $2\ell+1\in\mathbb{N}\cup\{0\}$

In this case, we should calculate

$$
\int_0^{\infty}\phi_n(\ell,\beta;r)\,\frac{1}{r}\,\lim_{\epsilon\to 0}W_{-i\eta(Z),\ell+\epsilon+\frac{1}{2}}(-2ikr)\,dr,
\tag{A.5a}
$$

but we solve instead

$$
\begin{aligned}
I_W^{(+)}(\ell)\ &:=\lim_{\epsilon\to 0}I_W(\ell+\epsilon)\\
&=\lim_{\epsilon\to 0}\int_0^{\infty}\phi_n(\ell,\beta;r)\,\frac{1}{r}\,W_{-i\eta(Z),\ell+\epsilon+\frac{1}{2}}(-2ikr)\,dr.
\end{aligned}
\tag{A.5b}
$$

It is convenient to start with (A.3) and use the following artifices for which we just sketch the ideas.

1. A split of the second series

$$\sum_{q=0}^{\infty}=\sum_{q=0}^{2\ell}+\sum_{q=2\ell+1}^{\infty}=\sum_{q=0}^{2\ell}+\sum_{Q=0}^{\infty} \tag{A.6}$$

where in the second equality we made a change of summation index $Q=q-(2\ell+1)$, and we use a transformation of one of the ${}_2F_1$ function[14]

$${}_2F_1(a,b,c;z)=(1-z)^{-b}\ {}_2F_1\left(b,c-a,c;\frac{z}{z-1}\right). \tag{A.7}$$

2. From $\Gamma(1-z)\Gamma(z)=\dfrac{\pi}{\sin(\pi z)}$ and $\sin[\pi(\alpha+1)]=-\sin(\pi\alpha)$ we obtain

$$\frac{\pi}{\sin[\pi(2\ell+2\epsilon+1)]}=-\Gamma(1-2\ell-2\epsilon)\Gamma(2\ell+2\epsilon). \tag{A.8}$$

Collecting these two results, the limit (A.5b) becomes

$$\begin{aligned}
I_W^{(+)}(\ell)=&\ \frac{(2\beta)^{\ell+1}}{N_{n,\ell}\Gamma(2\ell+2)}\frac{(-2ik)^{-\ell}}{\beta-ik}\frac{\Gamma(2\ell+1)}{\Gamma(\ell+1+i\eta(Z))}\\
&\times\sum_{q=0}^{2\ell}\frac{(1)_q(-\ell+i\eta(Z))_q}{(-2\ell)_q}\frac{Y^q}{q!}\ {}_2F_1(-n,1+q,2\ell+2;X)\\
&+\frac{(2\beta)^{\ell+1}}{N_{n,\ell}\Gamma(2\ell+2)}\frac{(-2ik)^{-\ell}}{\beta-ik}(1-X)^nY^{2\ell+1}\\
&\times\lim_{\epsilon\to0}-\frac{\pi}{\sin[\pi(2\ell+2\epsilon)]}\frac{1}{\Gamma(\ell+\epsilon+1+i\eta(Z))\,\Gamma(-\ell-\epsilon+i\eta(Z))}\\
&\times\left\{-Y^{2\epsilon}\sum_{q=0}^{\infty}\Gamma(\ell+\epsilon+1+i\eta(Z)+q)\frac{Y^q}{q!}\right.\\
&\times{}_2F_1\left(-n,-q,2\ell+2\epsilon+2;\frac{X}{X-1}\right)\\
&+\sum_{q=0}^{\infty}\frac{\Gamma(\ell-\epsilon+1+i\eta(Z)+q)\,\Gamma(q+1)}{\Gamma(q+1-2\epsilon)}\frac{Y^q}{q!}\\
&\left.\times{}_2F_1\left(-n,2\epsilon-q,2\ell+2\epsilon+2;\frac{X}{X-1}\right)\right\}.
\end{aligned}$$

The next step consists in performing the Taylor series (about $\epsilon = 0$). This implies, among other calculations, evaluating the derivative of hypergeometric functions ${}_2F_1$ with respect to their parameters; for that we use the results presented in Ref. 21 making a two variable hypergeometric function ${}_2\Theta_1^{(1)}$ appear. The calculations are rather simple and will be omitted here. The closed form finally obtained for the integral in the case of a nonnegative integer value of $2\ell + 1$ is the one presented in Section 3, formula (27).

REFERENCES

1. Alhaidari, A. D.; Heller, E. J.; Yamani, H. A.; Abdelmonem, M. S. *The J- Matrix Method, Development and Applications*. Springer: Berlin, 2008.
2. Heller, E. J.; Yamani, H. A. New L^2 Approach to Quantum Scattering: Theory. *Phys. Rev. A* **1974**, *9*, 1201.
3. Heller, E. J.; Yamani, H. A. J-Matrix Method: Application to s-Wave Electron-Hydrogen Scattering. *Phys. Rev. A* **1974**, *9*, 1209.
4. Yamani, H. A.; Fishman, L. J. J-Matrix Method: Extensions to Arbitrary Angular Momentum and to Coulomb Scattering. *Math. Phys.* **1975**, *16*, 410.
5. Broad, J.; Reinhardt, W. J-Matrix Method: Multichannel Scattering and Photoionization. *J. Phys. B: At. Mol. Phys.* 1976, *9*, 1491.
6. Bray, I. On Convergence of the Close-Coupling Method for Calculating Electron-Hydrogen Ionization. *J. Phys. B: At. Mol. Phys.* 2003, *36*, 2203.
7. Buchholz, H. *The Confluent Hypergeometric Function*. Springer: New York, 1969.
8. Del Punta, J. A. *Mathematical Methods in Atomic Physics* (Ph.D. thesis); Metz, France/ Bahía Blanca, Argentina: Université de Lorraine/Universidad Nacional del Sur.
9. Yamani, H. A.; Reinhardt, W. P. L^2 Discretizations of the Continuum: Radial Kinetic Energy and Coulomb Hamiltonian. *Phys. Rev. A* **1975**, *11*, 1144.
10. Broad, J. T. Gauss Quadrature Generated by Diagonalization of H in Finite L^2 Bases. *Phys. Rev. A* **1978**, *18*, 1012.
11. Broad, J. T. Calculation of Two-Photon Processes in Hydrogen With an L^2 Basis. *Phys. Rev. A* **1985**, *31*, 1494.
12. Bank, E.; Ismail, M. E. H. The Attractive Coulomb Potential Polynomials. *Constr. Approx.* **1985**, *1*, 103.
13. Joachain, C. J. *Quantum Collision Theory*. North-Holland Publishing Company: Amsterdam, 1975.
14. Abramowitz, M.; Stegun, I. A. *Handbook of Mathematical Functions*. Dover: New York, 1972.
15. Boersma, J. Expansions for Coulomb Wave Functions. *Math. Comput.* **1969**, *23*, 51.
16. Morse, P. M.; Feshbach, H. *Methods of Theoretical Physics, Part I and II*. McGraw-Hill Book Company: United States of America, 1953.
17. Nikiforov, A. F.; Uvarov, V. B. *Special Functions of Mathematical Physics*. Birkhauser: Germany, 1988.
18. Szegö, G. *Orthogonal Polynomials*. American Mathematical Society Colloquium Publications: Providence, 1939.
19. Erdelyi, A.; Magnus, W.; Oberhettinger, F.; Tricomi, F. G. *Higher Transcendental Functions Vol. I, II and III*. McGraw-Hill: New York, 1953.
20. Appell, P. Kampé de Fériet, J. *Fonctions Hypergeomitriques et Hyperspheriques: Polynômes d'Hermite*. Gauthier-Villars: Paris, 1926.

21. Ancarani, L. U.; Gasaneo, G. Derivatives of any Order of the Gaussian Hypergeometric Function ${}_2F_1(a, b, c; z)$ With Respect to the Parameters a, b and c. *J. Phys. A: Math. Theor.* 2009, *42*, 395208.
22. Szegö, G. On Certain Special Sets of Orthogonal Polynomials. *Proc. Am. Math. Soc.* **1950**, *1*, 731.
23. Pollaczek, M. F. Sur une Famille de Polynômes Orthogonaux qui Contient les Polynômes d'Hermite et de Laguerre Comme cas Limites. *CR Acad. Sci. Paris* **1950**, *230*, 1563.
24. Koekoek, R.; Lesky, P. A.; Swarttouw, R. F. *Hypergeometric Orthogonal Polynomials and Their q-Analogues. Springer Monographs in Mathematics* Springer-Verlag: Berlin, Heidelberg, 2010.
25. Gasaneo, G.; Colavecchia, F. D. Two-Body Coulomb Wave Functions as Kernel for Alternative Integral Transformations. *J. Phys. A: Math. Gen.* **2003**, *36*, 8443.
26. Saad, N.; Hall, R. Integrals Containing Confluent Hypergeometric Functions With Applications to Perturbed Singular Potentials. *J. Phys. A* **2003**, *36*, 7771.

CHAPTER FIVE

Numerical Hartree–Fock and Many-Body Calculations for Diatomic Molecules

John C. Morrison[*,1], **Jacek Kobus**[†]
[*]University of Louisville, Louisville, KY, United States
[†]Institute of Physics, Faculty of Physics, Astronomy, and Informatics, Nicolaus Copernicus University, Torun, Poland
[1]Corresponding author: e-mail address: johnc@erdos.math.louisville.edu

Contents

1. Introduction 104
   1.1 The Hartree–Fock Equations for Atoms 104
   1.2 The Hartree–Fock Equations for Diatomic Molecule 104
2. The Domain Decomposition Algorithm 105
3. A Model Problem 107
4. Many-Body Calculations of the Correlation Energy 108
5. Numerical Solutions of the Orbital and Pair Equations 112
   5.1 The Single-Electron Equation 112
   5.2 The Pair Equation for Two Noninteracting Electrons 113
   5.3 The First-Order Pair Equation 114
References 116

Abstract

The Hartree–Fock theory for diatomic molecules and a theoretical approach for performing many-body calculations are described. Using single-electron wave functions and energies produced by a numerical Hartree–Fock program, the Goldstone diagrams that arise in a perturbation expansion of the energy are evaluated by expressing the Goldstone diagrams in terms of pair functions that are the solution of first-order pair equations. The relevant pair equations are discretized and solved using the spline collocation method with a basis of third-order Hermite splines. Both the Hartree–Fock theory and many-body theory are more complex for diatomic molecules than they are for atoms. While the Hartree–Fock equations for atoms involve a single radial variable and the two-electron pair equation for atoms involve two radial variables, the Hartree–Fock equations for diatomic molecules involve two independent variables and the pair equation for diatomic molecules involves five independent variables. To deal with these problems of higher-dimensionality, we have developed numerical methods for dividing the variable space into smaller subregions in which the equations can be solved independently. This domain decomposition theory is described and numerical results are

Advances in Quantum Chemistry, Volume 76
ISSN 0065-3276
https://doi.org/10.1016/bs.aiq.2017.06.001

given for a single-electron model problem and for many-body calculations for diatomic molecules. Because the long-range goal of our work is to develop an extensive program for doing numerical coupled-cluster calculations on molecules, we will take special care to show how each part of our numerical approach is tested.

1. INTRODUCTION

The Hartree–Fock (HF) equations provide an approximate description of atoms and molecules and serve as the starting point of more accurate calculations. These equations are obtained in each case by requiring that the energy of the atom or molecule be stationary with respect to small changes in the form of the wave functions describing the electrons.

1.1 The Hartree–Fock Equations for Atoms

For an atom with a single-electronic configuration, the Hartree–Fock equations in atomic units can be written

$$\left[-\frac{1}{2}\nabla^2 - \frac{Z}{r} + \sum_b (2J_b - K_b)\right]\psi_a(\mathbf{r}) = \epsilon\psi_a(\mathbf{r}), \tag{1}$$

where Z is the charge of the nucleus in atomic units and r is the distance of the electron from the nucleus. The operator J_b represents the direct interaction of an election with another electron b and the operator K_b represents the exchange interaction between the electrons.

The solutions of the Hartree–Fock equations are of the general form

$$\psi(r,\theta,\phi) = \frac{P_{nl}(r)}{r} Y_{lm}(\theta,\phi), \tag{2}$$

where r is the distance of the electron from the nucleus and θ and ϕ are spherical coordinates.

Accurate numerical solutions of the Hartree–Fock equations for atoms have been reported by Fischer.[3,4]

1.2 The Hartree–Fock Equations for Diatomic Molecule

The Hartree–Fock equations for a diatomic molecule in atomic units can be written (cf. Ref. 5)

$$\left[-\frac{1}{2}\nabla^2 - \frac{Z_A}{r_A} - \frac{Z_B}{r_B} + \sum_b (2J_b - K_b)\right]\psi_a(\mathbf{r}) = \epsilon\psi_a(\mathbf{r}), \tag{3}$$

where r_A and r_B denote the distances of the electron from the two nuclei separated by R atomic units, and the charges of the two nuclei are denoted by Z_A and Z_B. The Coulomb interaction among the electrons is described by the direct and exchange potentials, J_b and K_b

Diatomic molecules are usually described in spheroidal coordinates

$$\begin{aligned} &\xi = (r_a + r_b)/R, && \text{where } 1 \leq \xi < \infty \\ &\eta = (r_a - r_b)/R, && \text{where } -1 \leq \eta \leq 1 \\ &\phi, && \text{where } 0 \leq \phi \leq 2\pi. \end{aligned} \tag{4}$$

One can ensure that the grid points are clustered near the nuclei where the wave functions change most rapidly and the normal derivatives of the wave function are zero on the boundaries where $\eta = \pm 1$ and $\xi = +1$ by making a variable transformation

$$\begin{aligned} &\eta = \cos\nu \quad \text{for } 0 \leq \nu \leq \pi, \\ &\xi = \cosh\mu \quad \text{for } 0 \leq \mu \leq \infty. \end{aligned} \tag{5}$$

The Hartree–Fock equations may then be written as

$$\begin{aligned} &-\frac{\partial^2 u_a}{\partial \nu^2} - (\cos\nu/\sin\nu)\frac{\partial u_a}{\partial \nu} - \frac{\partial^2 u_a}{\partial \mu^2} - (\cosh\mu/\sinh\mu)\frac{\partial u_a}{\partial \mu} \\ &\quad + \left(\frac{m^2}{\xi^2 - 1} + \frac{m^2}{1 - \eta^2}\right) u_a + (-RZ\xi + R\Delta\eta) u_a \\ &\quad + \left[\frac{R^2}{2}(\xi^2 - \eta^2) \sum_b (2J_b - K_b)\right] u_a = \frac{R^2}{2}(\xi^2 - \eta^2)\epsilon u_a, \end{aligned} \tag{6}$$

where $Z = Z_A + Z_B$ and $\Delta = Z_A - Z_B$.

Accurate numerical solutions of the Hartree–Fock equations for diatomic have been reported by McCullough,[6] Laaksonen,[7] and Kobus.[8]

2. THE DOMAIN DECOMPOSITION ALGORITHM

The main difficulty one encounters in applying numerical methods to realistic problems in quantum chemistry is that the time required rises precipitously as the number of grid points increases. The domain decomposition method enables one to reduce the time necessary to solve PDEs by breaking the entire domain into smaller regions where the equations are solved independently. We now describe how the algorithm can to applied for the simple problem of a region divided into two subregions by a vertical

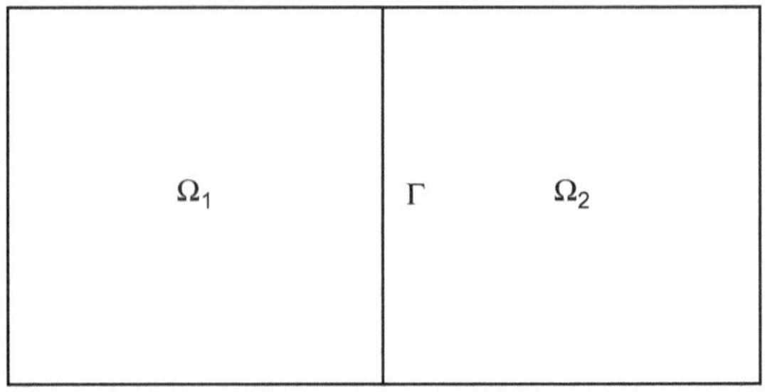

Fig. 1 Region divided into two subregions by interface. *Taken from Morrison, J.; Boyd, S.; Marsano, L.; Bialecki, B.; Ericsson, T.; Santos, J. P. Numerical Methods for Solving the Hartree-Fock Equations of Diatomic Molecules. Commun. Comput. Phys. 2009, 5, 959.*

interface. A derivation of the algorithm for this simple case can be found in the article[5] and in other articles on the spline collocation method.[1,2] The more general case of a two-dimensional region divided into subregions by an arbitrary number of planes parallel to the two coordinate axes can be found in a more recent article.[9]

A region that is divided into two regions by a vertical interface is illustrated in Fig. 1. We denote a PDE by

$$\mathbf{L}u(x,y)=f(x,y) \quad \text{for } a\leq x\leq b, \quad c\leq y\leq d, \tag{7}$$

where $\mathbf{L}$ is a differential operator and $u(x, y)$ and $f(x, y)$ are functions of two variables. As shown in Fig. 1, we shall denote the two regions by Ω_1 and Ω_2 and the interface between the two regions by Γ. The solutions in these two regions will be denoted by $u_1(x, y)$ and $u_2(x, y)$, and the values of the grid points on the interface between the two region will be denoted by Γ.

An equation for the solution in these two subregions can be obtained by resolving the function u into two functions $\hat{u}$ and $\tilde{u}$, which satisfy the same conditions as u along the exterior boundary

$$u(x,y)=\hat{u}(x,y)+\tilde{u}(x,y). \tag{8}$$

The properties of $\hat{u}$ and $\tilde{u}$ are complimentary with $\hat{u}$ satisfying the equation

$$\mathbf{L}\hat{u}_i(x,y)=f(x,y) \quad \text{for } i=1,2, \tag{9}$$

and $\tilde{u}$ satisfying the homogeneous equation

$$\mathbf{L}\tilde{u}_i(x,y)=0 \quad \text{for } i=1,2, \tag{10}$$

The functions $\hat{u}$ are equal to zero on the interface between the two regions, while $\tilde{u}$ is equal to the solution of the partial differential equation on the interface Γ. The functions, $\hat{u}_1$ and $\hat{u}_2$, which are equal to zero on the boundary between the two regions, may be regarded as known functions because the partial differential equation (9) with well-defined boundary condition may be solved for these two functions.

To obtain the solution u along the interface, we define an interface operator K by the equation

$$Kv(y) = W_\Gamma, \tag{11}$$

where the function W_Γ is defined by the equation

$$W_\Gamma(\xi) = \mathcal{D}_x V_1(\gamma, \xi) - \mathcal{D}_x V_2(\gamma, \xi), \quad \text{for } \xi \in \mathcal{G}. \tag{12}$$

Here ξ is a member of the set of grid points on the interface. The function, V_1 satisfies the same equation as $\tilde{u}_1$ in the region Ω_1, while the function V_2 satisfies the same equation as $\tilde{u}_2$ in the region Ω_2. The function, V_1 and V_2 are both equal to the spline function $v(y)$ on the interface between the two regions.

For a particular spline function $v(y)$ defined for $c \leq y \leq d$, the boundary conditions for the functions $V_1(x, y)$ and $V_2(x, y)$ are well defined and the function $W_\Gamma(\xi)$ is given by Eq. (12). As we have shown in Ref. 5, if the function $v(y)$ is equal to the solution of Eq. (7) along the interface, then the interface equation may be written

$$Ku_\Gamma = \mathcal{D}_x \hat{u}_2(\gamma, y) - \mathcal{D}_x \hat{u}_1(\gamma, y), \tag{13}$$

and we may define the following algorithm for solving the PDE (7).

Step 1. Using $\hat{u}_1$ and $\hat{u}_2$ obtained by solving Eq. (9) in the two regions, compute the right-hand side of Eq. (13).
Step 2. Solve Eq. (13) to find the solution of Eq. (7) along the interface.
Step 3. Solve Eq. (10) in the two regions to obtain the functions $\tilde{u}_1$ and $\tilde{u}_2$ and add these functions to $\hat{u}_1$ and $\hat{u}_2$ to obtain the solution u in regions Ω_1 and Ω_2.

3. A MODEL PROBLEM

The domain decomposition methods can be tested for a single electron by using the states of a single-electron atom or molecular ion. Equations for these states can be obtained by setting the direct and exchange terms equal to zero in Eq. (6). A simple model problem is provided by the singly ionized helium ion He^+ whose wave functions can be expressed analytically. For He^+, Z_A can be taken to be two and Z_B can be taken to be zero ($Z = Z_A + Z_B = 2$ and $\Delta = Z_A - Z_B = 2$). For diatomic molecules, the interatomic distance R is typically in the range between 2 and 6 atomic units. The spheroidal coordinate system may thus be used to describe atoms and diatomic molecules in a unified way.

Using spheroidal coordinates defined by Eq. (4) and the coordinate transformation (5), the wave function for the ground state of a He^+ ion centered at A can be written

$$u(\nu,\mu) = \left(\frac{Z_A^3}{\pi}\right)^{1/2} \exp\left[-\frac{1}{2}RZ_A(\cos\nu + \cosh\mu)\right]. \tag{14}$$

This wave function satisfies homogeneous Neumann boundary conditions on the vertical sides and on the lower boundary while satisfying homogeneous Dirichlet boundary condition on a sufficiently high upper boundary.

We tested the accuracy of the domain decomposition theory for solving PDEs by evaluating the right-hand side of Eq. (6) at the grid points and comparing the solution of the equation with the analytical solution (14). Table 1 gives the maximum difference of the numerical solution and analytic function in the each of 24 subregions obtained by dividing the original region into 4 pieces for the first variable and 6 pieces for the second variable.

4. MANY-BODY CALCULATIONS OF THE CORRELATION ENERGY

We shall now consider calculations of the correlation energy using many-body perturbation theory. A description of the methods we shall use can be found in the book by Lindgren and Morrison (cf. Ref. 10). In a perturbation calculation, the Hamiltonian is divided into a term H_0 representing the energy of electrons moving independently in an average potential and a perturbation V which is the difference of the Coulomb interaction and the interaction between the electrons with the average potential

$$H = H_0 + V. \tag{15}$$

H_0 is given by the equation

$$H_0 = \sum_{i=1}^{N}\left[-\frac{1}{2}\nabla_i^2 - \frac{Z_A}{r_{iA}} - \frac{Z_B}{r_{iB}} + \sum_i u_i\right] \tag{16}$$

Table 1 The Maximum Error of the Function u and Its Partial Derivatives u_ν and u_μ in a Region Divided Into 4 × 6 Subregions by Vertical and Horizontal Interfaces

k_μ	k_ν	Error (u)	Error (u_ν)	Error (u_μ)
1	1	2.6×10^{-7}	1.5×10^{-7}	9.6×10^{-8}
1	2	5.0×10^{-7}	8.2×10^{-7}	2.8×10^{-7}
1	3	2.1×10^{-6}	3.8×10^{-6}	5.9×10^{-7}
1	4	7.3×10^{-6}	4.6×10^{-6}	2.9×10^{-6}
2	1	1.4×10^{-7}	2.2×10^{-7}	2.9×10^{-7}
2	2	3.0×10^{-7}	7.3×10^{-7}	5.1×10^{-7}
2	3	8.4×10^{-7}	1.1×10^{-6}	1.1×10^{-6}
2	4	1.2×10^{-6}	7.1×10^{-7}	2.8×10^{-6}
3	1	1.5×10^{-7}	2.6×10^{-7}	3.7×10^{-7}
3	2	3.2×10^{-7}	3.3×10^{-7}	7.7×10^{-7}
3	3	5.6×10^{-7}	4.4×10^{-7}	1.5×10^{-6}
3	4	1.2×10^{-6}	7.0×10^{-7}	2.2×10^{-6}
4	1	1.6×10^{-7}	1.5×10^{-7}	5.3×10^{-7}
4	2	1.5×10^{-7}	1.8×10^{-7}	7.7×10^{-7}
4	3	4.0×10^{-7}	4.2×10^{-7}	1.6×10^{-6}
4	4	6.0×10^{-7}	4.2×10^{-7}	2.3×10^{-6}
5	1	1.0×10^{-7}	8.5×10^{-8}	5.3×10^{-7}
5	2	9.1×10^{-8}	9.5×10^{-8}	6.4×10^{-7}
5	3	2.3×10^{-7}	2.3×10^{-7}	1.6×10^{-6}
5	4	3.5×10^{-7}	2.3×10^{-7}	2.3×10^{-6}
6	1	7.1×10^{-8}	3.7×10^{-8}	4.7×10^{-7}
6	2	6.3×10^{-8}	6.9×10^{-8}	4.7×10^{-7}
6	3	1.6×10^{-7}	1.6×10^{-7}	1.2×10^{-6}
6	4	2.5×10^{-7}	1.6×10^{-7}	1.8×10^{-6}

The grid used for the calculation consisted of 48 intervals in each direction.

where r_{iA} and r_{iB} denote the distances of the i-th electron from the two nuclei separated by a distant R, and the charges of the two nuclei are denoted by Z_A and Z_B. V is given by the equation

$$V = \sum_{i<j} \frac{1}{r_{ij}} - \sum_i u_i. \tag{17}$$

The contributions to the correlation energy may be represented by Goldstone diagrams with the 2 second-order diagrams for the energy shown in Fig. 2 and the 12 third-order diagrams shown in Fig. 3.

To give some idea of how Goldstone diagrams are evaluated, we consider the first second-order diagram shown in Fig. 2, which corresponds to the mathematical expression,

$$\frac{1}{2} \sum_{abrs} \frac{\langle ab|1/r_{12}|rs\rangle \langle rs|/1r_{12}|ab\rangle}{\epsilon_a + \epsilon_b - \epsilon_r - \epsilon_s}. \tag{18}$$

This diagram, for which the orbitals, a and b, are in the same order in the two matrix elements, is called the *direct* diagram, while the second diagram in

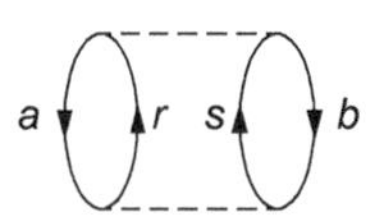

$$= \frac{1}{2} \sum_{abrs} \frac{\langle ab|\, r_{12}^{-1}\, |rs\rangle \langle rs|\, r_{12}^{-1}\, |ab\rangle}{\varepsilon_a + \varepsilon_b - \varepsilon_r - \varepsilon_s}$$

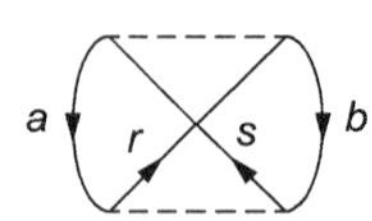

$$= -\frac{1}{2} \sum_{abrs} \frac{\langle ab|\, r_{12}^{-1}\, |sr\rangle \langle rs|\, r_{12}^{-1}\, |ab\rangle}{\varepsilon_a + \varepsilon_b - \varepsilon_r - \varepsilon_s}.$$

Fig. 2 Second-order Goldstone diagrams.

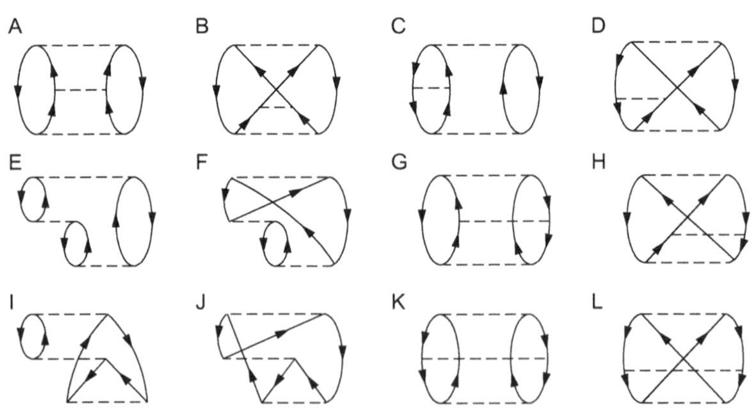

Fig. 3 Twelve third-order diagrams for the energy.

Fig. 2 for which orbital a is joined to the orbital b is called the *exchange* diagram. Beginning at the bottom of the each diagram and moving upward, there is a Coulomb matrix element for each dotted line and for every Coulomb interaction except the last there is an energy denominator that can be obtained by drawing a horizontal line above the dotted line and including a positive energy for each line directed downward and a negative energy for every line moving upward.

The first many-body calculations for atoms were performed by Kelly.[11] Some years later Morrison and Rajnak[12] carried out a calculation of the effective interaction of the two $4f$ electrons in the rare earth ion Pr^{3+} by using the numerical methods developed by Kelly together with an effective operator formulation of many-body perturbation theory. The difficulties offered by such many-body calculations is apparent from Eq. (18). Since the sum in this equation runs over all excited states—whether bound or free continuum states—one has to sum over all of the bound states and integrate over the continuum states. This is difficult because different grids must be used for the bound states and for the states deep in the continuum which oscillate very rapidly.

A means of overcoming these difficulties appeared in the literature in 1970 when Winter and his collaborators solved numerically a two-electron pair equation for helium.[13] With the publication of this work, it became apparent one could solve an inhomogeneous equation for the particular linear combinations of excited states that contribute to the Feynman diagrams. Adopting this approach, Morrison wrote a computer code to solve a single-particle inhomogeneous equation for the single-particle excitations of a rare earth ion[14] and a pair program to evaluate the effect of pair excitations.[15] This is the approach we shall adopt in this work.

Notice that the mathematical expression (18) for the direct diagram considered above can be written in terms of the two-electrons function

$$\rho(ab \to rs) = \frac{1}{2}\sum_{r,s} \frac{|rs\rangle\langle rs|1/r_{12}|ab\rangle}{\epsilon_a + \epsilon_b - \epsilon_r - \epsilon_s} \tag{19}$$

with the direct diagram being equal to

$$\frac{1}{2}\langle ab|1/r_{12}|\rho(ab \to rs)\rangle.$$

One can derive a first-order equation for the function $\rho(ab \to rs)$ by multiplying Eq. (19) from the left with $[\epsilon_a + \epsilon_b - h_0(1) - h_0(2)]$ and using the equations

$$h_0\psi_r = \epsilon_r\psi_r$$
$$h_0\psi_s = \epsilon_s\psi_s$$

to obtain

$$[\epsilon_a + \epsilon_b - h_0(1) - h_0(2)]\rho(ab \to rs) = \sum_{r,s} |rs\rangle\langle rs|1/r_{12}|ab\rangle, \tag{20}$$

where the sum on the right-hand side going over the excited states and has the effect of forcing the orthogonality of the right-hand side with respect to the occupied orbitals.

5. NUMERICAL SOLUTIONS OF THE ORBITAL AND PAIR EQUATIONS

5.1 The Single-Electron Equation

For perturbative calculations, we use the *spine collocation method* to discretize partial differential equations and eigenvalue problems. For problems involving a single variable x, a partition of points $x_0, x_1, \ldots, x_N$ is defined with the separation between adjacent points equal to $x_i - x_{i-1} = h$. As describe in previous articles (cf. Refs. 5 and 9), a basis of Hermite spines, v_i and s_i can be defined with respect to this partition. The solution of an ordinary differential equation can then be expressed as a linear combination of these basis splines

$$u(x) = \sum_{i=1}^{N} [\alpha_i v_i(x) + \beta_i s_i(x)]. \tag{21}$$

Gauss quadrature points, ξ_{i1} and ξ_{i2}, are defined within each interval $[x_{i-1}, x_i]$, and it can be shown that the vector $\mathbf{u}_G$ consisting of the values of the approximate solution at the Gauss points can be written as the product of a matrix $\mathbf{B}$ times a vector consisting of the spline coefficients

$$\mathbf{u}_G = \mathbf{B}\mathbf{u}.$$

The $\mathbf{B}$ matrix has 2×4 blocks running down its diagonal and has $2N + 2$ columns and $2N$ rows. Similarly, the first derivative may be represented by a matrix $\mathbf{C}$ with the same block structure and the negative of the second derivative is represented by a matrix $\mathbf{A}$. In the case of homogenous Dirichlet or Neumann boundary conditions, the boundary conditions can be imposed by removing from the matrices $\mathbf{B}$, $\mathbf{C}$, and $\mathbf{A}$ the columns corresponding to the zero value of the functions or its derivatives. The matrices then have the same number of rows and columns.

As we have seen, the wave function for a diatomic molecule depends upon two variables that serve to locate an electron in a plain that contains the axis of the molecule. The matrices for a diatomic molecule may be formed by taking the tensor product of the **B**, **C**, and **A** that correspond to the two variable, ν and μ. In particular, the tensor product $\mathbf{B}_\nu \otimes \mathbf{B}_\mu$ times an ordered array of the spline coefficients with respect to the ν and μ variables gives a ordered array of the values of the function at the Gauss points. We shall denote the tensor product of the two **B** matrices by **BBmat** and the collocation matrix for the molecule by **Amat**. The eigenvalue equation for a single electron in the molecule can thus be written

$$[\mathbf{Amat}][u] = \epsilon[\mathbf{BBmat}][u]. \tag{22}$$

At the beginning of each or our computer runs, the Hartree–Fock potential and wave functions are imported into our program and Eq. (22) is solved using LAPACK routines. If the grid used to solve Eq. (22) has a sufficient number of grid points, the solutions of Eq. (22) agree with the Hartree orbitals inputted into our program to a high level of accuracy.

5.2 The Pair Equation for Two Noninteracting Electrons

The pair equation for two noninteracting electrons is

$$[h_0(1) + h_0(2)]\psi(1,2) = (\epsilon_1 + \epsilon_2)\psi(1,2) \tag{23}$$

with the discrete representation

$$[\mathbf{Amat}(1) \otimes \mathbf{I}(2) + \mathbf{I}(1) \otimes \mathbf{Amat}(2)][u(1,2])] = (\epsilon_1 + \epsilon_2)[\psi(1,2])] \tag{24}$$

The immediate difficulty one encounters for solving the pair equation for a diatomic molecule is that the matrices are so large. The matrix obtained when one solves the single-electron equation using a 49 point grid with spine collocation has $4 \times 24 \times 24 = 2304$ rows and columns and the pair equation has $(2304)^2 = 5{,}308{,}416$ rows and columns. We will eventually use domain decomposition to overcome this difficulty but have first avoided forming the large tensor products by using the Slicot library.

For each right-hand side vector, **g**, the pair equation may be solved for a vector **x**. A set of vectors **g** and corresponding vectors **x** may be used to form the columns of the square matrices X and G, which may be shown to satisfy the equation

$$AX + XA^T = G$$

This last equation is a special case of the Sylvester equation

$$AX + XB = C \tag{25}$$

which may be solved by the Slicot library of routines without ever forming the huge Kronecker products, $A(1) \otimes I(2)$ and $I(1) \otimes A(2)$.

In the following table, we give the computer times necessary for solving a pair equation with the Slicot library and different numbers of grid points.

Times Using the Slicot Routines

n_x	n_y	Time (s)
16	16	13.3
24	24	140.1
32	32	916.0
40	40	3,462.0
48	48	11,037.4
4	48	43.0
4	96	53.3

We have found that to a very high level of accuracy, the solution of the pair equation for a noninteracting pair of electrons is equal to the corresponding product function—an indication that our numerical methods are correct.

5.3 The First-Order Pair Equation

The discrete form of the first-order pair equation is

$$([\mathbf{c1mat}] \otimes \mathbf{I} + \mathbf{I} \otimes [\mathbf{c1mat}])[u(1,2)] = \mathbf{G}, \tag{26}$$

where

$$[\mathbf{c1mat}] = \epsilon[\mathbf{BBmat}] - [\mathbf{Amat}]. \tag{27}$$

With a sparse grid having 25 points in the two coordinate directions, the collocation matrix for the single-electron problem has $4 \times 12 \times 12 = 576$ rows and columns, and the matrix for the pair equation has $572^2 = 331{,}776$ rows and columns and a band width of 50, 113. The memory required to store the pair matrix is the number of rows and columns times the band width times 8 for the word size—a little more the 133 GB—well within the capacity of our computers. For a 25 × 25 grid, we have solved the first-order pair equation using the Slicot library and also Gaussian elimination obtaining results that agree with each other to a high level of accuracy. In this context, we should say that we have previously solved the first-order pair equation for the sigma excitations of neutral helium using successive over-relaxation[16] obtaining the results shown in the following table.

Energy of the Ground State of the Helium Atom Including Excitations to *s* States With $(25 \otimes 25)^2$ Grid

E_{HF}	−2.861 680
E_2	−0.0181
$E_{\mathrm{HF}} + E_2$	−2.879 780
S-limit	2.879 022

The value of E_2 shown in the table, which was obtained by using a first-order pair function obtained by successive over-relaxation to evaluating the second-order contribution to the energy, is compared with an S-limit, which was calculated using a version of the Morrison–Lindgren pair program. The agreement is quite good. Before going forward, however, we must ensure that the results we have obtained with Gaussian elimination, with the Slicot library, and with successive over-relaxation are entirely consistent with each other. As any person who has developed a major computer code knows, one must check ones results very carefully at each stage of development.

The next step in developing our programs is to apply the technique of domain decomposition. As one can immediately see from the table showing times using the Slicot library, if one were to divide a region with 48 subinterval or 96 points by vertical interfaces into 12 subregions, it would only take 43.0 s to solve the pair equation in each subregion and 12 × 43 s for the entire region. The most time-intensive step in applying the domain decomposition method is solving the interface set of equations. Once the values of the solution on the interface is known, then the equation in each of the smaller subregions can be found independently. Fortunately, the first-order pair equation (20) is separable which means that the interface matrix for the first-order pair equation is the tensor product of the interface matrix for each electron.

REFERENCES

1. Bialecki, B.; Fairweather, G. Matrix Decomposition Algorithms in Orthogonal Spline Collocation for Separable Elliptic Boundary Value Problems. *SIAM J. Sci. Comput.* **1995**, *16*, 330.
2. Bialecki, B. Convergence Analysis of Orthogonal Spline Collocation for Elliptic Boundary Value Problems. *SIAM J. Numer. Anal.* **1998**, *35*, 617.
3. Fischer, C. F. Self-Consistent-Field (SCF) and Multiconfiguration (MC) Hartree-Fock (HF) Methods in Atomic Calculations: Numerical Integration Approaches. *Comput. Phys. Rep.* **1986**, *3*, 273.
4. Gaigalas, G.; Fischer, C. F. Extension of the HF Program to Partially Filled f-Subshells. *Comput. Phys. Commun.* **1996**, *98*, 255.
5. Morrison, J.; Boyd, S.; Marsano, L.; Bialecki, B.; Ericsson, T.; Santos, J. P. Numerical Methods for Solving the Hartree-Fock Equations of Diatomic Molecules. *Commun. Comput. Phys.* **2009**, *5*, 959.
6. McCullough Jr., E. A. Numerical Hartree-Fock Methods for Diatomic Molecules: A Partial-Wave Expansion Approach. *Comput. Phys. Rep.* **1986**, *4*, 265.
7. Laaksonen, L.; Pyykkö, P.; Sundholm, D. Fully Numerical Hartree-Fock Methods for Molecules. *Comput. Phys. Rep.* **1986**, *4*, 313.
8. Kobus, J. A finite difference Hartree-Fock program for atoms and diatomic molecules. *Comp. Phys. Commun.* **2013**, *184*, 799.
9. Morrison, J.; Steffen, K.; Pantoja, B.; Nagaiya, A.; Kobus, J.; Ericsson, T. Numerical Methods for Solving the Hartree-Fock Equations of Diatomic Molecules II. *Commun. Comput. Phys.* **2016**, *19*, 632.
10. Lingren, I.; Morrison, J. *Atomic Many-Body Theory, Springer Series on Atoms and Plasmas*, Vol. 3. Springer-Verlag: Berlin, 1986.
11. Kelly, H. F. Correlation Effects in Atoms. *Phys. Rev.* **1986**, *131*, 684.
12. Morrison, J.; Rajnak, K. Many-Body Calculations for the Heavy Atoms. *Phys. Rev. A* **1971**, *4*, 536.
13. Winter, N. W.; Laferriere, A.; McKoy, V. Numerical Solution of the Two-Electron Schrödinger Equation. *Phys. Rev. A* **1970**, *2*, 49.
14. Morrison, J. Effect of Core Polarization Upon the f-f Interactions of Rare-Earth and Actinide Ions. *Phys. Rev. A* **1972**, *6*, 643.
15. Morrison, J. Many-Body Calculations for the Heavy Atoms. III. Pair Correlations. *J. Phys. B* **1973**, *6*, 2205.
16. Boodoor, K.; Kobus, J.; Morrison, J. A Numerical Solution of the Pair Equation of a Model Two-Electron Diatomic System. *Int. J. Quantum Chem.* **2015**, *10*, 868.

CHAPTER SIX

Local Effective Hartree–Fock Potentials Obtained by the Depurated Inversion Method

Alejandra M.P. Mendez, Dario M. Mitnik[1], Jorge E. Miraglia
Instituto de Astronomía y Física del Espacio (IAFE), CONICET-Universidad de Buenos Aires, Buenos Aires, Argentina
[1]Corresponding author: e-mail address: dmitnik@df.uba.ar

Contents

1. Introduction 118
2. Theory 119
   2.1 HF Nodes and Derivatives 119
   2.2 Inverted Effective Potentials 122
   2.3 The Depuration Method 126
3. Results 127
4. Conclusions 128
Acknowledgment 130
Appendix 130
References 131

Abstract

In this work we show the results of a numerical experiment performed on the Hartree–Fock (HF) wave functions in order to understand the relationship between the positions of the orbital nodes and the inflection points (zeros of their second derivative). This analysis is equivalent to investigating the existence of a physical one-electron local potential representing the interactions between the electrons. We found that with successive improvements in the quality of the numerical methods, the nodes and the inflection points systematically become closer. When the nodes coincide exactly with the inflection points, the existence of an effective local potential would be proven. However, this requirement cannot be fulfilled unless an explicit constraint (missing in the standard method) is incorporated into the HF procedure. The depurated inversion method (DIM) was devised to obtain detailed *nl*-orbital potentials for atoms and molecules. The method is based on the inversion of Kohn–Sham-type equations, followed by a further careful optimization which eliminates singularities and also ensures the fulfillment of the appropriate boundary conditions. The orbitals resulting from these potentials have their internal inflection points located exactly at the nodes. In this way, the DIM can be employed to obtain effective potentials that accurately reproduce the HF orbitals.

Advances in Quantum Chemistry, Volume 76
ISSN 0065-3276
https://doi.org/10.1016/bs.aiq.2017.07.004

1. INTRODUCTION

The Hartree–Fock (HF) method is one of the best known and most commonly used approximation procedures to solve atomic, molecular, and solid systems because it greatly simplifies the problem of many electrons moving in a potential field. The method assumes, in accordance with the independent particle approximation and the Pauli exclusion principle, that the exact many-body wave function has the form of an antisymmetrized product of one-electron orbitals. After a self-consistent variational approach has been applied, a set of coupled equations for the orbitals is derived. Besides the minimization of the energies, the orbitals defining the wave function of the system are required to be orthonormal to each other. No further restrictions are imposed on the solutions.

In the present work, we will focus on the nodes of the HF orbitals and the radial location of their inflection points (the zeros of their second derivative). We will examine whether the orbital nodes can also be inflection points. Our investigation is based on a numerical experiment that studies in detail the influence of the methods employed for solving the HF equations on the radial location of the nodes and inflection points of the orbitals. The relative positions of these zeros are very important, since if they were not exactly at the same place, it would be impossible to obtain, through inversion, physical effective local potentials representing the interaction between the electrons. Based on our findings, we are tempted to postulate that although it is not explicitly required in any HF procedure, the nodes of the orbitals must also be turning points.

In a previous study,[1] a depuration inversion method (DIM) was devised. The method is based on the inversion of one-electron Kohn–Sham (KS) equations, whose solutions have been replaced by HF orbitals and energies. A further optimization procedure allows well-behaved effective potentials to ensue which, in turn, reproduce the original HF solutions very accurately. However, the new solutions have the advantage of having their internal inflection points located exactly at the same place as their nodes. Thus, our method can be considered as a means to introduce this additional constraint into the HF orbitals.

The chapter is organized as follows. In Section 2.1 the relative radial positions of the nodes and inflection points of the HF orbitals and their relationship with the accuracy of the computational calculations are analyzed. The origin of unphysical singularities in the effective potentials, due to the inversion procedure, is described in Section 2.2. The characteristics of these features are also analyzed in terms of the accuracy of the numerical calculations. In Section 2.3 the DIM method, conceived for handling the numerical problems encountered in the inversion procedure, is briefly formulated. Section 3 shows an example of how we are able to obtain local effective potentials that accurately replicate the HF wave functions by applying the DIM method. Atomic units are assumed throughout, unless stated otherwise.

2. THEORY

2.1 HF Nodes and Derivatives

In this section we study the nodes of HF orbitals and their inflection points. The analysis is supported by using an example (the 2*s* orbital of the magnesium atom), however, the conclusions are valid for all the atoms, for any level. The orbital u_{2s}^{HF} and its second derivative $u_{2s}^{\mathrm{HF}''}$ (multiplied by 0.1) are shown in Fig. 1 with solid and dashed lines, respectively. The orbital has one node and two inflection points. One of them, the external, is placed at $r \approx 0.9$ a.u. corresponding to the classical turning point. Beyond that point, the orbital decreases exponentially. The internal inflection point ($r \approx 0.18$ a.u.) is located, at first glance, at the same radius as the orbital node.

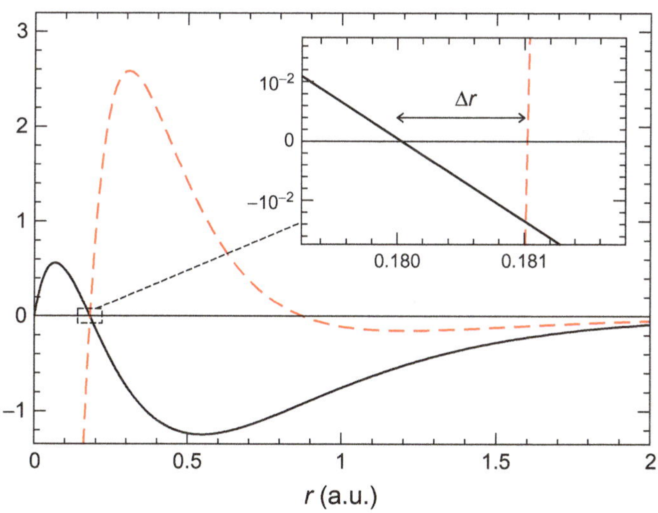

Fig. 1 Hartree–Fock orbital u_{2s}^{HF} (*solid line*) and scaled (×0.1) second derivative $u_{2s}^{\mathrm{HF}''}$ (*dashed line*) corresponding to the ground state of magnesium atom.

A question arises concerning the coincidence of the location of these points: is this the consequence of a restrictive condition of the HF method or is it just a fortuitous outcome? The later possibility is discarded since the same results are found systematically in all the orbitals of any atom. However, if this is a strict requirement, one should expect an explicit constraint in the HF procedure, forcing that. Moreover, a more meticulous inspection of the calculated orbital (see inset of Fig. 1) shows that the inflection point (dashed line) is not located exactly at the node position. Defining Δr as the distance between the node of the orbital $u_{nl}^{\mathrm{HF}}(r)$ and the closest zero of the corresponding second derivative $u_{nl}^{\mathrm{HF}''}(r)$, there is a distance of $\Delta r = 10^{-3}$ a.u. between these zeros in the 2*s* orbital example. We found about the same Δr in many other cases. Of course this distance can be considered negligible but there is no constraint within the HF procedure, imposing the exact coincidence of both positions.

In spite of that, one could expect that both zeros (orbital and second derivative) must coincide exactly, i.e., the nodes must also be internal inflection points. More precisely, one may be inclined to think that Δr should be exactly zero. If this hypothesis is proven to be true, it could be added as a constraint when performing HF calculations. The numerical experiment designed to scrutinize this hypothesis consists in performing variousapproximations with successive improvements of their accuracy, examining the behavior of the corresponding Δr. The quality of the numerical methods for solving the HF equations can be evaluated through the variation of the numerical algorithm accuracy order and by changing the density of points of the numerical grids. In the present work we used the linear multistep Adams–Moulton method for the differential equations and the Lagrangian differentiation method for the derivatives. The methodology proposed is implemented by modifying the NRHF code by Johnson (see Ref. 2 for further details), which uses eighth-order approximations as default. Nevertheless, the same results and conclusions were obtained with other codes, for example, the HF code by Fischer.[3]

Fig. 2 shows the HF orbitals u_{2s}^{HF} (solid lines) and their second derivative $u_{2s}^{\mathrm{HF}''}$ (dashed lines) in the proximity of the node. The least accurate calculation, in which all numerical methods are taken at their first order, is shown in Fig. 2A. In this calculation we used a numerical grid with 200 points—the minimum number needed to achieve convergence. This gave the biggest value for $\Delta r = 8 \times 10^{-3}$. Increasing the grid density to 400 points, this distance decreases significantly, $\Delta r = 4 \times 10^{-3}$, as shown in Fig. 2B. The best

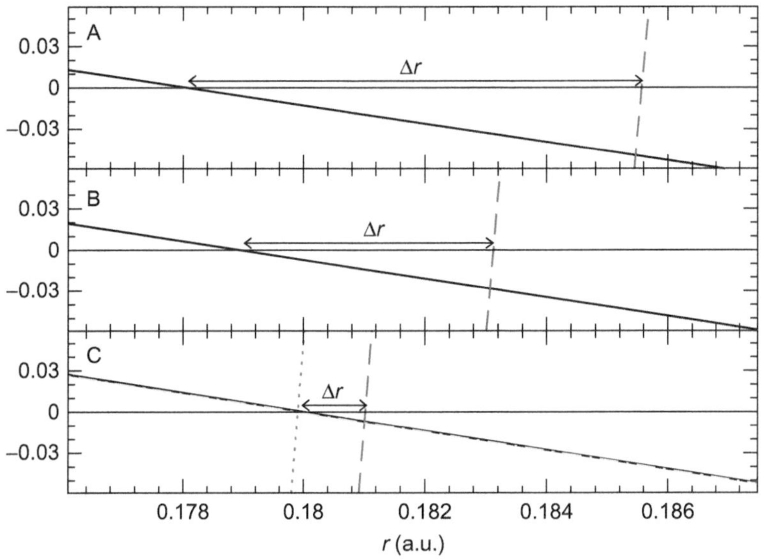

Fig. 2 Hartree–Fock orbital u_{2s}^{HF} and $u_{2s}^{HF''}$ (*solid* and *dashed lines*) when using (A) 200, (B) 400 points grid, and first-order approximations; (C) 1000 points grid and eighth-order approximation, u_{2s}^{OEP} and $u_{2s}^{OEP''}$ (*dot-dashed* and *dotted line*).

calculation performed is shown in Fig. 2C. In this case, an eighth-order approximation was considered in all the computational methods and a numerical grid of 1000 points was used. This arrangement led to the smallest difference $\Delta r = 1 \times 10^{-3}$. We were unable to obtain further improvements by increasing the number of points.

We performed an additional calculation using the optimized effective potential (OEP) method,[4,5] developed by Talman.[6] Fig. 2C also depicts the orbital u_{2s}^{OEP} near the node with dot-dashed line. However, because of the local character of the potential, its second derivative $u_{2s}^{OEP''}$—dotted line—is zero at the node.

We have recently developed the DIM[1] that enables one to acquire effective potentials which, in turn, reproduce the HF wave functions and energies very accurately. As we will see in the following section, the DIM potentials are physical, i.e., they have no singularities and they have the appropriate boundary conditions. The resulting orbitals have nodes that are also inflection points, a characteristic missing from the original HF orbitals. Unlike the OEP, the DIM effective potentials are orbital specific. This allows to obtain orbital inner shells energies much more accurately.

2.2 Inverted Effective Potentials

There are several methods to obtain effective potentials that describe a many-electron system. In principle, there is a direct way of calculating such potentials when the wave function is known, by inversion of the corresponding Schrödinger equations. The inversion scheme has been applied in different frameworks, from the density functional theory[7–9] to exact soluble models.[10] The procedure has been also applied in the calculation of atomic polarizability[11,12] and photoionization processes of atoms[13,14] and molecules.[15–17]

In our methodology,[1] we assume that the HF is turned into a set of KS-type equations, whose solutions are the HF orbitals and energies:

$$\left[-\frac{1}{2}\frac{d^2}{dr^2}+\frac{l(l+1)}{2r^2}+V_{nl}^{\mathrm{HF}}(r)\right]u_{nl}^{\mathrm{HF}}(r)=\varepsilon_{nl}^{\mathrm{HF}}\,u_{nl}^{\mathrm{HF}}(r)\,. \tag{1}$$

The direct inversion of Eq. (1) leads to effective *nl*-specific inverted potentials,

$$V_{nl}^{\mathrm{HF}}(r)=-\frac{1}{2}\frac{u_{nl}^{\mathrm{HF}''}(r)}{u_{nl}^{\mathrm{HF}}(r)}+\frac{l(l+1)}{2r^2}-\varepsilon_{nl}^{\mathrm{HF}}\,. \tag{2}$$

The direct computation of this equation is known to pose serious numerical problems.[18] The presence of genuine nodes in the orbitals can lead to the existence of poles and unphysical features around them. From Eq. (2) it is clear that the requirement of $\Delta r=0$ is an unavoidable imposition in order to allow the existence of an effective potential free of singularities. An additional difficulty arises at large radii. If the exponential decay of the orbitals is not rigorously reproduced by the second derivatives, the effective inverted potentials may diverge.

Both problems are illustrated in Fig. 3. In this figure we show the effective potential $V_{2s}^{\mathrm{HF}}(r)$ corresponding to the 2*s* orbital of Mg (in dashed line). The figure shows that the inverted potential exhibits a singularity at $r\approx 0.18$ a.u., which corresponds to the genuine node of u_{2s}^{HF}. As discussed above, the radial position of the inflection point is not located exactly at the node of the orbital (they are separated by Δr). This distance remains nonzero, even within a calculation at the highest order of accuracy, preventing complete cancellation of the node in the denominator of Eq. (2), and is responsible for the unphysical pole. The other unphysical divergence of the potential at $r>3$ a.u. (inset) is a consequence of the asymptotic exponential decay of the orbital, unbalanced by its second derivative.

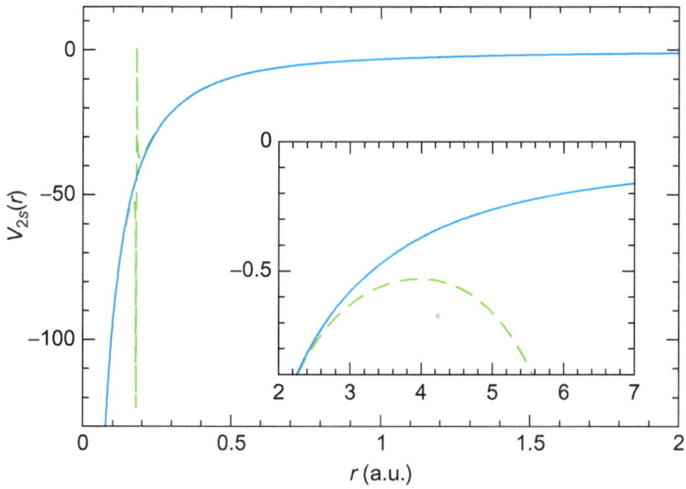

Fig. 3 Inverted V_{2s}^{HF} (*dashed line*) and depurated inverted V_{2s}^{DIM} (*solid line*) potential of the $2s$ orbital of magnesium atom.

A physical effective potential will allow generation of the corresponding structure of the system in a simple way. For example, for scattering process calculations it is convenient that the bound and continuum states should be computed on an equal footing. Nonetheless, requiring the existence of such a physical effective potential is not merely a demand of the atomic collision community. This requirement would be equivalent to imposing a $\Delta r = 0$ constraint in the HF procedure.

In the following, we will repeat the analysis performed above, but now the accuracy of the numerical methods involved in the calculations will be related to the appearance of the singularities of the effective potentials. To avoid handling the natural divergence due to the nucleus at the origin, we defined an *effective inverted charge*,

$$Z_{nl}^{\mathrm{HF}}(r) = -rV_{nl}^{\mathrm{HF}}(r)\,. \tag{3}$$

The effective charge around the node located at $r \approx 0.18$ a.u. is shown in Fig. 4. The least accurate calculation, using only first-order approximations and 200 points in the numerical grid (dotted line), produces a relatively wide pole (its width is about 0.3 a.u.). By increasing the accuracy of the numerical methods and the density of points in the grid to eighth order and 1000 points (dashed line), respectively, we are able to reduce this width by an order of magnitude. A pole-less physical potential would allow to obtain, by solving

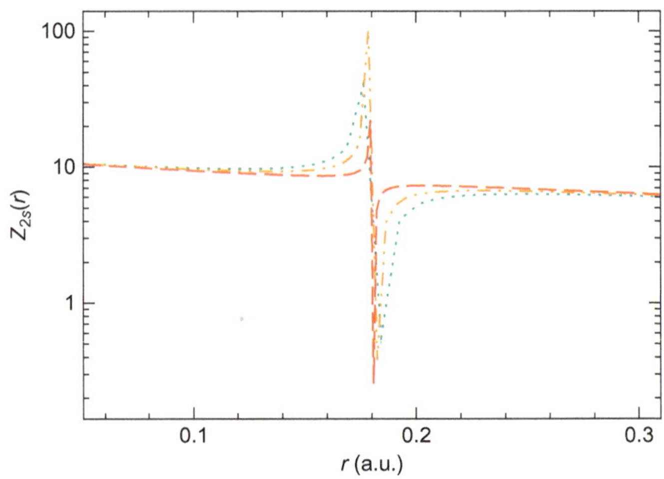

Fig. 4 Effective inverted charges around the node of u_{2s}^{HF} of Mg when different orders in the numerical method and grid densities are used.

the corresponding Schrödinger equation, orbitals having the internal inflection points at exactly the same location as its nodes.

The other feature of the potential considered above, related to its unphysical divergences at large distances (as a consequence of the different decreasing behavior between the wave functions and its second derivatives), can also be analyzed in terms of the accuracy of the numerical algorithms. In Fig. 5 we show the three calculations discussed previously. The least accurate calculation, corresponding to the first order in the numerical approximations (dotted line), shows a starting divergence at $r \approx 2$ a.u. Still, the best results for the effective inverted charge (dashed line) begin to diverge beyond 3 a.u.

We noticed an additional problematic feature in the HF solutions: the solutions may have oscillations (and therefore, spurious nodes) at the large r or "tail" region of the functions. These oscillations have already been pointed out by Fischer.[3] This failure to properly represent the orbitals is not caused by the numerical procedure but it is inherent to the method. Probably, due to the nonlocal character of the HF wave functions in which the behavior of a particular orbital depends on all others, these nodes may be attributed to surviving long-range exchange effects. Noticeably, we found the same spurious nodes at the same radial location even in calculations performed by using different numerical codes. As a general rule, the spurious nodes appear at very long distances, in regions where the amplitude of the orbital is very small. Therefore, their existence has no practical consequences and they can be effectively ignored in general.

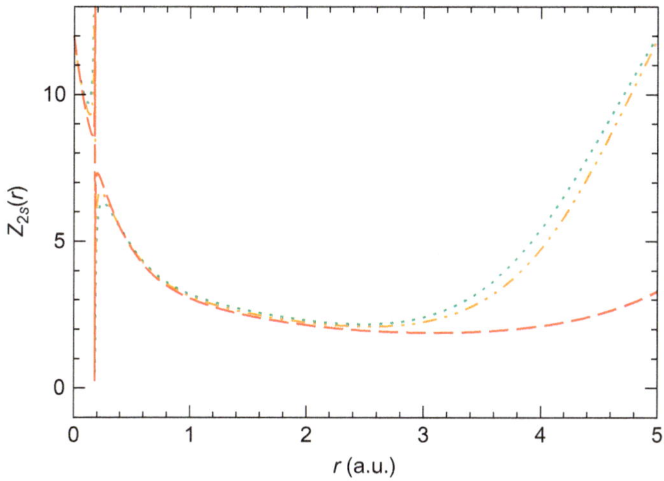

Fig. 5 Effective inverted charges computed with different orders in the numerical method and grid densities in the u_{2s}^{HF} of Mg atom.

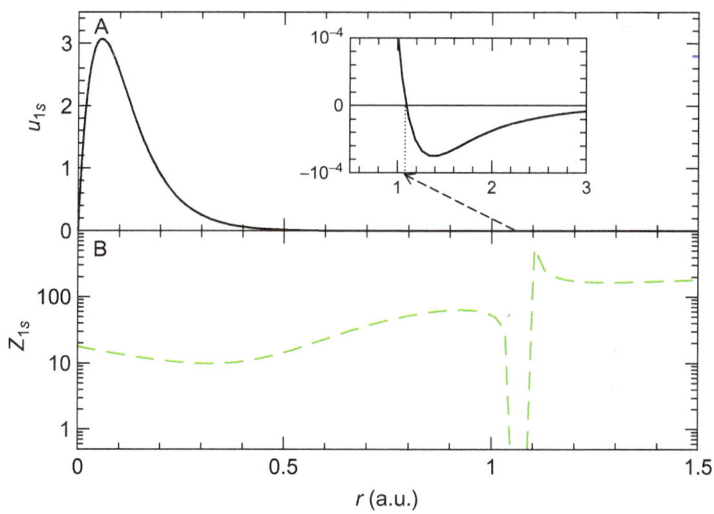

Fig. 6 (A) Hartree–Fock orbital u_{1s}^{HF} and (B) effective inverted charge Z_{1s}^{HF} of the 1*s* orbital of argon.

For example, the 1*s* orbital of the argon atom is nodeless, and it becomes practically zero for $r \geq 0.5$ a.u. However, a closer inspection shows a spurious node in the very large *r* region (at $r \approx 1.08$ a.u. as seen in Fig. 6A). Since the amplitude of the oscillation (see inset) is of the order of 10^{-4}, it passes generally unnoticed. Nonetheless, within the inversion scheme, the spurious node produces devastating effects. Evidence of this is the huge peak of the effective charge $Z_{1s}^{\mathrm{HF}}(r)$, as seen in Fig. 6B. As a matter of fact, the singularity is so big that it affects a broad range and causes the abrupt divergence of the effective charge for $r \geq 0.3$ a.u. This is really spectacular since a priori, there is no reason to suspect that a negligible oscillation in the tail of the wave function would produce such a big drawback at small distances. As a rule of thumb, we found that atoms with nuclear charge $Z_N \geq 13$ have spurious nodes, generally, in their most inner shells.

2.3 The Depuration Method

In order to deal with the problems of the inversion, we developed a complementary procedure. The depuration method consists, first, in erasing the unphysical poles and divergences from the inverted charges. Then, a parametric ad hoc formula is proposed in order to produce an analytic approximation to the charges. The physically correct boundary conditions are imposed through the proposed formula. The resulting effective potentials are then well behaved, without any singularity or divergence. A further optimization of the depurated potential is carried out. By solving iteratively the corresponding Schrödinger equation, orbitals u_{nl}^{DIM}, and energies $\varepsilon_{nl}^{\mathrm{DIM}}$ are obtained. Meanwhile, the fitting parameters are adjusted until the potential V_{nl}^{DIM} reproduces the original HF solutions with high accuracy.

We managed to constrain the potentials to have the right-boundary conditions by forcing the depurated inverted charge to behave as follows:

$$Z_{nl}^{\mathrm{DIM}}(r) \equiv -rV_{nl}^{\mathrm{DIM}}(r) \rightarrow \begin{cases} Z_N & \text{as } r \rightarrow 0 \\ 1 & \text{as } r \rightarrow \infty \end{cases} \tag{4}$$

where Z_N is the nuclear charge. Once the charge is determined at the boundaries, we can obtain a smooth analytic expression for $Z_{nl}^{\mathrm{DIM}}(r)$ by fitting $Z_{nl}^{\mathrm{HF}}(r)$ in the largest possible range, except at the regions where the charge has unphysical behavior (see Appendix for more details). This can be accomplished by enforcing the effective DIM charge to adjust the following analytical expression:

$$Z_{nl}^{\mathrm{DIM}}(r) = \sum_j Z_j e^{-\alpha_j r} + 1\,. \tag{5}$$

The asymptotic behavior defined by Eq. (4) is naturally fulfilled. In order to comply with the boundary condition at the origin, a further condition, $\Sigma Z_j = Z_N - 1$, is required for the fitting parameters.

The minimization of the adjustment parameters, Z_j and α_j, must be done carefully. Special attention should be paid to the adjustment region and the initial seeds chosen for the parameters. The fitting region is manually determined, that is, many points belonging to the original HF wave function should be discarded during the inversion procedure. Only the regions of the inverted charge with physical behavior must be preserved and considered, in such a way that Z_{nl}^{DIM} overlaps the inverted Z_{nl}^{HF} over a wide range. The unphysical behavior of the singularities and divergences are discarded. As the amplitude of the wave function becomes too small, the inversion must stop. Otherwise, the procedure diverges.

The most used optimization quantity in variational density functional approximation methods is the energy (some others consider the density). Although the minimization of the energy is a very important criterion, it is only one of the many observables that characterizes a quantum state. It has been proven that different functions (with different shapes) can produce, through a variational procedure, the same final energy. For instance Bartschat et al.[19,20] show a case in which two different potentials (one having exchange, the other neglecting it) led to very similar and accurate energies of the Rydberg series in several quasi-one-electron systems. However, a further examination of these potentials shows large discrepancies in scattering calculations.[21] Therefore, in addition to the energy, we have added the mean values of the orbitals as parameters in our variational procedure. This is achieved by optimizing the mean values $\langle 1/r \rangle$ (which characterize the quality of the wave function near the origin), and $\langle r \rangle$ (probing it at longer distances).

3. RESULTS

We denoted as u_{nl}^{DIM} and $\varepsilon_{nl}^{\mathrm{DIM}}$ the solutions obtained when the Schrödinger equation is solved with the depurated inverted potential V_{nl}^{DIM}.

In previous work,[1] the DIM was applied to ground state HF orbitals of several noble gases (He, Ne, Ar, and Kr) and the nitrogen atom. The potentials V_{nl}^{DIM} obtained reproduce wave functions and energies that agree with the HF ones with great accuracy. As a result of applying the depurated procedure, the singularities and divergences of the inverted potentials have been completely removed, and the V_{nl}^{DIM} potentials attained give a physically correct description of the problem.

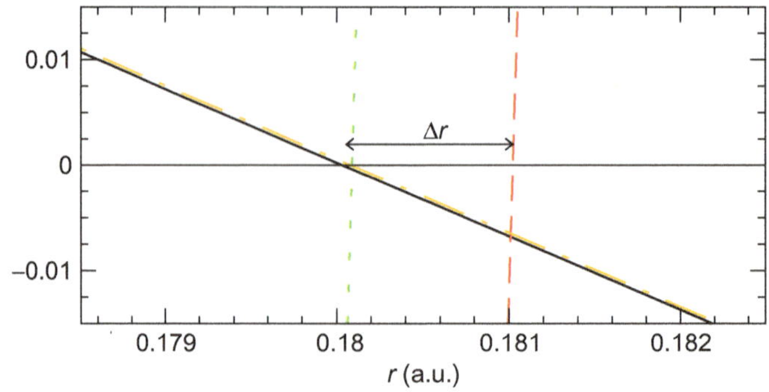

Fig. 7 Hartree–Fock orbital u^{HF} and corresponding second derivative $u^{HF''}$ (*solid* and *dashed lines*), DIM solved u^{DIM} and $u^{DIM''}$ (*dot-dashed* and *dotted lines*) for 2*s* orbital of magnesium.

Continuing with the example discussed in the previous section, a comparison between the DIM u_{2s}^{DIM} and the HF u_{2s}^{HF} orbitals of magnesium (and their corresponding second derivatives) is given in Fig. 7. The figure reproduces the inset of Fig. 1, additionally showing the DIM results (the corresponding V_{2s}^{DIM} potential is depicted in Fig. 3 with solid line). As stated above, the distance Δr, computed with eighth-order numerical approximations, between the nodes of u_{2s}^{HF}—dotted line—and the zero of its second derivative $u_{2s}^{HF''}$—dashed line—was found to be 1×10^{-3}. On the other hand, by using the V_{2s}^{DIM} potential, we obtained that the DIM second derivative $u_{2s}^{DIM''}$—dotted line—is zero exactly at the radius where the wave function u_{2s}^{DIM} —dash-dotted line—has a node.

As indicated previously, the conclusions reached here are general. They are valid for any other atom and for any level. We present the results for two additional atoms, boron and magnesium, in Table 1. The fitting parameters Z_j and α_j, defining the DIM charges (Eq. 5), allow to obtain orbital-specific potentials that, by solving the corresponding Schrödinger equation, replicate the original HF solutions within 1% and fulfill the $\Delta r = 0$ constraint.

4. CONCLUSIONS

The nodes and inflection points of the HF orbitals have been studied in detail. The influence of the quality of the numerical method used in the HF procedure was investigated. This was evaluated through the variation of the accuracy of the numerical algorithms and by changing the density of points in the numerical grids. The study showed that as the numerical quality

Table 1 Fitting Parameters for the Effective Charges of B and Mg Atoms for the Analytic Expression (5)

	nl	*Z*	α
B	1*s*	0.753400	6.14048
		1.43216	0.867257
		1.81444	0.905245
	2*s*	1.86154	1.10868
		0.153402	0.112430
		1.77713	2.40745
		0.207928	8.91487
	2*p*	1.69748	0.708184
		2.26098	3.54786
		0.041540	0.111029
Mg	1*s*	7.17402	1.31135
		3.38548	5.56370
		0.440500	1.26701
	2*s*	8.05058	3.30677
		2.74598	0.437365
		0.203440	22.3945
	2*p*	7.48607	2.78041
		1.28841	9.84139
		2.22552	0.475696
	3*s*	4.07701	6.01398
		6.18574	1.78888
		0.73725	0.665093

is improved, the distance Δr between the nodes of the HF orbitals and their inflection points became smaller. Further studies, considering higher orders in the numerical approximations, will allow one to perform an even more exhaustive inspection of these distances.

Therefore, one may propose the following hypothesis regarding these points: the nodes must be also inflection points. The similarities between the OEP and the HF orbitals would suggest a way to endorse this premise. The conjecture may or may not prove to be true. Subsequent work should be done in order to settle this, one way or the other. If the conjecture is true, it could be added as a constraint when performing the HF calculations. This would also imply that there exists a local potential that generates the HF orbitals.

The DIM, briefly reviewed here, allows effective potentials to be obtained that, by solving the corresponding Schrödinger equation, generate orbitals and energies that agree with the HF solutions with great accuracy. The method consists in considering the direct inversion of KS-type equations with HF solutions, and a further optimization. The depuration method imposes physically correct boundary conditions and a specific analytical expression for the inverted potentials.

We asserted that there is a direct relationship correlating the distances between the nodes and the inflection points and the unphysical behavior of the inverted potentials. We concluded that by generating the physically DIM effective potentials, we are able to obtain (after its diagonalization) orbitals that comply the hypothesis and reproduce the HF solutions very accurately.

ACKNOWLEDGMENT

This work was supported with PIP N°11220130100607 of CONICET, Argentina.

APPENDIX

The adjusting procedure consists in minimizing the function

$$\beta(x_i) = y(x_i) - f(x_i, \lambda_1, \lambda_2, \ldots, \lambda_m), \tag{A.1}$$

which is the difference between the curve to be adjusted, $y(r)$, and the adjusting function, $f(r)$. In the depurated inversion scheme, $y(r) = Z_{nl}^{\text{HF}}(r)$ corresponds to the inverted charge, and $f(r) = Z_{nl}^{\text{DIM}}(r)$ corresponds to the analytical form we have set up for it (Eq. 5). In order to minimize $\beta(x_i)$ respect to the m parameters λ_j that determine f, we define the matrix elements A_{ij},

$$A_{ij} \equiv \frac{d\beta(x_i)}{d\lambda_j} = \frac{df(x_i, \lambda_1, \lambda_2, \ldots, \lambda_m)}{d\lambda_j} \tag{A.2}$$

Then, we obtain the system of equations

$$\begin{bmatrix} d\beta(x_1) \\ d\beta(x_2) \\ \vdots \\ d\beta(x_n) \end{bmatrix} = \begin{bmatrix} A_{11} & A_{12} & \cdots & A_{1m} \\ A_{21} & A_{22} & \cdots & A_{2m} \\ \vdots & \vdots & \ddots & \vdots \\ A_{n1} & A_{n2} & \cdots & A_{nm} \end{bmatrix} \begin{bmatrix} d\lambda_1 \\ d\lambda_2 \\ \vdots \\ d\lambda_m \end{bmatrix}. \tag{A.3}$$

By multiplying both sides of the equation with transpose matrix $[A]^T$,

$$[A]^T[A][d\lambda] = [A]^T[d\beta], \tag{A.4}$$

we are able to get a system of equations that can be solved with standard numerical routines. The solution $[d\lambda]$ allows one to obtain the best parameters that minimizes $[\beta]$.

REFERENCES

1. Mendez, A. M. P.; Mitnik, D. M.; Miraglia, J. E. Depurated Inversion Method for Orbital-Specific Exchange Potentials. *Int. J. Quantum Chem.* **2016**, *116*, 1882.
2. Johnson, W. R. *Atomic Structure Theory: Lectures on Atomic Physics*. Springer-Verlag: Berlin, Germany, 2007.
3. Froese Fischer, C.; Brage, T.; Jönsson, P. *Computational Atomic Structure: An MCHF Approach*. Institute of Physics Publishing: Bristol, UK, 1997.
4. Sharp, R. T.; Horton, G. K. A Variational Approach to the Unipotential Many-Electron Problem. *Phys. Rev.* **1953**, *90*, 317.
5. Talman, J. D.; Shadwick, W. F. Optimized Effective Atomic Central Potential. *Phys. Rev. A* **1976**, *14*, 36.
6. Talman, J. D. A Program to Compute Variationally Optimized Effective Atomic Potentials. *Comput. Phys. Commun.* **1989**, *54*, 85.
7. Wu, Q.; Yang, W. A Direct Optimization Method for Calculating Density Functionals and Exchange-Correlation Potentials From Electron Densities. *J. Chem. Phys.* **2003**, *118*, 2498.
8. Gaiduk, A. P.; Ryabinkin, I. G.; Staroverov, V. N. Removal of Basis-Set Artifacts in Kohn-Sham Potentials Recovered From Electron Densities. *J. Chem. Theory Comput.* 2013, *9*, 3959.
9. Ryabinkin, I. G.; Kohut, S. V.; Staroverov, V. N. Reduction of Electronic Wave Functions to Kohn-Sham Effective Potentials. *Phys. Rev. Lett.* **2015**, *115*, 083001.
10. Filippi, C.; Umrigar, C. J.; Taut, M. Comparison of Exact and Approximate Density Functionals for an Exactly Soluble Model. *J. Chem. Phys.* **1994**, *100*, 1290.
11. Sternheimer, R. M. Electronic Polarizabilities of Ions From the Hartree-Fock Wave Functions. *Phys. Rev.* **1954**, *96*, 951.
12. Dalgarno, A.; Parkinson, D. The Polarizabilities of Atoms From Boron to Neon. *Proc. R. Soc. Lond. A* **1959**, *250*, 422.
13. Hilton, P. R.; Nordholm, S.; Hush, N. S. The Ground State Inversion Potential Method: Application to the Calculation of Photoionization Cross Sections. *J. Chem. Phys.* **1977**, *67*, 5213.
14. Süzer, S.; Hilton, P. R.; Hush, N. S.; Nordholm, S. UV Subshell Photoionisation Cross-Section of Atomic Zn, Cd, and Hg: Experiment and Theory. *J. Electron Spectrosc. Relat. Phenom.* 1977, *12*, 357.

15. Hilton, P. R.; Nordholm, S.; Hush, N. S. Photoionization Cross Section of Water by an Atomic Extrapolation Method. *Chem. Phys. Lett.* **1979**, *64*, 515.
16. Hilton, P. R.; Nordholm, S.; Hush, N. S. Document Ground-State Inversion Method Applied to Calculation of Molecular Photoionization Cross-Sections by Atomic Extrapolation: Interference Effects at Low Energies. *J. Electron Spectrosc. Relat. Phenom.* **1980**, *18*, 101.
17. Crljen, Ž.; Wendin, G. Many-Body Theory of Effective Local Potentials for Electronic Excitations. III. Application to Giant Dipole Resonances. *Phys. Rev. A* **1987**, *35*, 1571.
18. Mura, M. E.; Knowles, P. J.; Reynolds, C. A. Accurate Numerical Determination of Kohn-Sham Potentials From Electronic Densities: I. Two-Electron Systems. *J. Chem. Phys.* **1997**, *106*, 9659.
19. Albright, B. J.; Bartschat, K.; Flicek, P. R. Core Potentials for Quasi-One-Electron Systems. *J. Phys. B* **1993**, *26*, 337.
20. Bartschat, K. *Computational Atomic Physics*. Springer-Verlag: New York, USA, 1996.
21. Bartschat, K.; Bray, I. Local Versus Non-Local Core Potentials in Electron Scattering From Sodium Atoms. *J. Phys. B* **1996**, *29*, 271.

CHAPTER SEVEN

4-Center STO Interelectron Repulsion Integrals With Coulomb Sturmians

James E. Avery*[,1], John S. Avery†
*Niels Bohr Institute, University of Copenhagen, Copenhagen, Denmark
†H.C. Ørsted Institute, University of Copenhagen, Copenhagen, Denmark
[1]Corresponding author: e-mail address: avery@diku.dk

Contents

1. Introduction 133
   1.1 Prior Work 134
2. Coulomb Sturmian Expansions of STO 134
   2.1 Definitions 134
   2.2 Expanding Radial Functions 136
   2.3 Expanding Three-Dimensional Orbitals 139
3. STO Electron Repulsion Integrals 139
4. Checks of Convergence and Accuracy 141
   4.1 Checks Using the Exact 1-Center Case 141
   4.2 Checks Using the Exact 2-Center Case 143
5. Discussion 145
References 146

Abstract

We present a method for evaluating 4-center electron repulsion integrals (ERI) for Slater-type orbitals by way of expansions in terms of Coulomb Sturmians. The ERIs can then be evaluated using our previously published methods for rapid evaluation of Coulomb Sturmians through hyperspherical harmonics. Numerical investigations are made of the efficiency in 1- and 2-center cases where the exact integrals can be evaluated.

1. INTRODUCTION

Despite exponential-type orbitals (ETOs) being better able to represent both the cusps and the long-distance behavior of electrons in molecular orbitals, computational quantum chemistry is still dominated by Gaussian

Advances in Quantum Chemistry, Volume 76
ISSN 0065-3276
https://doi.org/10.1016/bs.aiq.2017.07.005

technology. This is due to the ease with which many-center interelectron repulsion integrals may be evaluated when Gaussians are used as a basis.

In the present chapter, as in a number of our previous papers, we hope to contribute to a growing body of work whose aim is to increase the speed, ease of use, and popularity of ETO calculations. Success in this effort will open the doors to more accurate molecular calculations than those that can achieved using Gaussians as a basis.

1.1 Prior Work

Because ETOs much better represent electronic wave functions than Gaussians, many quantum theorists have worked to overcome the difficult mathematical problems associated with evaluating many-center interelectron repulsion integrals in ETO bases. Early pioneers in this field include Harris, Michels, Steinborn, Weniger, Weatherford, Jones, and others.[1–4] An excellent review of the history and state of the art of molecular ETO integral calculation can be found in Ref. 5. A number of software packages have been developed for quantum chemistry calculations using Slater-type orbitals (STOs), the most prominent being the STOP package,[6] developed by Hoggan et al., SMILES,[7] developed by Rico, López, et al. at the Universidad Autónoma de Madrid, ADF,[8] by E.J. Baerends, J.G. Snijders, et al. in Amsterdam.

In our previous work, we have shown that all required molecular integrals for Coulomb Sturmians can be rapidly evaluated by way of hyperspherical harmonics,[9,10] and that the N-electron Schrödinger equation takes on a simpler form when they are used as a basis: the kinetic energy term vanishes, and all the integrals become independent of the exponent k.[11,12] The present paper continues the work begun in Ref. 13 of extending this method to molecular integrals for STOs by expanding them in terms of Coulomb Sturmians.

2. COULOMB STURMIAN EXPANSIONS OF STO

2.1 Definitions

We will write STOs as

$$\Phi_{\overline{\mu}}(\mathbf{x}) \equiv \mathcal{R}_{n,\zeta}(r)\, Y_{l,m}(\hat{\mathbf{x}}) \text{ with } \overline{\mu} \equiv (\zeta, n, l, m) \tag{1}$$

where the radial part is a monomial times a decaying exponential:

$$\mathcal{R}_{n,\zeta}(r) = N_{n,\zeta} \quad r^{n-1} e^{-\zeta r} \tag{2}$$

$N_{n,\zeta} = \frac{(2\zeta)^{n+\frac{1}{2}}}{\sqrt{(2n)!}}$ is the STO normalizing constant. We will expand the STO in terms of Coulomb Sturmians (CSO), which are similarly defined by

$$\chi_{n,l,m}(k,\mathbf{x}) = R_{n,l}(k,r)Y_{l,m}(\hat{\mathbf{x}}) \tag{3}$$

The radial function is a decaying exponential times an orthogonal polynomial, defined in terms of the confluent hypergeometric function

$$R_{n,l}(k,r) \equiv \mathcal{N}_{nl}e^{-kr}(2kr)^l F(l+1-n|2l+2|2kr) \tag{4}$$

We refer to, e.g., Ref. 10 for an in-depth review of the many useful properties of Coulomb Sturmians. Here we only note that the difficult 4-center electron repulsion integrals can be solved for Coulomb Sturmians by using their relationship to hyperspherical harmonics, as described in Ref. 9.

Where possible, we will work in the scaled coordinates $s = kr$, as the equations simplify. For this purpose, we define $\rho = \frac{\zeta}{k}$, such that the radial parts become

$$\begin{aligned}\mathcal{R}_{n,\zeta}(r) &= e^{-\zeta r}r^{n-1}\frac{(2\zeta)^{n+\frac{1}{2}}}{\sqrt{(2n)!}} \\ &= e^{-\rho s}s^{n-1}\frac{(2\rho)^{n+\frac{1}{2}}k^{n+\frac{1}{2}}}{\sqrt{(2n)!}\,k^{n-1}} = k^{3/2}\mathcal{R}_{n,\rho}(s)\end{aligned} \tag{5}$$

and similarly

$$R_{n,l}(k,r) = k^{3/2}R_{n,l}(1,s) \equiv k^{3/2}R_{n,l}(s) \tag{6}$$

for example, $R_{1,0}(s) = 2e^{-s}$, $R_{2,0}(s) = 2(1-s)e^{-s}$, $R_{2,1}(s) = \frac{2}{\sqrt{3}}se^{-s}$, and so on. The extra factor $k^{3/2}$ is absorbed in the relevant three-dimensional integrals by the change of coordinate from $\mathbf{x}$ to $\mathbf{s} = k\mathbf{x}$, so that the integrals become k-independent. In particular, the potential weighted orthonormality relation for Coulomb Sturmians becomes

$$\delta_{\mu',\mu} = \int d\mathbf{x}\ \chi^*_{\mu'}(k,\mathbf{x})\frac{n}{kr}\chi_\mu(k,\mathbf{x}) = \int d\mathbf{s}\ \chi^*_{\mu'}(\mathbf{s})\frac{n}{s}\chi_\mu(\mathbf{s}) \tag{7}$$

and for the radial functions

$$\delta_{n',n} = n\int_0^\infty ds\ s\ R_{n',l}(s)R_{n,l}(s) \tag{8}$$

2.2 Expanding Radial Functions

Due to the completeness properties of Coulomb Sturmians, we can write any square-integrable function $f(s)$ as:

$$f(s) = \sum_{n=l+1}^{\infty} R_{n,l}(s) c_{n,l} \tag{9}$$

for fixed $l \geq 0$. By Eq. (8), the expansion coefficients are

$$c_{n,l} = n \int_0^{\infty} ds \;\; s \;\; R_{n,l}(s) f(s) \tag{10}$$

If we now set $f(s) = s^l e^{-\rho s}$, i.e., if we consider the expansion

$$s^l e^{-\rho s} = \sum_{n=l+1}^{\infty} R_{n,l}(s) c_{n,l}(\rho) \tag{11}$$

we obtain

$$c_{n,l}(\rho) = n \int_0^{\infty} ds \;\; s^{l+1} \;\; R_{n,l}(s) e^{-\rho s} \tag{12}$$

This integral can be evaluated analytically when $n \geq l + 1$, and the result is

$$c_{n,l}(\rho) = 2^{l+1} n \sqrt{\frac{(n+l)!}{n(n-1-l)!}} \frac{(\rho-1)^{n-1-l}}{(\rho+1)^{n+1+l}} \tag{13}$$

In the special case where $\rho = 1$, corresponding to $\zeta = k$, the expansion is exact, and only the first term in the series is nonzero. It has the value

$$c_{n,n-1}(1) = \frac{\sqrt{(2n)!}}{2^{n+\frac{1}{2}}} = \frac{1}{N_{n,1}} \tag{14}$$

whereby Eq. (11) reduces to a single term

$$s^{n-1} \;\; e^{-s} = \frac{1}{N_{n,1}} R_{n,n-1}(s), \tag{15}$$

and in particular, $\Phi_{\zeta,n,l,m}(\mathbf{x}) = \chi_{n,l,m}(\zeta, \mathbf{x})$ when $l = n - 1$. When $\rho \neq 1$, the sum is infinite, and we must consider how fast its truncations

$$s^l e^{-\rho s} \simeq \sum_{n=l+1}^{l+N} R_{n,l}(s) c_{n,l}(\rho), \tag{16}$$

converge. Table 1 shows the rate of convergence as ρ grows away from one. It shows the L^2-error between the STO radial functions $\mathcal{R}_{n,\rho}(s)$ and the

Table 1 Convergence Dependence on ρ and N

$N\backslash\rho$	$\mathcal{R}_{1,\rho}(s)=N_{1,\rho}\ e^{-\rho s}$				
	1.1	**1.5**	**2**	**3**	**4**
5	10^{-6}	10^{-3}	10^{-2}	10^{-1}	10^{0}
10	10^{-12}	10^{-6}	10^{-4}	10^{-2}	10^{-1}
15	10^{-19}	10^{-9}	10^{-6}	10^{-3}	10^{-2}
20	10^{-25}	10^{-13}	10^{-8}	10^{-5}	10^{-3}
25	10^{-32}	10^{-16}	10^{-11}	10^{-6}	10^{-4}
30	10^{-38}	10^{-20}	10^{-13}	10^{-8}	10^{-5}
40	10^{-51}	10^{-26}	10^{-18}	10^{-11}	10^{-7}

$N\backslash\rho$	$\mathcal{R}_{2,\rho}(s)=N_{2,\rho}\ s\ e^{-\rho s}$				
	1.1	**1.5**	**2**	**3**	**4**
5	10^{-6}	10^{-2}	10^{-1}	10^{-1}	10^{0}
10	10^{-12}	10^{-6}	10^{-3}	10^{-2}	10^{-1}
15	10^{-18}	10^{-9}	10^{-5}	10^{-3}	10^{-2}
20	10^{-24}	10^{-12}	10^{-8}	10^{-4}	10^{-3}
25	10^{-31}	10^{-15}	10^{-10}	10^{-5}	10^{-4}
30	10^{-37}	10^{-19}	10^{-12}	10^{-7}	10^{-5}
40	10^{-50}	10^{-25}	10^{-17}	10^{-10}	10^{-7}

$N\backslash\rho$	$\mathcal{R}_{3,\rho}(s)=N_{3,\rho}\ s^2\ e^{-\rho s}$				
	1.1	**1.5**	**2**	**3**	**4**
5	10^{-6}	10^{-2}	10^{-2}	10^{0}	10^{0}
10	10^{-11}	10^{-5}	10^{-3}	10^{-1}	10^{-1}
15	10^{-17}	10^{-8}	10^{-5}	10^{-2}	10^{-1}
20	10^{-24}	10^{-11}	10^{-7}	10^{-4}	10^{-2}
25	10^{-30}	10^{-15}	10^{-9}	10^{-5}	10^{-3}
30	10^{-37}	10^{-18}	10^{-11}	10^{-6}	10^{-4}
40	10^{-49}	10^{-24}	10^{-16}	10^{-9}	10^{-6}

Continued

Table 1 Convergence Dependence on ρ and N—cont'd

$\mathcal{R}_{4,\rho}(s) = N_{4,\rho}\ s^3\ e^{-\rho s}$

$N \backslash \rho$	1.1	1.5	2	3	4
5	10^{-5}	10^{-2}	10^{-1}	10^{0}	10^{0}
10	10^{-11}	10^{-5}	10^{-3}	10^{-1}	10^{-1}
15	10^{-17}	10^{-8}	10^{-4}	10^{-2}	10^{-1}
20	10^{-23}	10^{-11}	10^{-6}	10^{-3}	10^{-2}
25	10^{-30}	10^{-14}	10^{-9}	10^{-4}	10^{-3}
30	10^{-36}	10^{-17}	10^{-11}	10^{-6}	10^{-3}
40	10^{-49}	10^{-24}	10^{-15}	10^{-8}	10^{-5}

The table shows L^2 errors of synthesized STO radial functions compared to exact STOs for $n = 1, 2, 3,$ and 4, rounded to nearest order of magnitude.

N-term synthesis $\bar{\mathcal{R}}_{n,\rho}^{(N)}(s)$, approximated with radial Coulomb Sturmians:

$$\left(\int_0^\infty ds\ s^2\ \|\mathcal{R}_{n,\rho}(s) - \bar{\mathcal{R}}_{n,\rho}^{(N)}(s)\|^2 \right)^{1/2} \tag{17}$$

Note that the error is with respect to the STO radial functions, normalized to a total radial integral of 1, so the absolute L^2-error can at the same time be seen as a relative error compared to the total s^2-weighted radial integral from zero to infinity.

We can see that the convergence rate is highly dependent on the ratio between the synthesis exponent k and the STO exponent ζ. $\rho = 1$ is exact; for $\rho = 1.1$, a total L^2-error limited to 10^{-6} results from 5 terms (and we can get the error down to $\approx 10^{-50}$ with 40 terms); for $\rho = 2$, we need 15–20 terms to get to 10^{-6}, depending on the STOs monomial power; for $\rho = 3$, we need 25–30 terms; and for $\rho = 4$, we need 35–40 terms. Hence, the efficiency of the present method hinges on being able to choose a common exponent k for the electron repulsion integral to be evaluated that is well balanced with respect to the four STO exponents. But, due to the completeness, we can always achieve any desired accuracy; and by precalculating similar tables to Table 1, we can even predict how many terms to use in the expansion to achieve the accuracy we want.

2.3 Expanding Three-Dimensional Orbitals

Consider now the expansion of a full STO orbital

$$\Phi_{\rho,n,l,m}(\mathbf{s}) = \mathcal{R}_{n,\rho}(s) Y_{l,m}(\hat{\mathbf{x}}) = Y_{l,m}(\hat{\mathbf{x}}) \sum_{n'=l+1}^{\infty} R_{n',l}(s)\ \mathcal{B}^{n'}_{n,l}(\rho) = \sum_{n'=l+1}^{\infty} \chi_{n,l,m}(s)\ \mathcal{B}^{n'}_{n,l}(\rho) \tag{18}$$

The expansion coefficients are derived as

$$\begin{aligned} \mathcal{B}^{n'}_{n,l}(\rho) &= \int d\mathbf{s}\ \chi^*_{n',l,m}(\mathbf{s}) \frac{n}{s} \Phi_{\rho,n,l,m}(\mathbf{s}) \\ &= \int_0^{\infty} ds\ \ s\ \ R_{n',l}(s) \mathcal{R}_{n,\rho}(s) \\ &= N_{n,\rho} \left(-\frac{\partial}{\partial \rho}\right)^{n-l-1} c_{n',l}(\rho) \end{aligned} \tag{19}$$

where we have used, in turn, Eq. (7), the spherical harmonic orthogonality, and the relationship

$$s^{l+j} e^{-\rho s} = \left(-\frac{\partial}{\partial \rho}\right)^{j} s^l e^{-\rho s} \tag{20}$$

An exact expression for these coefficients was recently derived by Morales in Ref. 14:

$$\begin{aligned} \mathcal{B}^{n'}_{n,l}(\rho) = {} & N_{n,\rho} \mathcal{N}_{n',l}\ n' 2^l (l+n)! \\ & (1+\rho)^{-(n+l+1)}\ {}_2F_1\left(l+1-n', n+l+1 \middle| 2+2l \middle| \frac{2}{1+\rho}\right) \end{aligned} \tag{21}$$

where $\mathcal{N}_{n',l} = \frac{2}{(2l+1)!}\sqrt{\frac{(l+n)!}{n(n-l-1)!}}$ is the Coulomb Sturmian normalization factor for $k = 1$. Because $l + 1 - n'$ is a negative integer, the ${}_2F_1$-hypergeometric function is just a polynomial of degree $n' - l - 1$ in $\frac{2}{1+\rho}$.

3. STO ELECTRON REPULSION INTEGRALS

In Refs. 9 and 10, we developed rapid methods for evaluating many-center molecular interelectron repulsion integrals when Coulomb Sturmians located on the atomic centers are used as a basis. Since we can express a 2-center density based on STOs in terms of Coulomb Sturmians by Eqs. (18), (19), and (21), we can transform many-center interelectron repulsion integrals from a CSO basis to an STO basis. We use the Fourier transform representation of $1/r_{12}$

$$\frac{1}{|\mathbf{x}_1 - \mathbf{x}_2|} = \frac{1}{2\pi^2}\int d\mathbf{p}\ \frac{1}{p^2}\ e^{i\mathbf{p}\cdot(\mathbf{x}_1-\mathbf{x}_2)} \tag{22}$$

If $\rho_1(\mathbf{x}_1)$ and $\rho_2(\mathbf{x}_2)$ are 2-center densities, the STO electron repulsion integrals take on the form

$$\begin{aligned}
&\int d\mathbf{x}_1 \int d\mathbf{x}_2\, \Phi^*_{\overline{\mu}_1}(\mathbf{x}_1 - \mathbf{X}_1)\Phi_{\overline{\mu}_2}(\mathbf{x}_1 - \mathbf{X}_2)\frac{1}{|\mathbf{x}_1 - \mathbf{x}_2|}\\
&\Phi^*_{\overline{\mu}_3}(\mathbf{x}_2 - \mathbf{X}_3)\Phi_{\overline{\mu}_4}(\mathbf{x}_2 - \mathbf{X}_4)\\
&\equiv \int d\mathbf{x}_1 \int d\mathbf{x}_2\ \rho_1(\mathbf{x}_1)\ \frac{1}{|\mathbf{x}_1 - \mathbf{x}_2|}\ \rho_2(\mathbf{x}_2)\\
&= \frac{1}{2\pi^2}\int d\mathbf{p}\ \frac{1}{p^2}\int d\mathbf{x}_1 \int d\mathbf{x}_2\ \rho_1(\mathbf{x}_1)\ e^{i\mathbf{p}\cdot(\mathbf{x}_1-\mathbf{x}_2)}\ \rho_2(\mathbf{x}_2)\\
&= 4\pi \int d\mathbf{p}\ \frac{1}{p^2}\ \rho_1^t(-\mathbf{p})\ \rho_2^t(\mathbf{p})
\end{aligned} \tag{23}$$

where $\mathbf{X}_1 \neq \mathbf{X}_2$ and $\mathbf{X}_3 \neq \mathbf{X}_4$ are the four nuclear centers, and

$$\rho_j^t(\mathbf{p}) \equiv \frac{1}{(2\pi)^{3/2}}\int d\mathbf{x}_j\ e^{-i\mathbf{p}\cdot\mathbf{x}_j}\rho_j(\mathbf{x}_j) \tag{24}$$

Thus we need to evaluate the Fourier transforms of STO 2-center densities. We do this by expressing the 2-center densities in terms of Coulomb Sturmians, for which we can compute the 2-center density Fourier transforms using the method described in Ref. 9.

$$\begin{aligned}
\rho_1^t(-\mathbf{p}) &= \frac{1}{(2\pi)^{3/2}}\int d\mathbf{x}_1\ e^{i\mathbf{p}\cdot\mathbf{x}_1}\Phi^*_{\overline{\mu}_1}(\mathbf{x}_1 - \mathbf{X}_1)\Phi_{\overline{\mu}_2}(\mathbf{x}_1 - \mathbf{X}_2)\\
&= \sum_{n>l_1}\sum_{n'>l_2}\mathcal{B}^n_{n_1,l_1}(\zeta_1/k)\mathcal{B}^{n'}_{n_2,l_2}(\zeta_2/k)\\
&\times \frac{1}{(2\pi)^{3/2}}\int d\mathbf{x}_1\ e^{i\mathbf{p}\cdot\mathbf{x}_j}\ \chi^*_{n,l_1,m_1}(\mathbf{x}_1 - \mathbf{X}_1)\chi_{n',l_2,m_2}(\mathbf{x}_1 - \mathbf{X}_2)\\
\rho_2^t(\mathbf{p}) &= \frac{1}{(2\pi)^{3/2}}\int d\mathbf{x}_2\ e^{-i\mathbf{p}\cdot\mathbf{x}_2}\Phi^*_{\overline{\mu}_3}(\mathbf{x}_2 - \mathbf{X}_3)\Phi_{\overline{\mu}_4}(\mathbf{x}_2 - \mathbf{X}_4)\\
&= \sum_{n''>l_3}\sum_{n'''>l_4}\mathcal{B}^{n''}_{n_3,l_3}(\zeta_3/k)\mathcal{B}^{n'''}_{n_4,l_4}(\zeta_4/k)\\
&\times \frac{1}{(2\pi)^{3/2}}\int d\mathbf{x}_2\ e^{-i\mathbf{p}\cdot\mathbf{x}_2}\ \chi^*_{n'',l_3,m_3}(\mathbf{x}_2 - \mathbf{X}_3)\chi_{n''',l_4,m_4}(\mathbf{x}_2 - \mathbf{X}_4)
\end{aligned} \tag{25}$$

Substituting the Fourier-transformed 2-center STO densities of Eq. (25) into Eq. (23), we obtain 4-center STO interelectron repulsion integrals. Combining these relationships yields

$$\begin{aligned}&\int d\mathbf{x}_1 \int d\mathbf{x}_2 \;\; \Phi^*_{\overline{\mu}_1}(\mathbf{x}_1-\mathbf{X}_1)\Phi_{\overline{\mu}_2}(\mathbf{x}_1-\mathbf{X}_2)\frac{1}{|\mathbf{x}_1-\mathbf{x}_2|}\Phi^*_{\overline{\mu}_3}(\mathbf{x}_2-\mathbf{X}_3)\Phi_{\overline{\mu}_4}(\mathbf{x}_2-\mathbf{X}_4)\\ &=\sum_{n>l_1}\sum_{n'>l_2}\sum_{n''>l_3}\sum_{n'''>l_4} \mathcal{B}^n_{n_1,l_1}(\zeta_1/k)\mathcal{B}^{n'}_{n_2,l_2}(\zeta_2/k)\mathcal{B}^{n''}_{n_3,l_3}(\zeta_3/k)\mathcal{B}^{n'''}_{n_4,l_4}(\zeta_4/k)\\ &\times\int d\mathbf{x}_1\int d\mathbf{x}_2 \;\; \chi^*_{n',l_1,m_1}(\mathbf{x}_1-\mathbf{X}_1)\chi_{n,l_2,m_2}(\mathbf{x}_1-\mathbf{X}_2)\frac{1}{|\mathbf{x}_1-\mathbf{x}_2|}\\ &\times\chi^*_{n'',l_3,m_3}(\mathbf{x}_2-\mathbf{X}_3)\chi_{n''',l_4,m_4}(\mathbf{x}_2-\mathbf{X}_4)\end{aligned} \tag{26}$$

We next need to choose an appropriate value of k in such a way that the convergence of the expansions will be rapid. In general, multicenter repulsion integrals with STOs involve four different values of the Slater exponents, but for our Coulomb Sturmian methods to work, a common value of k must be used. For evaluating one-particle integrals such as the overlap, which involve only two Slater exponents, our experience has shown the convergence is most rapid if we let

$$k=\sqrt{\zeta_1\zeta_2} \tag{27}$$

This suggests that for 4-center interelectron repulsion integrals we should let

$$k=(\zeta_1\zeta_2\zeta_3\zeta_4)^{1/4} \tag{28}$$

but of course, other schemes can be tried to investigate which choice of k converges most rapidly.

4. CHECKS OF CONVERGENCE AND ACCURACY

4.1 Checks Using the Exact 1-Center Case

In the case of a single center (i.e., for atoms), interelectron repulsion integrals can be evaluated exactly both for an STO basis and when Coulomb Sturmians are used as a basis. Therefore we can use this case to test the rate of convergence of the expansion shown in Eq. (26). Table 2 shows the distinct ERIs for a simple carbon calculation with two STOs: 1s with $\zeta = 57/10$, and 2s with $\zeta = 13/8 = 1.625$. The exact repulsion integrals are shown together with the relative differences (i.e., the number of accurate

Table 2 This Table Illustrates the Rate of Convergence for the Synthesis of Interelectron Repulsion Integrals in Terms of Coulomb Sturmians Given a Variety of ρ-Values

	ζ_1	ζ_2	ζ_3	ζ_4		Rel. Error for N-Term Expansion			
	ρ_1	ρ_2	ρ_3	ρ_4	Exact ERI	$N = 5$	$N = 10$	$N = 20$	$N = 40$
$(1s^2\|1s^2)$	5.70 1	5.70 1	5.70 1	5.70 1	3.5625	0	0	0	0
$(2s^2\|2s^2)$	1.625 1	1.625 1	1.625 1	1.625 1	0.59033203125000...	0	0	0	0
$(1s^2\|1s2s)$	5.70 1.37	5.70 1.37	5.70 1.37	1.625 0.39	0.47667864389933...	10^{-2}	10^{-5}	10^{-9}	10^{-20}
$(1s^2\|2s^2)$	5.70 1.87	5.70 1.87	1.625 0.53	1.625 0.53	0.80746939159423...	10^{-3}	10^{-7}	10^{-13}	10^{-26}
$(1s2s\|1s2s)$	5.70 1.87	1.625 0.53	5.7 1.87	1.625 0.53	0.08160009547519...	10^{-2}	10^{-6}	10^{-12}	10^{-25}
$(1s2s\|2s^2)$	5.70 2.56	1.625 0.73	1.625 0.73	1.625 0.73	0.17203279275189...	10^{-2}	10^{-6}	10^{-12}	10^{-26}

The example used are all the distinct single-center ERI (which we can evaluate exactly) for a simple carbon calculation with the STOs 1s, $\zeta = 57/10$ and 2s, $\zeta = 13/8 = 1.625$. We have used $k = (\zeta_1\zeta_2\zeta_3\zeta_4)^{1/4}$.

digits) for the N-term expansions for various N, rounded to nearest order of magnitude. We see that the relative errors resemble those for the individual STO expansions, shown in Table 1, when selecting the worst of the $\rho_1, \ldots, \rho_4$. This suggests that the convergence of the final integral can be estimated well by the expansion accuracy, and that precomputed tables similar to Table 1 can be used to automatically select expansion sizes appropriate for choosing a desired accuracy. We expect the errors introduced by the truncated expansion in the multicenter case to be similar to or smaller than for the single-center case, as is suggested by the relative ERI errors as a function of R, shown in Fig. 1 in the next subsection. More numerical work will be done to investigate both of these conjectures in a rigorous manner.

4.2 Checks Using the Exact 2-Center Case

STO interelectron repulsion integrals between two 1-center densities located on different centers, X_1 and X_2, may be evaluated easily and exactly, provided that $\zeta_1 + \zeta_2 = \zeta_3 + \zeta_4$, as they can be expressed exactly as Coulomb Sturmians, and we can calculate the exact result by using our Sturmian programs. As an example, the special case where $n_1 = n_2 = n_3 = n_4 = 1$ and $l_1 = l_2 = l_3 = l_4 = 0$ yields

$$
\begin{aligned}
&J^{\text{STO}}_{1s,1s,1s,1s}(R)\\
&= 4\pi \int d\mathbf{p} \frac{1}{p^2}\, e^{-i\mathbf{p}\cdot\mathbf{R}} \left[\Phi^*_{\zeta_1,1,0,0}\Phi_{\zeta_2,1,0,0}\right]^t(-\mathbf{p}) \left[\Phi^*_{\zeta_3,1,0,0}\Phi_{\zeta_4,1,0,0}\right]^t(\mathbf{p})\\
&= \frac{4\left(e^{-R(\zeta_1+\zeta_2)}\left(-R(\zeta_1+\zeta_2)(R(\zeta_1+\zeta_2)(R(\zeta_1+\zeta_2)+9)+33)-48\right)+48\right)(\zeta_1\zeta_2\zeta_3\zeta_4)^{3/2}}{3R(\zeta_1+\zeta_2)^6}
\end{aligned}
\tag{29}
$$

where $\mathbf{R} = \mathbf{X}_2 - \mathbf{X}_1$ and $R = |\mathbf{R}|$. When $R \to 0$, the integral in Eq. (29) reduces to

$$
J^{\text{STO}}_{1s,1s,1s,1s}(0) = \frac{20(\zeta_1\zeta_2\zeta_3\zeta_4)^{3/2}}{(\zeta_1+\zeta_2)^5} \tag{30}
$$

(and in the special case when $\zeta_1 = \cdots = \zeta_4 = k$, it becomes simply $\frac{5k}{8}$) while for large values of R,

$$
J^{\text{STO}}_{1s,1s,1s,1s}(R) \to \frac{64(\zeta_1\zeta_2\zeta_3\zeta_4)^{3/2}}{R(\zeta_1+\zeta_2)^6} \tag{31}
$$

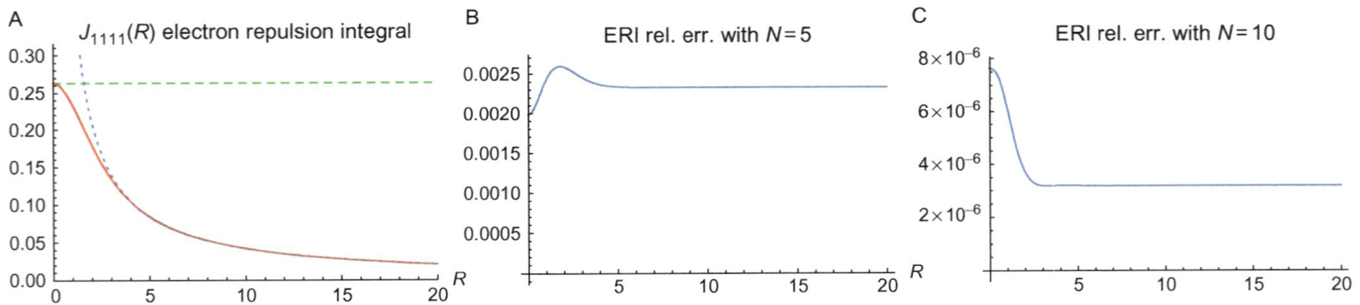

Fig. 1 (A) The exact electron repulsion integral of Eq. (29) as a function of nuclear separation distance R, together with the $R=0$-limit and the large-R asymptote. The approximate integrals with $N=5$ and $N=10$ terms are shown together with the exact curve, both indistinguishable by eye. (B) The relative error of the $N=5$-expansion as a function of R (roughly 3-digit accuracy). (C) The relative error of the $N=10$-expansion as a function of R (roughly 6-digit accuracy).

Similar exact expressions for the electron repulsion as functions of R, as well as the exact asymptotic expression for large R, can be derived in the same way for other integrals when $2k = \zeta_1 + \zeta_2 = \zeta_3 + \zeta_4$, allowing us to calculate the error compared to the generally applicable CSO expansion with $k = (\zeta_1\zeta_2\zeta_3\zeta_4)^{1/4}$.

When $\zeta_1 + \zeta_2 \neq \zeta_3 + \zeta_4$, the integral leads to the Harris W-function.

$$W_{i,i'}^{l,j} = \frac{2}{\pi}\int_0^\infty dp \; \frac{p^{l+2j}}{(p^2+z_1^2)^i (p^2+z_2^2)^{i'}} j_l(pR) \tag{32}$$

This function may be generated using recursion relations, but it is easier to evaluate it numerically.

Fig. 1 shows the exact electron repulsion integral of Eq. (29) for the case where $\zeta_1 = 1/2$, $\zeta_2 = 3/2$, $\zeta_3 = 3/2$, and $\zeta_4 = 1/2$, as a function of nuclear separation R. The limit for $R = 0$ (Eq. 30) and the large-R asymptote (Eq. 31) are also shown, as are syntheses using $N = 5$ and $N = 10$ Coulomb Sturmians. Neither of the truncated expansions can be distinguished from the exact ERI by eye, but the $N = 5$-expansion has a maximum relative error of $\sim 10^{-3}$, and the $N = 10$-expansion has maximum relative error of $\sim 10^{-5}$. Table 2 can be used to determine the number of terms to include in the expansion to obtain the desired accuracy. The convergence appears to be linear in N, in the sense that doubling the number of terms roughly doubles the number of correct digits, but further study is needed in order to verify this.

5. DISCUSSION

In our previous books and papers,[10,11,15,16] we have developed methods by which the theory of hyperspherical harmonics may be used to evaluate molecular integrals rapidly and accurately when Coulomb Sturmians are used as a basis. This includes many-center interelectron repulsion integrals, whose evaluation has long been regarded as difficult and slow when ETOs are used as a basis. Having developed these methods for Coulomb Sturmians, it is desirable to extend them to STOs, and we present one possible scheme for this here. Our method makes use of an expansion of STOs in terms of Coulomb Sturmians. The convergence of this expansion is very rapid when the Slater exponents ζ_j differ from each other only by a factor of 2 or 3, but when the Slater exponents differ more than that, convergence becomes slower, as is illustrated by the test cases shown in Section 4.

We hope that this work will contribute to the growing ETO movement in the quantum chemistry community, which is increasingly offering an alternative to Gaussian technology.

REFERENCES

1. Harris, F. E.; Michels, H. H. The Evaluation of Molecular Integrals for Slater-Type Orbitals. *Adv. Chem. Phys* **1967**, *13*, 205–266.
2. Steinborn, E. O. On the Evaluation of Exponential (Slater) Type Integrals. In: *Methods in Computational Molecular Physics*, Dierksen, G. H. F. ; Wilson, S., Eds.; Reidel: Dordrecht, 1983.
3. Weniger, E. A.; Steinborn, E. O. The Fourier Transforms of Some Exponential-Type Basis Functions and Their Relevance for Multicenter Problems. *J. Chem. Phys.* **1983**, *78*, 6121.
4. Weatherford, C. A.; Jones, H. W. (Eds.). *International Conference on ETO Multicenter Integrals*, Tallahassee, USA, 1981. Reidel: Dordrecht, 1982.
5. Hoggan, P. E.; Ruiz, M. B.; Ozdogan, T. Molecular Integrals Over Slater-Type Orbitals. From Pioneers to Recent Progress. In: *Quantum Frontiers of Atoms and Molecules in Physics, Chemistry, and Biology*, Putz, M. V. Ed.; NOVA Science Publishers: New York, 2011; pp. 61–90.
6. Bouferguene, A.; Fares, M.; Hoggan, P. E. STOP: A Slater-Type Orbital Package for Molecular Electronic Structure Determination. *Int. J. Quantum Chem.* **1996**, *57*, 801–810.
7. Fernandez Rico, J.; Lopez, R.; Ema, I.; Ramirez, G. SMILES: A Reference Program for Molecular Calculations With Slater-Type Orbitals. *J. Comput. Chem.* **1998**, *19*, 1284–1293.
8. *SCM, Theoretical Chemistry*, Vrije Universiteit: Amsterdam, The Netherlands. ADF2012: Amsterdam Density Functional, 2012. http://www.scm.com.
9. Avery, J. E. Fast Electron Repulsion Integrals for Molecular Coulomb Sturmians. In: *Advances in Quantum Chemistry, Advances in Quantum Chemistry*; Hoggan, P. E. Ed.; Vol. 67. Academic Press: Cambridge, MA, 2013; pp. 129–151, https://doi.org/10.1016/B978-0-12-411544-6.00006-6.
10. Avery, J. E.; Avery, J. S. Molecular Integrals for Exponential-Type Orbitals Using Hyperspherical Harmonics. *Adv. Quantum Chem.* **2015**, *70*, 265–324, https://doi.org/10.1016/bs.aiq.2014.07.004.
11. Avery, J. E.; Avery, J. S. Can Coulomb Sturmians be Used as a Basis for N-Electron Molecular Calculations? *J. Phys. Chem. A* **2009**, *113*, 14565–14572, https://doi.org/10.1021/jp9040502.
12. Avery, J. S.; Avery, J. E. Coulomb Sturmians as a Basis for Molecular Calculations. *Mol. Phys.* **2012**, *110* (15), 1593–1608. https://doi.org/10.1080/00268976.2012.658876.
13. Avery, J. E.; Avery, J. S. Molecular Integrals for Slater Type Orbitals Using Coulomb Sturmians. *J. Math. Chem.* **2014**, *52* (1), 301–312. https://doi.org/10.1007/s10910-013-0264-2.
14. Morales, D. A. On the Evaluation of Integrals With Coulomb Sturmian Radial Functions. *J. Math. Chem.* **2016**, *54* (3), 682–689, https://doi.org/10.1007/s10910-015-0588-1.
15. Avery, J. S.; Rettrup, S.; Avery, J. E. *Symmetry-Adapted Basis Sets: Automatic Generation for Problems in Chemistry and Physics*. World Scientific: Singapore, 2012.
16. Avery, J. E. New Computational Methods in the Quantum Theory of Nanostructures (Ph.D. thesis), University of Copenhagen, 2011.

CHAPTER EIGHT

Recurrence Relations for Four-Electron Integrals Over Gaussian Basis Functions

Giuseppe M.J. Barca*, Pierre-François Loos[1,*,†]
*Research School of Chemistry, Australian National University, Canberra, ACT, Australia
†Laboratoire de Chimie et Physique Quantiques, Université de Toulouse, CNRS, UPS, Toulouse, France
[1]Corresponding author: e-mail address: loos@irsamc.ups-tlse.fr

Contents

1. Introduction 147
2. Four-Electron Integrals 149
   2.1 Four-Electron Operators 150
3. Fundamental Integrals 152
4. Recurrence Relations 154
   4.1 Vertical Recurrence Relations 154
   4.2 Transfer Recurrence Relations 157
   4.3 Horizontal Recurrence Relations 158
5. Algorithm 158
6. Concluding Remarks 162
Acknowledgments 162
References 162

Abstract

In the spirit of the Head-Gordon–Pople algorithm, we report vertical, transfer, and horizontal recurrence relations for the efficient and accurate computation of four-electron integrals over Gaussian basis functions. Our recursive approach is a generalization of our algorithm for three-electron integrals (Barca et al., 2016). The RRs derived in the present study can be applied to a general class of multiplicative four-electron operators. In particular, we consider various types of four-electron integrals that may arise in explicitly correlated F12 methods.

1. INTRODUCTION

In 1985, starting from the Hylleraas functional[1–3] and using the interelectronic distance $r_{12} = |\boldsymbol{r}_1 - \boldsymbol{r}_2|$ as a correlation factor, Kutzelnigg derived

Advances in Quantum Chemistry, Volume 76
ISSN 0065-3276
https://doi.org/10.1016/bs.aiq.2017.03.004

the seminal form of the MP2-R12 equations.[4] This explicitly correlated method (more formally stated together with Klopper in 1987[5]) was later extended to higher level of theory and more accurate correlation factors $f_{12} = f(r_{12})$,[6–15] such as Gaussian geminals

$$f_{12} = \exp\left(-\lambda r_{12}^2\right), \tag{1}$$

or Slater geminals

$$f_{12} = \exp\left(-\lambda r_{12}\right). \tag{2}$$

The resulting "F12 methods" achieve chemical accuracy for small organic molecules with relatively small Gaussian basis sets[16–21] and are quickly becoming the first-choice method for high accuracy.[19,20]

However, the inclusion of the correlation factor f_{12} dredged up an old problem: in addition to two-electron integrals (traditional ones and new ones), three-electron integrals over $f_{13}f_{23}$, $r_{12}^{-1}f_{13}$, and $r_{12}^{-1}f_{13}f_{23}$, as well as four-electron integrals over $r_{12}^{-1}f_{14}f_{23}$, $r_{12}^{-1}f_{13}f_{34}$, and $r_{12}^{-1}f_{13}f_{14}$ arise. These integrals are not known analytically except when Gaussian geminals are used,[22,23] and, at that time, the only way to evaluate them would have been via expensive Gauss–Legendre quadratures.[24,25] Additionally, citing Kutzelnigg and Klopper,[26] "*even if fast procedures for the evaluation of these integrals were available, one would have to face the problem of the large number of these integrals; while that of two-electron integrals is* $\sim N^4$, *there are* $\sim N^6$ *three-electron and* $\sim N^8$ *four-electron integrals. The storing and manipulating of these integrals could be handled only for extremely small basis sets.*"

Undoubtedly, in the late 1980s, the two-electron integrals technology was still in development.[27–34] Nowadays, though still challenging, these integrals could be computed much more effectively via judicious recursive schemes, designing the quadrature only to the fundamental integrals.[35] Another important remark is that the actual number of *significant* (i.e., greater than a given threshold) three- and four-electron integrals in a large system, is, at worst, $\mathcal{O}(N^3)$ or $\mathcal{O}(N^4)$. Such scaling results, e.g., by exploiting robust density fitting[36] or upper bound-based screening methods.[37]

Nevertheless, the success of the R12 method was due to the decision of avoiding three- and four-electron integrals entirely through the insertion of the *resolution of the identity* (RI)[17,26,38]

$$\hat{I} \approx \sum_{\mu}^{N_{\mathrm{RI}}} |\chi_\mu\rangle\langle\chi_\mu|. \tag{3}$$

In this way, three- and four-electron integrals are approximated as linear combinations of products of more conventional two-electron integrals. Of course, the accuracy of the RI approximation (3) relies entirely on the assumption that the auxiliary basis set is sufficiently large ($N_{\mathrm{RI}} \gg N$). Thus, in the general context of explicitly correlated methods, it is licit to ask: what is the suitable method to evaluate three- and four-electron integrals?

The use of RI as method of choice does not seem definitive to us. In fact, eschewing the RI approximation would offer at least two advantages:

(i) smaller one-electron basis as the larger auxiliary basis set would not be required anymore;

(ii) the three- and four-electron integrals would be computed exactly.

Moreover, one could avoid the commutator rearrangements involved in the computation of integrals over the kinetic energy operator.[39]

In a recent paper,[35] we reported recurrence relations (RRs) to compute three-electron integrals over Gaussian basis functions for multiplicative general operators of the form $f_{12}g_{13}h_{23}$. Here, we generalize our previous study to four-electron integrals.

The present chapter is organized as follows. In Section 2, we introduce notations and we define various key quantities. Section 3 explains how to calculate fundamental integrals required to start the recursive scheme. In Section 4, we reports vertical, transfer, and horizontal RRs for four-electron integrals. Finally, in Section 5, we propose a recursive scheme based on these RRs to calculate classes of four-electron integrals. Atomic units are used throughout.

2. FOUR-ELECTRON INTEGRALS

A primitive Gaussian-type function (PGF) is specified by an orbital exponent α, a center $\mathbf{A} = (A_x, A_y, A_z)$, and angular momentum $\boldsymbol{a} = (a_x, a_y, a_z)$:

$$\varphi_{\boldsymbol{a}}^{\mathbf{A}}(\boldsymbol{r}) = (x - A_x)^{a_x}(y - A_y)^{a_y}(z - A_z)^{a_z} e^{-\alpha|\boldsymbol{r}-\mathbf{A}|^2}. \tag{4}$$

A contracted Gaussian-type function (CGF) is defined as a normalized sum of PGFs

$$\psi_{a}^{\mathbf{A}}(\boldsymbol{r}) = \sum_{k=1}^{K_A} D_{ak}(x-A_x)^{a_x}(y-A_y)^{a_y}(z-A_z)^{a_z} e^{-\alpha_k|\boldsymbol{r}-\mathbf{A}|^2}, \tag{5}$$

where K_A is the degree of contraction and the D_{ak} are contraction coefficients. Throughout this chapter, we use physicist's notations, and we write the integral over a four-electron operator f_{1234} of CGFs as

$$\begin{aligned}
\langle \boldsymbol{a}_1\boldsymbol{a}_2\boldsymbol{a}_3\boldsymbol{a}_4|\boldsymbol{b}_1\boldsymbol{b}_2\boldsymbol{b}_3\boldsymbol{b}_4\rangle &\equiv \langle \boldsymbol{a}_1\boldsymbol{a}_2\boldsymbol{a}_3\boldsymbol{a}_4|f_{1234}|\boldsymbol{b}_1\boldsymbol{b}_2\boldsymbol{b}_3\boldsymbol{b}_4\rangle \\
&= \iiint \psi_{\boldsymbol{a}_1}^{\mathbf{A}_1}(\boldsymbol{r}_1)\psi_{\boldsymbol{a}_2}^{\mathbf{A}_2}(\boldsymbol{r}_2)\psi_{\boldsymbol{a}_3}^{\mathbf{A}_3}(\boldsymbol{r}_3)\psi_{\boldsymbol{a}_4}^{\mathbf{A}_4}(\boldsymbol{r}_4)f_{1234}\psi_{\boldsymbol{b}_1}^{\mathbf{B}_1}(\boldsymbol{r}_1) \\
&\quad \psi_{\boldsymbol{b}_2}^{\mathbf{B}_2}(\boldsymbol{r}_2)\psi_{\boldsymbol{b}_3}^{\mathbf{B}_3}(\boldsymbol{r}_3)\psi_{\boldsymbol{b}_4}^{\mathbf{B}_4}(\boldsymbol{r}_4)d\boldsymbol{r}_1 d\boldsymbol{r}_2 d\boldsymbol{r}_3 d\boldsymbol{r}_4.
\end{aligned} \tag{6}$$

Additionally, square-bracketed integrals denote integrals over PGFs:

$$\begin{aligned}
[\boldsymbol{a}_1\boldsymbol{a}_2\boldsymbol{a}_3\boldsymbol{a}_4|\boldsymbol{b}_1\boldsymbol{b}_2\boldsymbol{b}_3\boldsymbol{b}_4] &\equiv [\boldsymbol{a}_1\boldsymbol{a}_2\boldsymbol{a}_3\boldsymbol{a}_4|f_{1234}|\boldsymbol{b}_1\boldsymbol{b}_2\boldsymbol{b}_3\boldsymbol{b}_4] \\
&= \iiint \varphi_{\boldsymbol{a}_1}^{\mathbf{A}_1}(\boldsymbol{r}_1)\varphi_{\boldsymbol{a}_2}^{\mathbf{A}_2}(\boldsymbol{r}_2)\varphi_{\boldsymbol{a}_3}^{\mathbf{A}_3}(\boldsymbol{r}_3)\varphi_{\boldsymbol{a}_4}^{\mathbf{A}_4}(\boldsymbol{r}_4)f_{1234}\varphi_{\boldsymbol{b}_1}^{\mathbf{B}_1}(\boldsymbol{r}_1) \\
&\quad \varphi_{\boldsymbol{b}_2}^{\mathbf{B}_2}(\boldsymbol{r}_2)\varphi_{\boldsymbol{b}_3}^{\mathbf{B}_3}(\boldsymbol{r}_3)\varphi_{\boldsymbol{b}_4}^{\mathbf{B}_4}(\boldsymbol{r}_4)d\boldsymbol{r}_1 d\boldsymbol{r}_2 d\boldsymbol{r}_3 d\boldsymbol{r}_4.
\end{aligned} \tag{7}$$

The fundamental integral (i.e., the integral in which all eight basis functions are s-type PGFs) is defined as $[\mathbf{0}] \equiv [\mathbf{0000}|\mathbf{0000}]$ with $\mathbf{0} = (0,0,0)$. The Gaussian product rule reduces it from eight to four centers:

$$[\mathbf{0}] = S_1 S_2 S_3 S_4 \iiint \varphi_{\mathbf{0}}^{\mathbf{Z}_1}(\boldsymbol{r}_1)\varphi_{\mathbf{0}}^{\mathbf{Z}_2}(\boldsymbol{r}_2)\varphi_{\mathbf{0}}^{\mathbf{Z}_3}(\boldsymbol{r}_3)\varphi_{\mathbf{0}}^{\mathbf{Z}_4}(\boldsymbol{r}_4)f_{1234}d\boldsymbol{r}_1 d\boldsymbol{r}_2 d\boldsymbol{r}_3 d\boldsymbol{r}_4, \tag{8}$$

where

$$\zeta_i = \alpha_i + \beta_i, \quad \mathbf{Z}_i = \frac{\alpha_i \mathbf{A}_i + \beta_i \mathbf{B}_i}{\zeta_i}, \quad S_i = \exp\left(-\frac{\alpha_i \beta_i}{\zeta_i}|\mathbf{A}_i\mathbf{B}_i|^2\right), \tag{9}$$

with $\mathbf{A}_i\mathbf{B}_i = \mathbf{A}_i - \mathbf{B}_i$. For conciseness, we will adopt a notation in which missing indices represent s-type Gaussians. For example, $[\boldsymbol{a}_2\boldsymbol{a}_3]$ is a shorthand for $[\mathbf{0}\boldsymbol{a}_2\boldsymbol{a}_3\mathbf{0}|\mathbf{0000}]$. We will also use unbold indices, e.g., $[a_1a_2a_3a_4|b_1b_2b_3b_4]$ to indicate a complete class of integrals from a shell-octet.

2.1 Four-Electron Operators

In the present study, we are particularly interested in the four-electron operators $g_{13}h_{23}i_{34}$ (trident) and $f_{12}h_{23}i_{34}$ (four-electron chain or 4-chain) because they can be required in explicitly correlated methods such as F12 methods.[17–20] Explicitly correlated calculations may also require three-electron integrals over the (3-chain) $f_{12}h_{23}$ and (cyclic) $f_{12}g_{13}h_{23}$ operators, as well as two-electron integrals over f_{12}. However, we will eschew the study of the two-electron integrals here as they have been extensively studied in the past 25 years.[11,13,15,26,40–49] Note that the nuclear attraction integrals can be easily obtained by taking the large-exponent limit of a s-type shell-pair. We refer the interested reader to Refs. 34 and 50 for more details about the computation of nuclear attraction integrals.

The structure of these operators is illustrated in Fig. 1, where we have adopted a diagrammatic representation. Starting with the "pacman" operator $f_{12}g_{13}h_{23}i_{34}$, we are going to show that one can easily derive all the RRs required to compute two-, three-, and four-electron integrals following simple rules. Therefore, in the following, we will focus our analysis on this master "pacman" operator.

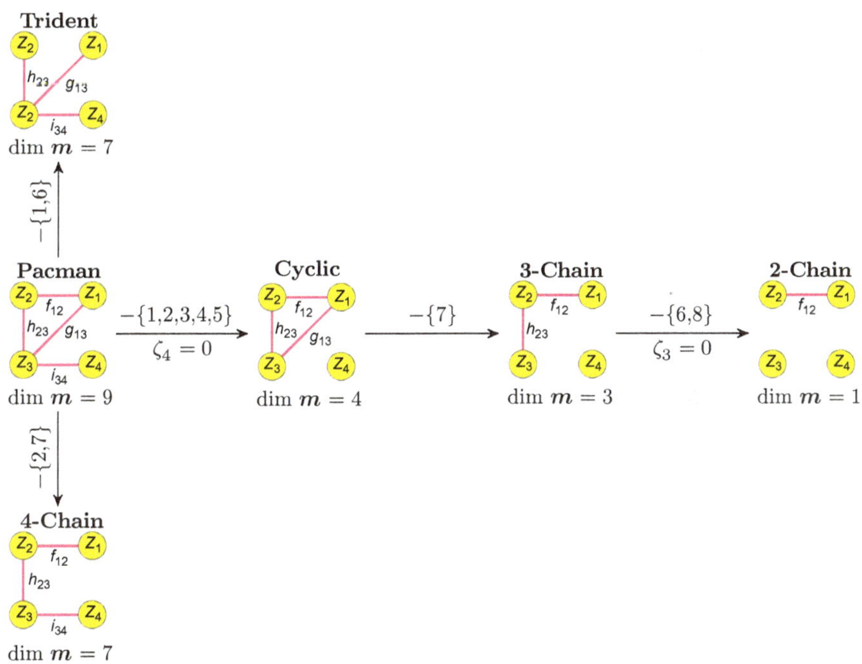

Fig. 1 Diagrammatic representation of various four-, three-, and two-electron integrals involved in explicitly correlated methods. dim $\boldsymbol{m}$ refers to the dimensionality of the auxiliary index $\boldsymbol{m}$ (see Eq. (15)). The values in the *curly brackets* indicate which components of the auxiliary index vector $\boldsymbol{m}$ must be removed.

3. FUNDAMENTAL INTEGRALS

The first step required to compute integrals of arbitrary angular momentum is the computation of the (momentumless) fundamental integrals $[\mathbf{0}]$. These are derived starting from Eq. (8) using the Gaussian integral representation of each two-electron operator. For instance, we have

$$f_{12} = \int_0^\infty F(t_{12}) \exp\left(-t_{12}^2 r_{12}^2\right) dt_{12}, \tag{10}$$

where $F(t_{12})$ is a Gaussian kernel. Table 1 contains kernels $F(t)$ for a variety of important two-electron operators f_{12}. From the formulas in Table 1, one can also easily deduce the kernels for related functions, such as f_{12}^2, f_{12}/r_{12}, and $\nabla^2 f_{12}$, which are of interest in explicitly correlated methods. More general kernels can be found in Ref. 35.

Next, the integration over $\boldsymbol{r}_1$, $\boldsymbol{r}_2$, $\boldsymbol{r}_3$, and $\boldsymbol{r}_4$ can be carried out, yielding

$$[\mathbf{0}] = S_1 S_2 S_3 S_4 \iiiint F(t_{12})\, G(t_{13})\, H(t_{23})\, I(t_{34})\, w_{\mathbf{0}}(\boldsymbol{t})\, d\boldsymbol{t}, \tag{11}$$

Table 1 Kernels $F(t)$ of the Gaussian Integral Representation for Various f_{12} Operators

f_{12}	$F(t)$
1	$\delta(t)$
r_{12}^{-1}	$2/\sqrt{\pi}$
r_{12}^{-2}	$2t$
$(r_{12}^2 + \lambda^2)^{-1/2}$	$(2/\sqrt{\pi}) \exp\left(-\lambda^2 t^2\right)$
$\exp\left(-\lambda\, r_{12}\right)$	$(\lambda\, t^{-2}/\pi) \exp\left(-\lambda^2 t^{-2}/4\right)$
$r_{12}^{-1} \exp\left(-\lambda\, r_{12}\right)$	$(2/\sqrt{\pi}) \exp\left(-\lambda^2 t^{-2}/4\right)$
$\exp\left(-\lambda^2 r_{12}^2\right)$	$\delta(t - \lambda)$
$r_{12}^{-1} \operatorname{erfc}(\lambda\, r_{12})$	$(2/\sqrt{\pi})\, \theta(t - \lambda)$
$r_{12}^{-1} \operatorname{erf}(\lambda\, r_{12})$	$(2/\sqrt{\pi})[1 - \theta(t - \lambda)]$

$\delta(x)$ and $\theta(x)$ are, respectively, the Dirac delta and Heaviside step functions, and $\operatorname{erf}(x)$ and $\operatorname{erfc}(x)$ are the error function and its complement version, respectively.[51]

where $\boldsymbol{t} = (t_{12}, t_{13}, t_{23}, t_{34})$ and

$$w_0(\boldsymbol{t}) = \left(\frac{\pi^4}{D(\boldsymbol{t})}\right)^{3/2} \exp\left[-\frac{N(\boldsymbol{t})}{D(\boldsymbol{t})}\right]. \tag{12}$$

Defining the following polynomials

$$s_1 = t_{12}^2, \quad s_2 = t_{13}^2, \quad s_3 = t_{23}^2, \tag{13a}$$

$$s_4 = t_{34}^2, \quad s_5 = t_{12}^2 t_{13}^2 + t_{12}^2 t_{23}^2 + t_{13}^2 t_{23}^2, \quad s_6 = t_{12}^2 t_{34}^2, \tag{13b}$$

$$s_7 = t_{13}^2 t_{34}^2, \; s_8 = t_{23}^2 t_{34}^2, \quad s_9 = t_{12}^2 t_{13}^2 t_{34}^2 + t_{12}^2 t_{23}^2 t_{34}^2 + t_{13}^2 t_{23}^2 t_{34}^2, \tag{13c}$$

we have

$$\begin{aligned} D(\boldsymbol{t}) = {} & \zeta_1\zeta_2\zeta_3\zeta_4 + (\zeta_1+\zeta_2)\zeta_3\zeta_4 s_1 + \zeta_2\zeta_4(\zeta_1+\zeta_3)s_2 \\ & + \zeta_1\zeta_4(\zeta_2+\zeta_3)s_3 + \zeta_1\zeta_2(\zeta_3+\zeta_4)s_4 + \zeta_4(\zeta_1+\zeta_2+\zeta_3)s_5 \\ & + (\zeta_1+\zeta_2)(\zeta_3+\zeta_4)s_6 + \zeta_2(\zeta_1+\zeta_3+\zeta_4)s_7 \\ & + \zeta_1(\zeta_2+\zeta_3+\zeta_4)s_8 + (\zeta_1+\zeta_2+\zeta_3+\zeta_4)s_9, \end{aligned} \tag{14a}$$

$$\begin{aligned} N(\boldsymbol{t}) = {} & \zeta_3\zeta_4\kappa_{12}s_1 + \zeta_2\zeta_4\kappa_{13}s_2 + \zeta_1\zeta_4\kappa_{23}s_3 \\ & + \zeta_1\zeta_2\kappa_{34}s_4 + \zeta_4(\kappa_{12}+\kappa_{13}+\kappa_{23})s_5 + [(\zeta_3+\zeta_4)\kappa_{12} + (\zeta_1+\zeta_2)\kappa_{34}]s_6 \\ & + \zeta_2(\kappa_{13}+\kappa_{14}+\kappa_{34})s_7 + \zeta_1(\kappa_{23}+\kappa_{24}+\kappa_{34})s_8 \\ & + (\kappa_{12}+\kappa_{13}+\kappa_{14}+\kappa_{23}+\kappa_{24}+\kappa_{34})s_9, \end{aligned} \tag{14b}$$

where $\kappa_{ij} = \zeta_i\zeta_j \, |\mathbf{Z}_i - \mathbf{Z}_j|^2$.

Following Obara and Saika,[32,52] vertical RRs (VRRs) are obtained by differentiation of Eq. (11) with respect to the center coordinates. Therefore, one can show that (11) has to be generalized to the following form:

$$[\mathbf{0}]^{\boldsymbol{m}} = S_1 S_2 S_3 S_4 \iiiint F(t_{12})\, G(t_{13})\, H(t_{23})\, I(t_{34})\, w_{\boldsymbol{m}}(\boldsymbol{t})\, d\boldsymbol{t}, \tag{15}$$

where $\boldsymbol{m} = (m_1, m_2, m_3, m_4, m_5, m_6, m_7, m_8, m_9)$ is, for the pacman operator, a nine-dimensional auxiliary index, and

$$w_{\boldsymbol{m}}(\boldsymbol{t}) = w_0(\boldsymbol{t}) \prod_{k=1}^{\dim \boldsymbol{m}} \left[\frac{s_k}{D(\boldsymbol{t})}\right]^{m_k}. \tag{16}$$

As reported in Fig. 1, while the fundamental integrals of the pacman operator contains 9 auxiliary indices (i.e., $\dim \boldsymbol{m} = 9$), the two interesting four-electron operators (trident and chain) contains 7. The cyclic and chain three-electron operators have only four and three, respectively, while the two-electron chain operator has a single m component (i.e., $\dim \boldsymbol{m} = 1$). These numbers are drastically reduced if one uses Gaussian geminals due to their factorization properties.[53]

4. RECURRENCE RELATIONS

In this section, we report vertical, transfer, and horizontal RRs for the computation of four-electron integrals of arbitrary angular momentum. In particular, we refer the interested readers to the appendix of Ref. 35 for more details about how to derive these VRRs following the late R. Ahlrichs.[54]

4.1 Vertical Recurrence Relations

To build angular momentum over center $\mathbf{A}_1$, we have derived the following 24-term VRR:

$$
\begin{aligned}
[a_1^+ a_2 a_3 a_4]^m = {} & \mathbf{Z}_1\mathbf{A}_1 [a_1 a_2 a_3 a_4]^m - \zeta_2\zeta_3\zeta_4 \mathbf{Z}_{12} [a_1 a_2 a_3 a_4]^{\{1\}} \\
& - \zeta_2\zeta_3\zeta_4 \mathbf{Z}_{13} [a_1 a_2 a_3 a_4]^{\{2\}} - \zeta_4(\zeta_2 \mathbf{Z}_{12} + \zeta_3 \mathbf{Z}_{13}) [a_1 a_2 a_3 a_4]^{\{5\}} \\
& - \zeta_2(\zeta_3 + \zeta_4) \mathbf{Z}_{12} [a_1 a_2 a_3 a_4]^{\{6\}} - \zeta_2(\zeta_3 \mathbf{Z}_{13} + \zeta_4 \mathbf{Z}_{14}) [a_1 a_2 a_3 a_4]^{\{7\}} \\
& - (\zeta_2 \mathbf{Z}_{12} + \zeta_3 \mathbf{Z}_{13} + \zeta_4 \mathbf{Z}_{14}) [a_1 a_2 a_3 a_4]^{\{9\}} \\
& + \frac{a_1}{2\zeta_1} \Big\{ [a_1^- a_2 a_3 a_4]^m - \zeta_2\zeta_3\zeta_4 \mathbf{Z}_{12} [a_1^- a_2 a_3 a_4]^{\{1\}} \\
& - \zeta_2\zeta_3\zeta_4 \mathbf{Z}_{13} [a_1^- a_2 a_3 a_4]^{\{2\}} - \zeta_4(\zeta_2 + \zeta_3) [a_1^- a_2 a_3 a_4]^{\{5\}} \\
& - \zeta_2(\zeta_3 + \zeta_4) [a_1^- a_2 a_3 a_4]^{\{6\}} - \zeta_2(\zeta_3 + \zeta_4) [a_1^- a_2 a_3 a_4]^{\{7\}} \\
& - (\zeta_2 + \zeta_3 + \zeta_4) [a_1^- a_2 a_3 a_4]^{\{9\}} \Big\} \\
& + \frac{a_2}{2} \Big\{ \zeta_3\zeta_4 [a_1 a_2^- a_3 a_4]^{\{1\}} + \zeta_4 [a_1 a_2^- a_3 a_4]^{\{5\}} \\
& + (\zeta_3 + \zeta_4) [a_1 a_2^- a_3 a_4]^{\{6\}} + [a_1 a_2^- a_3 a_4]^{\{9\}} \Big\} \\
& + \frac{a_3}{2} \Big\{ \zeta_2\zeta_4 [a_1 a_2 a_3^- a_4]^{\{2\}} + \zeta_4 [a_1 a_2 a_3^- a_4]^{\{5\}} \\
& + \zeta_2 [a_1 a_2 a_3^- a_4]^{\{7\}} + [a_1 a_2 a_3^- a_4]^{\{9\}} \Big\} \\
& + \frac{a_4}{2} \Big\{ \zeta_2 [a_1 a_2 a_3 a_4^-]^{\{7\}} + [a_1 a_2 a_3 a_4^-]^{\{9\}} \Big\},
\end{aligned}
\tag{17}
$$

where the superscript + or − denotes an increment or decrement of one unit of Cartesian angular momentum. (Thus, $\boldsymbol{a}^{\pm}$ is analogous to $\boldsymbol{a} \pm \mathbf{1}_i$ in the notation of Obara and Saika.) The value in the curly superscript indicates which component of the auxiliary index vector $\boldsymbol{m}$ is incremented.

Because Eq. (17) builds angular momentum over $\mathbf{A}_1$ and all four bra centers have nonzero angular momentum, we will call this expression $\text{VRR}_4^{\mathbf{A}_1}$. The VRRs used to obtain $[\boldsymbol{a}_1^+ \boldsymbol{a}_2 \boldsymbol{a}_3]^{\boldsymbol{m}}$, $[\boldsymbol{a}_1^+ \boldsymbol{a}_2]^{\boldsymbol{m}}$, and $[\boldsymbol{a}_1^+]^{\boldsymbol{m}}$ can be easily derived from Eq. (17) by setting successively $\boldsymbol{a}_4 = \mathbf{0}$, $\boldsymbol{a}_3 = \mathbf{0}$, and $\boldsymbol{a}_2 = \mathbf{0}$. These are, respectively, named $\text{VRR}_3^{\mathbf{A}_1}$, $\text{VRR}_2^{\mathbf{A}_1}$, and $\text{VRR}_1^{\mathbf{A}_1}$.

One can easily derive VRR_4 for the trident and 4-chain operators following the simple rules given in Fig. 1. We obtain the VRRs for the trident operator by removing the terms $\{1\}$ and $\{6\}$. Similarly, 4-chain VRRs are obtained by removing the terms $\{2\}$ and $\{7\}$. This yields an 18- and 17-term $\text{VRR}_4^{\mathbf{A}_1}$ for the trident and 4-chain operators, respectively.

$\text{VRR}_4^{\mathbf{A}_2}$, $\text{VRR}_4^{\mathbf{A}_3}$, and $\text{VRR}_4^{\mathbf{A}_4}$ are used to build angular momentum over $\mathbf{A}_2$, $\mathbf{A}_3$, and $\mathbf{A}_4$, respectively. Their expressions are

$$
\begin{aligned}
[\boldsymbol{a}_1 \boldsymbol{a}_2^+ \boldsymbol{a}_3 \boldsymbol{a}_4]^{\boldsymbol{m}} = {} & \mathbf{Z}_2\mathbf{A}_2 [\boldsymbol{a}_1 \boldsymbol{a}_2 \boldsymbol{a}_3 \boldsymbol{a}_4]^{\boldsymbol{m}} + \zeta_1\zeta_3\zeta_4 \mathbf{Z}_{12} [\boldsymbol{a}_1 \boldsymbol{a}_2 \boldsymbol{a}_3 \boldsymbol{a}_4]^{\{1\}} \\
& - \zeta_1\zeta_3\zeta_4 \mathbf{Z}_{23} [\boldsymbol{a}_1 \boldsymbol{a}_2 \boldsymbol{a}_3 \boldsymbol{a}_4]^{\{3\}} + \zeta_4(\zeta_1 \mathbf{Z}_{12} - \zeta_3 \mathbf{Z}_{23}) [\boldsymbol{a}_1 \boldsymbol{a}_2 \boldsymbol{a}_3 \boldsymbol{a}_4]^{\{5\}} \\
& + \zeta_1(\zeta_3 + \zeta_4) \mathbf{Z}_{12} [\boldsymbol{a}_1 \boldsymbol{a}_2 \boldsymbol{a}_3 \boldsymbol{a}_4]^{\{6\}} - \zeta_1(\zeta_3 \mathbf{Z}_{23} + \zeta_4 \mathbf{Z}_{24}) [\boldsymbol{a}_1 \boldsymbol{a}_2 \boldsymbol{a}_3 \boldsymbol{a}_4]^{\{8\}} \\
& + (\zeta_1 \mathbf{Z}_{12} - \zeta_3 \mathbf{Z}_{23} - \zeta_4 \mathbf{Z}_{24}) [\boldsymbol{a}_1 \boldsymbol{a}_2 \boldsymbol{a}_3 \boldsymbol{a}_4]^{\{9\}} \\
& + \frac{a_2}{2\zeta_2} \Big\{ [\boldsymbol{a}_1 \boldsymbol{a}_2^- \boldsymbol{a}_3 \boldsymbol{a}_4]^{\boldsymbol{m}} - \zeta_1\zeta_3\zeta_4 [\boldsymbol{a}_1 \boldsymbol{a}_2^- \boldsymbol{a}_3 \boldsymbol{a}_4]^{\{1\}} - \zeta_1\zeta_3\zeta_4 [\boldsymbol{a}_1 \boldsymbol{a}_2^- \boldsymbol{a}_3 \boldsymbol{a}_4]^{\{3\}} \\
& - \zeta_4(\zeta_1 - \zeta_3) [\boldsymbol{a}_1 \boldsymbol{a}_2^- \boldsymbol{a}_3 \boldsymbol{a}_4]^{\{5\}} + \zeta_1(\zeta_3 + \zeta_4) [\boldsymbol{a}_1 \boldsymbol{a}_2^- \boldsymbol{a}_3 \boldsymbol{a}_4]^{\{6\}} \\
& - \zeta_1(\zeta_3 + \zeta_4) [\boldsymbol{a}_1 \boldsymbol{a}_2^- \boldsymbol{a}_3 \boldsymbol{a}_4]^{\{8\}} - (\zeta_1 + \zeta_3 + \zeta_4) [\boldsymbol{a}_1 \boldsymbol{a}_2^- \boldsymbol{a}_3 \boldsymbol{a}_4]^{\{9\}} \Big\} \\
& + \frac{a_1}{2} \Big\{ \zeta_3\zeta_4 [\boldsymbol{a}_1^- \boldsymbol{a}_2 \boldsymbol{a}_3 \boldsymbol{a}_4]^{\{1\}} + \zeta_4 [\boldsymbol{a}_1^- \boldsymbol{a}_2 \boldsymbol{a}_3 \boldsymbol{a}_4]^{\{5\}} \\
& + (\zeta_3 + \zeta_4) [\boldsymbol{a}_1^- \boldsymbol{a}_2 \boldsymbol{a}_3 \boldsymbol{a}_4]^{\{6\}} + [\boldsymbol{a}_1^- \boldsymbol{a}_2 \boldsymbol{a}_3 \boldsymbol{a}_4]^{\{9\}} \Big\} \\
& + \frac{a_3}{2} \Big\{ \zeta_1\zeta_4 [\boldsymbol{a}_1 \boldsymbol{a}_2 \boldsymbol{a}_3^- \boldsymbol{a}_4]^{\{3\}} + \zeta_4 [\boldsymbol{a}_1 \boldsymbol{a}_2 \boldsymbol{a}_3^- \boldsymbol{a}_4]^{\{5\}} \\
& + \zeta_1 [\boldsymbol{a}_1 \boldsymbol{a}_2 \boldsymbol{a}_3^- \boldsymbol{a}_4]^{\{8\}} + [\boldsymbol{a}_1 \boldsymbol{a}_2 \boldsymbol{a}_3^- \boldsymbol{a}_4]^{\{9\}} \Big\} \\
& + \frac{a_4}{2} \Big\{ \zeta_1 [\boldsymbol{a}_1 \boldsymbol{a}_2 \boldsymbol{a}_3 \boldsymbol{a}_4^-]^{\{8\}} + [\boldsymbol{a}_1 \boldsymbol{a}_2 \boldsymbol{a}_3 \boldsymbol{a}_4^-]^{\{9\}} \Big\},
\end{aligned}
\tag{18}
$$

$$
\begin{aligned}
[a_1a_2a_3^+a_4]^m = {} & \mathbf{Z}_3\mathbf{A}_3[a_1a_2a_3a_4]^m \\
& + \zeta_1\zeta_2\zeta_4\mathbf{Z}_{13}[a_1a_2a_3a_4]^{\{2\}} + \zeta_1\zeta_2\zeta_4\mathbf{Z}_{23}[a_1a_2a_3a_4]^{\{3\}} \\
& + \zeta_1\zeta_2\zeta_4\mathbf{Z}_{34}[a_1a_2a_3a_4]^{\{4\}} + \zeta_4(\zeta_1\mathbf{Z}_{13} + \zeta_2\mathbf{Z}_{23})[a_1a_2a_3a_4]^{\{5\}} \\
& - \zeta_4(\zeta_1 + \zeta_2)\mathbf{Z}_{34}[a_1a_2a_3a_4]^{\{6\}} + \zeta_2(\zeta_1\mathbf{Z}_{13} - \zeta_4\mathbf{Z}_{34})[a_1a_2a_3a_4]^{\{7\}} \\
& + \zeta_1(\zeta_2\mathbf{Z}_{23} - \zeta_4\mathbf{Z}_{34})[a_1a_2a_3a_4]^{\{8\}} \\
& + (\zeta_1\mathbf{Z}_{13} + \zeta_2\mathbf{Z}_{23} - \zeta_4\mathbf{Z}_{34})[a_1a_2a_3a_4]^{\{9\}} \\
& + \frac{a_3}{2\zeta_3}\Big\{[a_1a_2a_3^-a_4]^m - \zeta_1\zeta_2\zeta_4[a_1a_2a_3^-a_4]^{\{2\}} - \zeta_1\zeta_2\zeta_4[a_1a_2a_3^-a_4]^{\{3\}} \\
& - \zeta_1\zeta_2\zeta_4[a_1a_2a_3^-a_4]^{\{4\}} - \zeta_4(\zeta_1 + \zeta_2)[a_1a_2a_3^-a_4]^{\{5\}} \\
& - \zeta_4(\zeta_1 + \zeta_2)[a_1a_2a_3^-a_4]^{\{6\}} - \zeta_2(\zeta_1 + \zeta_4)[a_1a_2a_3^-a_4]^{\{7\}} \\
& - \zeta_1(\zeta_2 + \zeta_4)[a_1a_2a_3^-a_4]^{\{8\}} - (\zeta_1 + \zeta_2 + \zeta_4)[a_1a_2a_3^-a_4]^{\{9\}}\Big\} \\
& + \frac{a_1}{2}\Big\{\zeta_2\zeta_4[a_1^-a_2a_3a_4]^{\{2\}} + \zeta_4[a_1^-a_2a_3a_4]^{\{5\}} \\
& + \zeta_2[a_1^-a_2a_3a_4]^{\{7\}} + [a_1^-a_2a_3a_4]^{\{9\}}\Big\} \\
& + \frac{a_2}{2}\Big\{\zeta_1\zeta_4[a_1a_2^-a_3a_4]^{\{3\}} + \zeta_4[a_1a_2^-a_3a_4]^{\{5\}} \\
& + \zeta_1[a_1a_2^-a_3a_4]^{\{8\}} + [a_1a_2^-a_3a_4]^{\{9\}}\Big\} \\
& + \frac{a_4}{2}\Big\{\zeta_1\zeta_2[a_1a_2a_3a_4^-]^{\{4\}} + (\zeta_1 + \zeta_2)[a_1a_2a_3a_4^-]^{\{6\}} \\
& + \zeta_2[a_1a_2a_3a_4^-]^{\{7\}} + \zeta_1[a_1a_2a_3a_4^-]^{\{8\}} + [a_1a_2a_3a_4^-]^{\{9\}}\Big\}
\end{aligned}
\tag{19}
$$

$$
\begin{aligned}
[a_1a_2a_3a_4^+]^m = {} & \mathbf{Z}_4\mathbf{A}_4[a_1a_2a_3a_4]^m + \zeta_1\zeta_2\zeta_3\mathbf{Z}_{34}[a_1a_2a_3a_4]^{\{4\}} \\
& + \zeta_3(\zeta_1 + \zeta_2)\mathbf{Z}_{34}[a_1a_2a_3a_4]^{\{6\}} + \zeta_2(\zeta_1\mathbf{Z}_{14} + \zeta_3\mathbf{Z}_{34})[a_1a_2a_3a_4]^{\{7\}} \\
& + \zeta_1(\zeta_2\mathbf{Z}_{24} + \zeta_3\mathbf{Z}_{34})[a_1a_2a_3a_4]^{\{8\}} \\
& + (\zeta_1\mathbf{Z}_{14} + \zeta_2\mathbf{Z}_{24} + \zeta_3\mathbf{Z}_{34})[a_1a_2a_3a_4]^{\{9\}} \\
& + \frac{a_4}{2\zeta_4}\Big\{[a_1a_2a_3a_4^-]^m - \zeta_1\zeta_2\zeta_3[a_1a_2a_3a_4^-]^{\{4\}} \\
& - \zeta_3(\zeta_1 + \zeta_2)[a_1a_2a_3a_4^-]^{\{6\}} - \zeta_2(\zeta_1 + \zeta_3)[a_1a_2a_3a_4^-]^{\{7\}} \\
& - \zeta_1(\zeta_2 + \zeta_3)[a_1a_2a_3a_4^-]^{\{8\}} - (\zeta_1 + \zeta_2 + \zeta_3)[a_1a_2a_3a_4^-]^{\{9\}}\Big\} \\
& + \frac{a_1}{2}\Big\{\zeta_2[a_1^-a_2a_3a_4]^{\{7\}} + [a_1^-a_2a_3a_4]^{\{9\}}\Big\} \\
& + \frac{a_2}{2}\Big\{\zeta_1[a_1a_2^-a_3a_4]^{\{8\}} + [a_1a_2^-a_3a_4]^{\{9\}}\Big\} \\
& + \frac{a_3}{2}\Big\{\zeta_1\zeta_2[a_1a_2a_3^-a_4]^{\{4\}} + (\zeta_1 + \zeta_2)[a_1a_2a_3^-a_4]^{\{6\}} \\
& + \zeta_2[a_1a_2a_3^-a_4]^{\{7\}} + \zeta_1[a_1a_2a_3^-a_4]^{\{8\}} + [a_1a_2a_3^-a_4]^{\{9\}}\Big\}
\end{aligned}
\tag{20}
$$

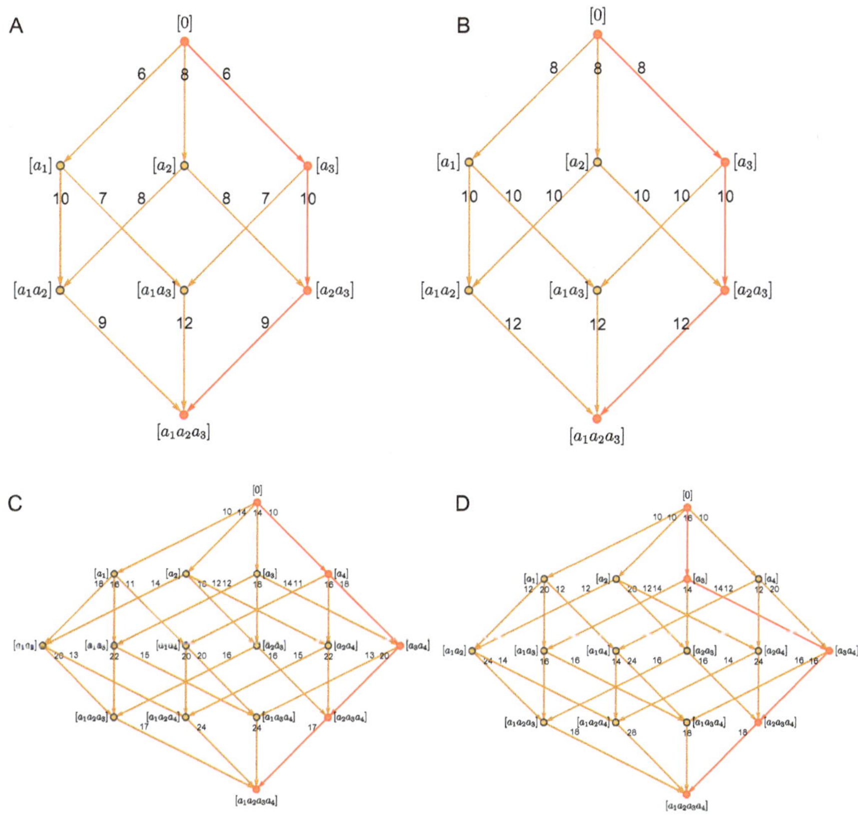

Fig. 2 Graph representation of the VRRs for the 3-chain $f_{12}h_{23}$ (*top left*), cyclic $f_{12}g_{13}h_{23}$ (*top right*), 4-chain $f_{12}h_{23}i_{34}$ (*bottom left*), and trident $g_{13}h_{23}i_{34}$ (*bottom right*) operators. The *edge label* gives the number of terms in the corresponding VRR. The *red path* corresponds to the algorithm generating the smallest number of intermediates.

Again, the corresponding expressions for VRR_1, VRR_2, and VRR_3 can be easily derived from Eqs. (18), (19), and (20). The number of terms for each of these VRRs is reported in Fig. 2 for the 3-chain $f_{12}h_{23}$ (top left), cyclic $f_{12}g_{13}h_{23}$ (top right), 4-chain $f_{12}h_{23}i_{34}$ (bottom left), and trident $g_{13}h_{23}i_{34}$ (bottom right) operators.

4.2 Transfer Recurrence Relations

TRRs redistribute angular momentum between centers hosting to different electrons. Using translational invariance, one can derive

$$\begin{aligned}[a_1^+ a_2 a_3 a_4] =& \frac{a_1}{2\zeta_1}[a_1^- a_2 a_3 a_4] + \frac{a_2}{2\zeta_1}[a_1 a_2^- a_3 a_4] \\ &+ \frac{a_3}{2\zeta_1}[a_1 a_2 a_3^- a_4] + \frac{a_4}{2\zeta_1}[a_1 a_2 a_3 a_4^-] \\ &- \frac{\zeta_2}{\zeta_1}[a_1 a_2^+ a_3 a_4] - \frac{\zeta_3}{\zeta_1}[a_1 a_2 a_3^+ a_4] - \frac{\zeta_4}{\zeta_1}[a_1 a_2 a_3 a_4^+] \\ &- \frac{\beta_1 \mathbf{A}_1\mathbf{B}_1 + \beta_2 \mathbf{A}_2\mathbf{B}_2 + \beta_3 \mathbf{A}_3\mathbf{B}_3 + \beta_4 \mathbf{A}_4\mathbf{B}_4}{\zeta_1}[a_1 a_2 a_3 a_4].\end{aligned} \tag{21}$$

4.3 Horizontal Recurrence Relations

The so-called HRRs enable to shift momentum between centers over the same electronic coordinate:

$$\langle a_1 a_2 a_3 a_4 | b_4^+ \rangle = \langle a_1 a_2 a_3 a_4^+ | b_4 \rangle + \mathbf{A}_4\mathbf{B}_4 \langle a_1 a_2 a_3 a_4 | b_4 \rangle, \tag{22a}$$

$$\langle a_1 a_2 a_3 a_4 | b_3^+ b_4 \rangle = \langle a_1 a_2 a_3^+ a_4 | b_3 b_4 \rangle + \mathbf{A}_3\mathbf{B}_3 \langle a_1 a_2 a_3 a_4 | b_3 b_4 \rangle, \tag{22b}$$

$$\langle a_1 a_2 a_3 a_4 | b_2^+ b_3 b_4 \rangle = \langle a_1 a_2^+ a_3 a_4 | b_2 b_3 b_4 \rangle + \mathbf{A}_2\mathbf{B}_2 \langle a_1 a_2 a_3 a_4 | b_2 b_3 b_4 \rangle, \tag{22c}$$

$$\langle a_1 a_2 a_3 a_4 | b_1^+ b_2 b_3 b_4 \rangle = \langle a_1^+ a_2 a_3 a_4 | b_1 b_2 b_3 b_4 \rangle + \mathbf{A}_1\mathbf{B}_1 \langle a_1 a_2 a_3 a_4 | b_1 b_2 b_3 b_4 \rangle. \tag{22d}$$

Note that HRRs can be applied to contracted integrals because they are independent of the contraction coefficients and exponents.

5. ALGORITHM

In this section, we describe a recursive scheme for the computation of three- and four-electron integrals based on the late-contraction path of the Head–Gordon–Pople (HGP) algorithm.[33] The general skeleton of the algorithm is shown in Fig. 3 for two representative examples: the 3-chain operator $f_{12}g_{23}$ (left) and the 4-chain operator $f_{12}h_{23}i_{34}$ (right). First, let us focus on the 3-chain operator.

To compute a class of three-electron integrals $\langle a_1 a_2 a_3 | b_1 b_2 b_3 \rangle$, starting from the fundamental integrals $[000]^{\boldsymbol{m}}$, we first build up angular momentum over center $\mathbf{A}_3$ with the 6-term VRR_1 to obtain $[00a_3]$. Then, we use the 10-term VRR_2 over $\mathbf{A}_2$ to obtain $[0a_2a_3]$. Finally, we build up momentum over the last bra center $\mathbf{A}_1$ using the 9-term VRR_3 to get $[a_1a_2a_3]$.

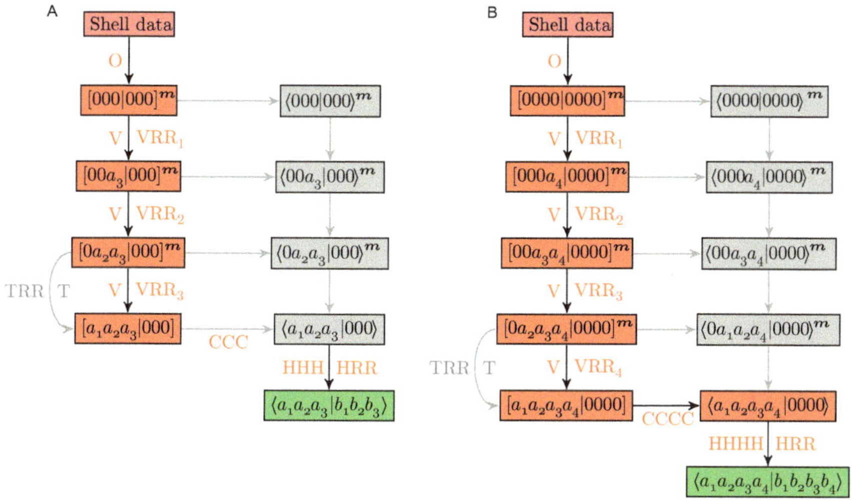

Fig. 3 Schematic representation of the algorithm used to compute three-electron integrals over the 3-chain operator $f_{12}g_{23}$ (*left*) and the four-electron integrals over the 4-chain operator $f_{12}h_{23}i_{34}$ (*right*). The three- and four-electron algorithms follow a OVVVCCCHHH and OVVVVCCCCHHHH path, respectively.

Note that, in our previous paper,[35] we claimed that it would be computationally cheaper to use the 6-term TRR instead of VRR_3 because the number of terms in the TRR is much smaller than in VRR_3. However, we have found here that the number of intermediates (i.e., the number of precomputed classes needed to calculate a given class) required by the paths involving the TRR is much larger (see Table 2). This is readily understood: note that a unit increase in momentum on the last center requires the same increase on all the other centers (as evidenced by the second term in the right-hand side of (21)). Hence, the TRR is computationally expensive for three- and four-electron integrals due to the large number of centers.

As illustrated in the top left graph of Fig. 2, other paths, corresponding to different VRRs, are possible. However, we have found that they do generate a larger number of intermediates, as reported in Table 2.

Similarly to the 3-chain operator, for the 4-chain operator, we get $[a_1a_2a_3a_4]$ by successively building up momentum over $\mathbf{A}_4$, $\mathbf{A}_3$, $\mathbf{A}_2$, and $\mathbf{A}_1$. The number of intermediates required by the other paths is gathered in Table 2. Again, the paths involving the 8-term TRR (reported in Eq. (21)) are much more expensive.

The last two steps of the algorithm are common to the three- and four-electron integral schemes. Following the HGP algorithm,[33] we contract the

Table 2 Number of Intermediates Required to Compute Various Integral Classes for Two-, Three-, and Four-Electron Operators

Integral	Type	Operator	Path	Number of Terms	Centers	Integral Class		
						$[p\ldots p]$	$[d\ldots d]$	$[f\ldots f]$
Two-electron		f_{12}	**VV**	**(4,6)**	$(\mathbf{A}_2,\mathbf{A}_1)$	**4**	**13**	**25**
			VT	(4,4)	$(\mathbf{A}_2,\mathbf{A}_1)$	7	19	37
Three-electron	Chain	$f_{12}h_{23}$	VVV	(6,7,12)	$(\mathbf{A}_3,\mathbf{A}_1,\mathbf{A}_2)$	34	230	881
			VVV	**(6,10,9)**	$(\mathbf{A}_3,\mathbf{A}_2,\mathbf{A}_1)$	**32**	**209**	**778**
			VVV	(8,8,9)	$(\mathbf{A}_2,\mathbf{A}_3,\mathbf{A}_1)$	32	212	801
			VVT	(6,7,6)	$(\mathbf{A}_3,\mathbf{A}_1,\mathbf{A}_2)$	38	246	873
			VVT	(6,10,6)	$(\mathbf{A}_3,\mathbf{A}_2,\mathbf{A}_1)$	43	314	1,256
			VVT	(8,8,6)	$(\mathbf{A}_2,\mathbf{A}_3,\mathbf{A}_1)$	40	260	923
	Cyclic	$f_{12}g_{13}h_{23}$	**VVV**	**(8,10,12)**	$(\mathbf{A}_3,\mathbf{A}_2,\mathbf{A}_1)$	**52**	**469**	**2,216**
			VVT	(8,10,6)	$(\mathbf{A}_3,\mathbf{A}_2,\mathbf{A}_1)$	61	539	2,426
Four-electron	Chain	$f_{12}h_{23}i_{34}$	VVVV	(10,11,20,24)	$(\mathbf{A}_4,\mathbf{A}_1,\mathbf{A}_3,\mathbf{A}_2)$	465	13,781	150,961
			VVVV	**(10,18,20,17)**	$(\mathbf{A}_4,\mathbf{A}_3,\mathbf{A}_2,\mathbf{A}_1)$	**436**	**12,535**	**133,891**
			VVVV	(14,14,20,17)	$(\mathbf{A}_3,\mathbf{A}_4,\mathbf{A}_2,\mathbf{A}_1)$	433	12,704	138,913
			VVVV	(14,18,16,17)	$(\mathbf{A}_3,\mathbf{A}_2,\mathbf{A}_4,\mathbf{A}_1)$	435	12,863	141,679
			VVVT	(10,11,20,8)	$(\mathbf{A}_4,\mathbf{A}_1,\mathbf{A}_3,\mathbf{A}_2)$	532	16,295	181,178

		VVVT	(10,18,20,8)	($\mathbf{A}_4$,$\mathbf{A}_3$,$\mathbf{A}_2$,$\mathbf{A}_1$)	560	17,029	188,242
		VVVT	(14,14,20,8)	($\mathbf{A}_3$,$\mathbf{A}_4$,$\mathbf{A}_2$,$\mathbf{A}_1$)	559	17,487	199,050
		VVVT	(14,18,16,8)	($\mathbf{A}_3$,$\mathbf{A}_2$,$\mathbf{A}_4$,$\mathbf{A}_1$)	543	16,612	185,869
Trident	$g_{13}h_{23}i_{34}$	VVVV	(10,12,14,28)	($\mathbf{A}_4$,$\mathbf{A}_2$,$\mathbf{A}_1$,$\mathbf{A}_3$)	445	13,139	143,619
		VVVV	(10,12,24,18)	($\mathbf{A}_4$,$\mathbf{A}_2$,$\mathbf{A}_3$,$\mathbf{A}_1$)	447	13,381	148,911
		VVVV	(10,20,16,18)	($\mathbf{A}_4$,$\mathbf{A}_3$,$\mathbf{A}_2$,$\mathbf{A}_1$)	418	12,447	133,853
		VVVV	**(16,14,16,18)**	($\mathbf{A}_3$,$\mathbf{A}_4$,$\mathbf{A}_2$,$\mathbf{A}_1$)	**418**	**12,054**	**129,322**
		VVVT	(10,12,14,8)	($\mathbf{A}_4$,$\mathbf{A}_2$,$\mathbf{A}_1$,$\mathbf{A}_3$)	470	13,306	136,584
		VVVT	(10,12,24,8)	($\mathbf{A}_4$,$\mathbf{A}_2$,$\mathbf{A}_3$,$\mathbf{A}_1$)	546	16,917	191,171
		VVVT	(10,20,16,8)	($\mathbf{A}_4$,$\mathbf{A}_3$,$\mathbf{A}_2$,$\mathbf{A}_1$)	521	15,515	168,958
		VVVT	(16,14,16,8)	($\mathbf{A}_3$,$\mathbf{A}_4$,$\mathbf{A}_2$,$\mathbf{A}_1$)	499	13,969	142,264

The path generating the minimum number of intermediates is highlighted in bold. The number of terms in the RRs and the associated incremental center is also reported.

integrals: $[\mathbf{a}_1\mathbf{a}_2\mathbf{a}_3\mathbf{a}_4|\mathbf{0000}]$ to form $\langle\mathbf{a}_1\mathbf{a}_2\mathbf{a}_3\mathbf{a}_4|\mathbf{0000}\rangle$ in the four-electron case, or $[\mathbf{a}_1\mathbf{a}_2\mathbf{a}_3|\mathbf{000}]$ to form $\langle\mathbf{a}_1\mathbf{a}_2\mathbf{a}_3|\mathbf{000}\rangle$ in the three-electron case. More details about the contraction step can be found in Ref. 34. The final step of the algorithm shifts momentum to the ket centers from the bra centers with the help of the 2-term HRRs reported in Section 4.3.

6. CONCLUDING REMARKS

In this study, we have reported RRs for the efficient and accurate computation of four-electron integrals over Gaussian basis functions and a general class of multiplicative four-electron operators of the form $f_{12}g_{13}h_{23}i_{34}$. Starting from this master operator, one can easily derive the RRs for various operators arising in explicitly correlated methods following simple diagrammatic rules (see Fig. 1).

Here, we have derived three types of RRs: (i) starting from the fundamental integrals, vertical RRs (VRRs) allow to increase the angular momentum over the bra centers; (ii) the transfer RR (TRR) redistributes angular momentum between centers hosting different electrons and can be used instead of the VRR on the last bra center; (iii) the horizontal RRs (HRRs) enable to shift momentum from the bra to the ket centers corresponding to the same electronic coordinate. Importantly, HRRs can be applied to contracted integrals.

Finally, after carefully studying the different paths one can follow to build up angular momentum (see Fig. 2), we have proposed a late-contraction recursive scheme which minimizes the number of intermediates to be computed (see Fig. 3). We believe our approach represents a major step toward a fast and accurate computational scheme for three- and four-electron integrals within explicitly correlated methods. It also paves the way to contraction-effective methods for these types of integrals.[55] In particular, an early contraction scheme would have significant computational benefits.

ACKNOWLEDGMENTS

P.F.L. thanks the NCI National Facility for generous grants of supercomputer time, and the Australian Research Council for a Discovery Project grant (DP140104071). The authors would like to thank the anonymous referee for helpful and constructive comments.

REFERENCES

1. Hylleraas, E. A. On the Ground State of the Helium Atom. *Z. Phys.* **1928**, *48*, 469.
2. Hylleraas, E. A. A New Calculation of the Energy of Helium in the Ground State as well as the Lowest Term of Ortho-helium. *Z. Phys.* **1929**, *54*, 347.

3. Sinanoğlu, O. Many-Electron Theory of Atoms and Molecules. I. Shells, Electron Pairs vs Many-Electron Correlations. *J. Chem. Phys.* **1962**, *36*, 3198.
4. Kutzelnigg, W. R12-Dependent Terms in the Wave Function as Closed Sums of Partial Wave Amplitudes for Large l. *Theor. Chim. Acta* **1985**, *68*, 445.
5. Klopper, W.; Kutzelnigg, W. Møller-Plesset Calculations Taking Care of the Correlation Cusp. *Chem. Phys. Lett.* **1987**, *134*, 17.
6. Noga, J.; Kutzelnigg, W. Coupled Cluster Theory That Takes Care of the Correlation Cusp by Inclusion of Linear Terms in the Interelectronic Coordinates. *J. Chem. Phys.* **1994**, *101*, 7738.
7. Klopper, W.; Kutzelnigg, W. Wave Functions With Terms Linear in the Interelectronic Coordinates to Take Care of the Correlation Cusp. III. Second-Order Møller-Plesset (MP2-R12) Calculations on Molecules of First Row Atoms. *J. Chem. Phys.* **1991**, *94*, 2020.
8. Klopper, W.; Rohse, R.; Kutzelnigg, W. CID and CEPA Calculations With Linear R12 Terms. *Chem. Phys. Lett.* **1991**, *178*, 455.
9. Termath, V.; Klopper, W.; Kutzelnigg, W. Wave Functions With Terms Linear in the Interelectronic Coordinates to Take Care of the Correlation Cusp. II. Second-Order Møller-Plesset (MP2-R12) Calculations on Closed-Shell Atoms. *J. Chem. Phys.* **1991**, *94*, 2002.
10. Persson, B. J.; Taylor, P. R. Accurate Quantum-Chemical Calculations: The Use of Gaussian-Type Geminal Functions in the Treatment of Electron Correlation. *J. Chem. Phys.* **1996**, *105*, 5915.
11. Persson, B. J.; Taylor, P. R. Molecular Integrals Over Gaussian-Type Geminal Basis Functions. *Theor. Chem. Acc.* **1997**, *97*, 240.
12. May, A. J.; Manby, F. R. An Explicitly Correlated Second Order Møller-Plesset Theory Using a Frozen Gaussian Geminal. *J. Chem. Phys.* **2004**, *121*, 4479.
13. Ten-no, S. Initiation of Explicitly Correlated Slater-type Geminal Theory. *Chem. Phys. Lett.* **2004**, *398*, 56.
14. Tew, D. P.; Klopper, W. New Correlation Factors for Explicitly Correlated Electronic Wave Functions. *J. Chem. Phys.* **2005**, *123*, 074101.
15. May, A. J.; Valeev, E. F.; Polly, R.; Manby, F. R. Analysis of the Errors in Explicitly Correlated Electronic Structure Theory. *Phys. Chem. Chem. Phys.* **2005**, *7*, 2710.
16. Klopper, W.; Manby, F. R.; Ten-no, S.; Valeev, E. F. R12 Methods in Explicitly Correlated Molecular Electronic Structure Theory. *Int. Rev. Phys. Chem.* **2006**, *25*, 427.
17. Hattig, C.; Klopper, W.; Kohn, A.; Tew, D. P. Explicitly Correlated Electrons in Molecules. *Chem. Rev.* **2012**, *112*, 4.
18. Kong, L.; Bischo, F. A.; Valeev, E. F. Explicitly Correlated R12/F12 Methods for Electronic Structure. *Chem. Rev.* **2012**, *112*, 75.
19. Ten-no, S. Explicitly Correlated Wave Functions: Summary and Perspective. *Theor. Chem. Acc.* **2012**, *131*, 1070.
20. Ten-no, S.; Noga, J. Explicitly Correlated Electronic Structure Theory From R12/F12 Ansatze. *WIREs Comput. Mol. Sci.* **2012**, *2*, 114.
21. Gruneis, A.; Hirata, S.; Ohnishi, Y.-Y.; Ten-no, S. Perspective: Explicitly Correlated Electronic Structure Theory for Complex Systems. *J. Chem. Phys.* **2017**, *146*, 080901.
22. Boys, S. F. The Integral Formulae for the Variational Solution of the Molecular Many-Electron Wave Equations in Terms of Gaussian Functions With Direct Electronic Correlation. *Proc. R. Soc. A* **1960**, *258*, 402.
23. Singer, K. The Use of Gaussian (Exponential Quadratic) Wave Functions in Molecular Problems. I. General Formulae for the Evaluation of Integrals. *Proc. R. Soc. A* **1960**, *258*, 412.
24. Preiskorn, A.; Zurawski, B. Many-Electron, Multicenter Integrals in Superposition of Correlated Configurations Method. I. One- and Two-Electron Integrals. *Int. J. Quantum Chem.* **1985**, *27*, 641.

25. Clementi, E. *Modern Techniques in Computational Chemistry: MOTECC-89*; ESCOM Science Publishers B.V.: New York, 1989.
26. Kutzelnigg, W.; Klopper, W. Wave Function With Terms Linear in the Interelectronic Coordinates to Take Care of the Correlation Cusp. I. General theory. *J. Chem. Phys.* **1991**, *94*, 1985.
27. McMurchie, L. E.; Davidson, E. R. One- and Two-Electron Integrals Over Cartesian Gaussian Functions. *J. Comput. Phys.* **1978**, *26*, 218.
28. Pople, J. A.; Hehre, W. J. Computation of Electron Repulsion Integrals Involving Contracted Gaussian Basis Functions. *J. Comput. Phys.* **1978**, *27*, 161.
29. King, H. F.; Dupuis, M. Numerical Integration Using RYS Polynomials. *J. Comput. Phys.* **1976**, *21*, 144.
30. Dupuis, M.; Rys, J.; King, H. F. Evaluation of Molecular Integrals Over Gaussian Basis Functions. *J. Chem. Phys.* **1976**, *65*, 111.
31. Rys, J.; Dupuis, M.; King, H. F. Computation of Electron Repulsion Integrals Using the RYS Quadrature Method. *J. Comput. Chem.* **1983**, *4*, 154.
32. Obara, S.; Saika, A. Efficient Recursive Computation of Molecular Integrals Over Cartesian Gaussian Functions. *J. Chem. Phys.* **1986**, *84*, 3963.
33. Head-Gordon, M.; Pople, J. A. A Method for Two-Electron Gaussian Integral and Integral Derivative Evaluation Using Recurrence Relations. *J. Chem. Phys.* **1988**, *89*, 5777.
34. Gill, P. M. W. Molecular Integrals Over Gaussian Basis Functions. *Adv. Quantum Chem.* **1994**, *25*, 141.
35. Barca, G. M. J.; Loos, P. F.; Gill, P. M. W. Many-Electron Integrals Over Gaussian Basis Functions. I. Recurrence Relations for Three-Electron Integrals. *J. Chem. Theory Comput.* **2016**, *12*, 1735.
36. Womack, J. C.; Manby, F. R. Density Fitting for Three-Electron Integrals in Explicitly Correlated Electronic Structure Theory. *J. Chem. Phys.* **2014**, *140*, 044118.
37. Barca, G. M. J.; Loos, P. F.; Gill, P. M. W. Many-Electron Integrals Over Gaussian Basis Functions. II. Upper Bounds. J. Chem. Theory Comput. (in preparation).
38. Werner, H. J.; Adler, T. B.; Manby, F. R. General Orbital Invariant MP2-F12 Theory. *J. Chem. Phys.* **2007**, *126*, 164102.
39. Rohse, R.; Klopper, W.; Kutzelnigg, W. Configuration Interaction Calculations With Terms Linear in the Interelectronic Coordinate for the Ground State of H_3^+. A Benchmark Study. *J. Chem. Phys.* **1993**, *99*, 8830.
40. Klopper, W.; Rohse, R. Computation of Some New Two-Electron Gaussian Integrals. *Theor. Chem. Acc.* **1992**, *83*, 441.
41. Klopper, W.; Samson, C. C. M. Explicitly Correlated Second-Order Møller-Plesset Methods With Auxiliary Basis Sets. *J. Chem. Phys.* **2002**, *116*, 6397.
42. Manby, F. R. Density Fitting in Second-Order Linear-R12 Møller-Plesset Perturbation Theory. *J. Chem. Phys.* **2003**, *119*, 4607.
43. Werner, H. J.; Manby, F. R.; Knowles, P. J. Fast Linear Scaling Second-Order Møller-Plesset Perturbation Theory (MP2) Using Local and Density Fitting Approximations. *J. Chem. Phys.* **2003**, *118*, 8149.
44. Klopper, W. A Hybrid Scheme for the Resolution-of-the-Identity Approximation in Second-Order Møller-Plesset Linear-R12 Perturbation Theory. *J. Chem. Phys.* **2004**, *120*, 10890.
45. Ten-no, S. Explicitly Correlated Second Order Perturbation Theory: Introduction of a Rational Generator and Numerical Quadratures. *J. Chem. Phys.* **2004**, *121*, 117.
46. Manby, F. R.; Werner, H. J.; Adler, T. B.; May, A. J. Explicitly Correlated Local Second-Order Perturbation Theory With a Frozen Geminal Correlation Factor. *J. Chem. Phys.* **2006**, *124*, 094103.
47. Ten-no, S. New Implementation of Second-Order Møller-Plesset Perturbation Theory With an Analytic Slater-type Geminal. *J. Chem. Phys.* **2007**, *126*, 014108.

48. Komornicki, A.; King, H. F. A General Formulation for the Efficient Evaluation of *n*-Electron Integrals Over Products of Gaussian Charge Distributions With Gaussian Geminal Functions. *J. Chem. Phys.* **2011**, *134*, 244115.
49. Reine, S.; Helgaker, T.; Lindh, R. Multi-Electron Integrals. *WIREs Comput. Mol. Sci.* **2012**, *2*, 290.
50. Gill, P. M. W.; Pople, J. A. The PRISM Algorithm for Two-Electron Integrals. *Int. J. Quantum Chem.* **1991**, *40*, 753.
51. Olver, F. W. J.; Lozier, D. W.; Boisvert, R. F.; Clark, C. W., Eds. *NIST Handbook of Mathematical Functions*; Cambridge University Press: New York, 2010.
52. Obara, S.; Saika, A. General Recurrence Formulas for Molecular Integrals Over Cartesian Gaussian Functions. *J. Chem. Phys.* **1988**, *89*, 1540.
53. Barca, G. M. J.; Loos, P. F. Three- and Four-Electron Integrals Involving Gaussian Geminals: Fundamental Integrals, Upper Bounds and Recurrence Relations. J. Chem. Phys. (submitted) arXiv:1704.08051.
54. Ahlrichs, R. A Simple Algebraic Derivation of the Obara-Saika Scheme for General Two-Electron Interaction Potentials. *Phys. Chem. Chem. Phys.* **2006**, *8*, 3072.
55. Barca, G. M. J.; Gill, P. M. W. Two-Electron Integrals Over Gaussian Geminals. *J. Chem. Theory Comput.* **2016**, *12*, 4915.

CHAPTER NINE

Ab Initio Molecular Local Nuclear Magnetic Shielding Tensors

Nabil Joudieh*, **Ali Bağcı**[†], **Philip E. Hoggan**[‡,1]
*University of Damascus, Damascus, Syrian Arab Republic
[†]University of Warsaw, Warsaw, Poland
[‡]Institute Pascal, UMR 6602 CNRS, Aubière Cedex, France
[1]Corresponding author: e-mail address: pehoggan@yahoo.com

Contents

1. Introduction 168
2. HFR Formalism for Field and Moments 169
3. HFR With Field and Moments: NDDO Level 171
4. HFR/NDDO London Approximation for Fields 172
5. HFR INDO Formalism for Field and Moments Within the London Approximation 173
6. Second-Order Energy Correction $E^2(\vec{B}, \vec{\mu})$ Using CPHF 174
7. Double Perturbed HF 176
8. Exchange of Perturbation in HF 177
9. HFR Exchange of Perturbations 177
10. Perturbation Exchange for Molecular Diamagnetism 178
11. Theorem of Exchange of Perturbation Directions 179
12. The Tensors χ_{he}, $\Phi_{\kappa,\,he}$, and $\sigma_{\kappa,\,he}$ 180
13. Discussion and Conclusion 181
References 181

Abstract

A formalism to evaluate susceptibility tensors in molecules χ and those of nuclear shielding σ_k is developed using GIAO (gauge-including AOs). It uses the coupled-perturbed Hartree–Fock formalism.

Originality resides in the definition of local susceptibilities.

An in-house MOPAC code provides an NDDO approximation to this molecular site approach which has also been used for chemical shift determination within the GAUSSIAN suite of programs.

Advances in Quantum Chemistry, Volume 76
ISSN 0065-3276
https://doi.org/10.1016/bs.aiq.2017.03.001

1. INTRODUCTION

The aim of this work is to provide a method of evaluating chemical shifts and tensorial magnetic properties of molecules. We have previously shown in Ref. 1 that the accuracy of an NDDO approximation implemented in MOPAC[2–4] is competitive with DFT in a large Gaussian basis[5] for dipolar nuclei, although the proton spectra are difficult to interpret. In this work, we provide an ab initio framework and compare it with the semiempirical approximations, using the molecular site approach and local tensors.[6]

Standard theory of molecular diamagnetism[7–10] gives the energy of a molecule, comprising nuclei possessing spin and with magnetic dipole moments $\vec{\mu}_k$, placed in a constant homogeneous field $\vec{B}$, in the following matrix notation:

$$E\left\{\vec{B},\vec{\mu}\right\} = E\{0,0\} - \frac{1}{2}B^{+}\chi B - \frac{1}{2}\sum_{\kappa=1}^{N_A}\mu_{\kappa}^{+}\Theta_{\kappa}\mu_{\kappa} + \sum_{\kappa=1}^{N_A}\mu_{\kappa}^{+}\sigma_{\kappa}B + \Delta E. \quad (1)$$

$E\{0,0\}$ denotes the zero field, zero moment molecular energy. In addition, the energy $E\left\{\vec{B},\vec{\mu}\right\}$ … is expanded in a Mac–Laurin–Taylor series about zero, in the 6-dimensional space spanned by $\{B_i,\mu_i\}$, $i = 1, 2, 3$:

$$\begin{aligned} E\left\{\vec{B},\vec{\mu}\right\} =& E\{0,0\} + \frac{1}{2}\sum_{i=1}^{3}\sum_{j=1}^{3}\left(\frac{\partial_2 E}{\partial B_i \partial B_j}\right)_{0,0} B_i B_j \\ &+\frac{1}{2}\sum_{\kappa=1}^{N_A}\sum_{i=1}^{3}\sum_{j=1}^{3}\left(\frac{\partial_2 E}{\partial \mu_{i\kappa}\partial \mu_{j\kappa}}\right)_{0,0} \mu_{i\kappa}\mu_{j\kappa} \\ &+\sum_{\kappa=1}^{N_A}\sum_{i=1}^{3}\sum_{j=1}^{3}\left(\frac{\partial_2 E}{\partial B_i \partial \mu_{j\kappa}}\right)_{0,0} B_i \mu_{j\kappa} + \Delta E. \end{aligned} \quad (2)$$

All terms in the expansion containing odd powers of the moments $B_i^q \cdot \mu_i^q$, $(q \geq 0, p+q = 2n+1, n \geq 0)$, must be zero, because of energy invariance with respect to reversal of field and moment orientation $E\left\{\vec{B},\vec{\mu}\right\}$. Comparing (1) and (2) allows the components of tensors χ, Θ, and σ to be defined

$$\chi_{ij} = -\left(\frac{\partial_2 E}{\partial B_i \partial B_j}\right)_{0,0};\quad \Theta_{\kappa,ij} = -\left(\frac{\partial_2 E}{\partial \mu_{i\kappa}\partial \mu_{j\kappa}}\right)_{0,0};\quad \sigma_{\kappa,ij} = \left(\frac{\partial_2 E}{\partial B_i \partial \mu_{j\kappa}}\right)_{0,0} \tag{3}$$

χ is the diamagnetic susceptibility tensor, whereas Θ_κ has not been reported in the literature. It is the local susceptibility tensor for the electron distribution under the local field due to moment $\vec{\mu}_\kappa$ of a nucleus with spin. The nuclear screening tensor σ_κ is the molecular response at nucleus κ. Average susceptibilities and screening constants $\overline{\chi}$ and $\overline{\sigma}\langle\rangle$ exist.

At the ab initio level for $\overline{\chi}$ Ditchfield[9] used AO including gauge GIAO/HFPT.

Perturbed Hartree–Fock Roothaan (HFR) formalisms[11] include field and moments for molecules, as listed below.

2. HFR FORMALISM FOR FIELD AND MOMENTS

The Hamiltonian for a molecule of $n = 2m$ electrons of moment $\vec{\mu}_\kappa$ in a constant homogeneous magnetic field is

$$H\left\{\vec{r}_i; \vec{B}, \vec{\mu}\right\} = \sum_{i=1}^{n} h_i\left(\vec{r}_i; \vec{B}, \vec{\mu}\right) + \sum_{i<j} \frac{e^{-2}}{r_{ij}} \tag{4}$$

$$h_i\left(\vec{r}_i; \vec{B}, \vec{\mu}\right) = \frac{1}{2m}\left(\vec{P}_i^{0} + \frac{e}{c}\vec{A}_i\right)^2 + \sum_{\kappa=1}^{N_A}\left(-\frac{Z_\kappa e^{-2}}{r_{\kappa i}}\right) \tag{5}$$

$$\vec{A}_i = \vec{A}_{iB} + \vec{A}_{i\mu} = \frac{1}{2}\left(\vec{B} \wedge \vec{r}_i\right) + \sum_{\kappa=1}^{N_A} \frac{\vec{\mu}_\kappa \wedge \vec{r}_{i\kappa}}{r_{i\kappa}^2} \tag{6}$$

A Slater determinant is built up of the m occupied field-dependent MOs that satisfy the variation principle for the moments present.

$$F\left(\vec{B}, \vec{\mu}\right)\Psi_h\left(\vec{B}, \vec{\mu}\right) = e_h\left(\vec{B}, \vec{\mu}\right)\Psi_h\left(\vec{B}, \vec{\mu}\right) \tag{7}$$

$$\left\langle \Psi_h\left(\vec{B}, \vec{\mu}\right) | \Psi_j\left(\vec{B}, \vec{\mu}\right)\right\rangle = \delta_{hj} \tag{8}$$

$$F\left(\vec{B}, \vec{\mu}\right) = h\left(\vec{B}, \vec{\mu}\right) + G\left(\vec{B}, \vec{\mu}\right) \tag{9}$$

$$h\left(\vec{B},\vec{\mu}\right)=\frac{1}{2m}\left(\vec{P}^0+\frac{e}{c}\vec{A}\right)^2 \tag{10}$$

$$G\left(\vec{B},\vec{\mu}\right)=\sum_{j=1}^{m}\left[2J_j\left(1;\vec{B},\vec{\mu}\right)-K_j\left(1;\vec{B},\vec{\mu}\right)\right] \tag{11}$$

$$J_j\left(1;\vec{B},\vec{\mu}\right)=\int_{D_2}\frac{e^{-2}}{r_{12}}\Psi_j\left(2;\vec{B},\vec{\mu}\right)\Psi_j^*\left(2;\vec{B},\vec{\mu}\right)d\tau_2 \tag{12}$$

$$K_j\left(1;\vec{B},\vec{\mu}\right)=\int_{D_2}\frac{e^{-2}}{r_{12}}\Psi_j\left(1;\vec{B},\vec{\mu}\right)\Psi_j^*\left(2;\vec{B},\vec{\mu}\right)d\tau_2P_{12} \tag{13}$$

To solve the HF equations, the LCAO approximation must be used and invariance to the choice of origin and axes is assured by using the GIAO.[12–21]

$$\Psi_h\left(\vec{r};\vec{B},\vec{\mu}\right)=\sum_{s=1}^{n}C_{hs}\left(\vec{B},\vec{\mu}\right);\quad \Phi_s\left(\vec{r};\vec{B}\right)=\Phi^+C_h \tag{14}$$

$$\Phi_s\left(\vec{r};\vec{B}\right)=\Gamma_s\left(\vec{r};\vec{B}\right)\varphi_s\left(\vec{r}_s\right) \tag{15}$$

$$\Gamma_s\left(\vec{r};\vec{B}\right)=\exp\left[-\frac{ie}{\hbar c}\vec{A}_s\cdot\vec{r}\right] \tag{16}$$

$$\vec{A}_s=\frac{1}{2}\left(\vec{B}_s\vec{d}_s\right) \tag{17}$$

Here $\vec{r}$ and $\vec{r}_s$ denote the position vector of an electron in the molecule, with respect to the origin and to nucleus s. The vector potential $\vec{A}_s$ is defined for nucleus s. Γ_s is the gauge factor for the GIAO. Thus, the HFR equation for fields and moments using GIAO become

$$FC_h=e_hSC_h \tag{18}$$

$$C_j^+SC_h=\delta_{jh} \tag{19}$$

$$F_{rs}=h_{rs}+G_{rs} \tag{20}$$

$$h_{rs}=\left\langle\Phi_r^*\left|\frac{1}{2m}\left(\vec{P}^0+\frac{e}{c}\vec{A}\right)^2+\sum_{\kappa=1}^{N_A}\left(-\frac{Z_\kappa e^{-2}}{r_\kappa}\right)\right|\Phi_s\right\rangle \tag{21}$$

$$G_{rs}=\sum_{t=1}^{n}\sum_{u=1}^{n}p_{tu}q_{rstu} \tag{22}$$

$$q_{rstu}=\langle\Phi_r^*\Phi_s|\Phi_t\Phi_u^*\rangle-\frac{1}{2}\langle\Phi_r^*\Phi_t|\Phi_s\Phi_u^*\rangle \tag{23}$$

$$E\left(\vec{B},\vec{\mu}\right)=\frac{1}{2}tr\left\{P^T\left(\vec{B},\vec{\mu}\right)\left[h\left(\vec{B},\vec{\mu}\right)+F\left(\vec{B},\vec{\mu}\right)\right]\right\} \quad (24)$$

$$p_{rs}\left(\vec{B},\vec{\mu}\right)=\sum_{h=1}^{m}2C_{rh}^{*}\left(\vec{B},\vec{\mu}\right)C_{sh}\left(\vec{B},\vec{\mu}\right) \quad (25)$$

3. HFR WITH FIELD AND MOMENTS: NDDO LEVEL

The NDDO approximation,[22,23] for GIAO bases satisfies[24,25]:

$$\Phi_{r(R)}^{*}\left(1;\vec{B}\right)\Phi_{s(S)}^{*}\left(1;\vec{B}\right)d\tau_1=\delta_{RS}\Phi_{r(R)}^{*}\left(1;\vec{B}\right)\Phi_{s(S)}^{*}\left(1;\vec{B}\right)d\tau_1 \quad (26)$$

For an *ns*, *np* valence basis the overlap matrix becomes $S=I$. The Roothaan secular equations are

$$h\Phi_s=\Gamma_s\left\{h^0+\frac{e}{mc}\left(\vec{A}-\vec{A}_s\right)\vec{P}^0+\frac{e^2}{2mc^2}\left(\vec{A}-\vec{A}_s\right)^2\right\}\varphi_s \quad (27)$$

gives for h_{rs} the relation:

$$h_{rs}=\overset{\sim 0}{h}_{rs}+\left\langle\Gamma_r^*\Gamma_s\varphi_r\left|\frac{e}{mc}\left(\vec{A}-\vec{A}_s\right)\vec{P}^0+\frac{e^2}{2mc^2}\left(\vec{A}-\vec{A}_s\right)^2\right|\varphi_s\right\rangle \quad (28)$$

$$\overset{\sim 0}{h}_{rs}=\left\langle\Gamma_r^*\Gamma_s\varphi_r\left|T^0+\sum_{\kappa=1}^{N_A}V_\kappa\right|\varphi_s\right\rangle;\quad V_\kappa=-\frac{Z_\kappa e^{-2}}{r_\kappa} \quad (29)$$

The product $\Gamma_r^*\Gamma_s$ in $\overset{\sim 0}{h}_{rs}$ does not allow NDDO to be used except for the G matrix, where the identity:

$$\sum_{r=1}^{n}\sum_{s=1}^{n}\chi_{rs}\equiv\sum_{R=1}^{N_A}\sum_{S=1}^{N_A}\sum_{r(S)}\sum_{s(S)}\chi_{rs} \quad (30)$$

applies to replace the sum over orbitals with that over atoms $R\in[1,N_A]$ of the molecule and OA $R;\varphi_{r(R)}$ and, assuming equal gauge factors:

$$\Gamma_{r(R)}\equiv\Gamma_{s(S)}\quad\forall r,s\in R\Rightarrow\Gamma_{r(R)}^{*}\Gamma_{s(R)}=1 \quad (31)$$

The G matrix for NDDO becomes

$$G_{r(R)s(S)}\left(\vec{B},\vec{\mu}\right)=\sum_{T\neq R}\sum_{t(T)}\sum_{u(T)}P_{tu}\langle\varphi_r\varphi_s|\varphi_t\varphi_u\rangle$$
$$+\sum_{t(R)}\sum_{u(R)}P_{tu}\left[\langle\varphi_r\varphi_s|\varphi_t\varphi_u\rangle-\frac{1}{2}\langle\varphi_r\varphi_t|\varphi_s\varphi_u\rangle\right] \tag{32}$$

$$G_{r(R)s(S)}\left(\vec{B},\vec{\mu}\right)=-\frac{1}{2}\sum_{t(R)}\sum_{u(S)}P_{tu}\langle\varphi_r\varphi_t|\varphi_s\varphi_u\rangle \tag{33}$$

These relations and the property in (31) imply the field is absent from the two-electron integrals and thus only implicitly present in the density matrix.

4. HFR/NDDO LONDON APPROXIMATION FOR FIELDS

The London approximation[26] simplifies the position vectors of electrons to those of the atoms in the axes defining the field:

$$\vec{r}=\frac{1}{2}\left(\vec{d}_r+\vec{d}_s\right)+\frac{1}{2}\left(\vec{r}_R+\vec{r}_S\right) \tag{34}$$

The London approximation gives $\vec{r}\simeq\frac{1}{2}\left(\vec{d}_r+\vec{d}_s\right)$, giving the product $\Gamma_r^*\Gamma_s$ the form:

$$\Gamma_r^*\Gamma_s=\lambda_{rs}=\exp\left[\frac{ie}{2\hbar c}\left(\vec{A}_r-\vec{A}_s\right)\left(\vec{d}_r+\vec{d}_s\right)\right] \tag{35}$$

Hermiticity of h is restored neglecting the antihermitic part $h: h_{rs}\rightarrow\frac{1}{2}\left(h_{rs}+h_{rs}^+\right)$. Giving:

$$h_{rs}=\lambda_{rs}\left\{h_{rs}^0+\frac{e}{4mc}\sum_{p=1}^{3}B_pI_{rs}^p+\frac{e^2}{16mc^2}\sum_{p=1}^{3}\sum_{q=1}^{3}B_pB_qJ_{rs}^{pq}\right.$$
$$+\frac{e}{mc}\sum_{N=1}^{N_A}\sum_{p=1}^{3}\mu_{p_N}K_{rs}^p+\frac{e^2}{2mc^2}\sum_{N=1}^{N_A}\sum_{p=1}^{3}\sum_{q=1}^{3}\mu_{p_N}\mu_{q_N}L_{rs}^{pq} \tag{36}$$
$$\left.+\frac{e^2}{4mc^2}\sum_{N=1}^{NA}\sum_{p=1}^{3}\sum_{q=1}^{3}B_p\mu_{q_N}M_{rs}^{pq}\right\}$$

$$I_{rs}^p=\left\langle\varphi_r\middle|\vec{u}_p\cdot\vec{L}_S\middle|\varphi_s\right\rangle-\left\langle\varphi_s\middle|\vec{u}_p\cdot\vec{L}_R\middle|\varphi_r\right\rangle \tag{37}$$

$$J_{rs}^{pq}=\langle\varphi_r|\left(\vec{u}_p\wedge\vec{r}_R\right)\left(\vec{u}_q\wedge\vec{r}_R\right)+\left(\vec{u}_p\wedge\vec{r_S}\right)\left(\vec{u}_q\wedge\vec{r_S}\right)|\varphi_s\rangle \tag{38}$$

$$K_{rs}^{p}=\left\langle \varphi_r \left| \frac{\vec{u}_p \vec{L}_N}{r_N^3} \right| \varphi_s \right\rangle - \left\langle \varphi_s \left| \frac{\vec{u}_p \vec{L}_N}{r_N^3} \right| \varphi_r \right\rangle \tag{39}$$

$$L_{rs}^{pq}=\left\langle \varphi_r \left| \frac{\left(\vec{u}_p \wedge \vec{r}_N\right)\left(\vec{u}_q \wedge \vec{r}_N\right)}{r_N^6} \right| \varphi_s \right\rangle \tag{40}$$

$$M_{rs}^{pq}=\left\langle \varphi_r \left| \frac{\left(\vec{u}_p \wedge \vec{r}_R\right)\left(\vec{u}_q \wedge \vec{r}_N\right)+\left(\vec{u}_p \wedge \vec{r}_S\right)\left(\vec{u}_q \wedge \vec{r}_N\right)}{r_N^3} \right| \varphi_s \right\rangle \tag{41}$$

$$\lambda_{rs}=\exp\left[\frac{ie}{2\hbar c}\sum_{p=1}^{3} B_p S_{rs}^{p}\right] \tag{42}$$

$$S_{rs}^{p}=\vec{u}_p\left(\vec{d}_r \wedge \vec{d}_s\right); \quad \vec{L}_x=\vec{r}_x \wedge \vec{P}^0; \quad \vec{P}^0=-i\hbar\vec{\nabla}; \quad x=N,R,S \tag{43}$$

$\vec{u}_p, p=1,2,3$ is the unit vector. The h_{rs}^0 unperturbed elements for NDDO are

$$h_{rs}^0=|1-(\delta_{rs}+\delta_{SR})|U_{rs}+\delta_{RS}\sum_{T\neq R}^{Na}\langle\varphi_r|V_T|\varphi_s\rangle \tag{44}$$

$$U_{r(R)s(R)}=\left\langle\varphi_{r(R)}|T^0+V_R|\varphi_{s(R)}\right\rangle \tag{45}$$

$$U_{r(R)s(S)}=\left\langle\varphi_{r(R)}|T^0+V_R+V_S|\varphi_{s(S)}\right\rangle \tag{46}$$

The NDDO elements G_{rs} are unaffected by the London approximation and given by (36).

5. HFR INDO FORMALISM FOR FIELD AND MOMENTS WITHIN THE LONDON APPROXIMATION

The ZDO approximation on a GIAO basis gives

$$\Phi_{r(R)}^{*}\left(\vec{r};\vec{B}\right)\Phi_{s(S)}\left(\vec{r};\vec{B}\right)d\tau=\delta_{RS}\delta_{rs}\Phi_{r(R)}\left(\vec{r};\vec{B}\right)\Phi_{s(S)}\left(\vec{r};\vec{B}\right) \tag{47}$$

ZDO is included in NDDO with approximate atomic overlaps. The core Hamiltonian $h_{rs}\left(\vec{B},\vec{\mu}\right)$ takes the same form in NDDO and INDO. In h_{rs}^0 ZDO requires that invariance to a change of axes be restored in h_{rs}^0 and G_{rs}. For h_{rs}^0 the approximation used is

$$\left\langle \varphi_{r(R)} \middle| V_T \middle| \varphi_{s(R)} \right\rangle = \delta_{rs} V_{RT} \quad \forall \varphi_r \in R \tag{48}$$

For invariant INDO h^0_{rs} is written

$$h^0_{rs} = |1 - (\delta_{rs} + \delta_{RS})| U_{rs} + \delta_{rs} \sum_{T \neq R}^{N_A} V_{RT} \tag{49}$$

Similarly, in the following approximation:

$$\left\langle \Phi^*_{r(R)} \Phi_{r(R)} \middle| \Phi^*_{t(T)} \Phi_{t(T)} \right\rangle \equiv \left\langle \varphi_{r(R)} \varphi_{r(R)} \middle| \varphi_{t(T)} \varphi_{t(T)} \right\rangle = \gamma_{RT} \tag{50}$$

preserves invariance of the INDO method. The G matrix for invariant INDO takes the form:

$$\begin{aligned} G_{r(R)s(R)} = {} & \delta rs \sum_{T \neq R} \sum_{t(T)} \gamma_{RT} P_{tt} \\ & + \sum_{t(R)} \sum_{u(R)} P_{tu} \left[\left\langle \varphi_{r(R)} \varphi_{s(R)} \middle| \varphi_{t(R)} \varphi_{u(R)} \right\rangle - \frac{1}{2} \left\langle \varphi_{r(R)} \varphi_{u(R)} \middle| \varphi_{t(R)} \varphi_{s(R)} \right\rangle \right] \end{aligned} \tag{51}$$

$$G_{r(R)s(S)} = -\frac{1}{2} P_{sr} \gamma_{RS} \tag{52}$$

6. SECOND-ORDER ENERGY CORRECTION $E^2\left(\vec{B}, \vec{\mu}\right)$ USING CPHF

Using coupled-perturbed Hartree–Fock (CPHF) the energy is correct to second order and is written

$$E^2 = tr\left\{ P^{0T} h^2 + \frac{1}{2} P^1 T h^1 \right\} \tag{53}$$

c^0_i is the i-th unperturbed eigenvector and $F^1_{ij} = \langle \Psi^0_i | F^1 | \Psi^0_j \rangle$. The corrections $\chi^1 = h^1$, G^1, F^1, P^1 can be determined by Maclaurin expansion and perturbation series for the matrices $\chi = h$, G, F, P perturbed by magnetic fields and moments:

$$\chi\left(\vec{B},\vec{\mu}\right)=\chi^0+\sum_{p=1}^{3}B_p\left(\frac{\partial\chi}{\partial B_p}\right)_{0,0}+\sum_{N=1}^{N_A}\sum_{p=1}^{3}\mu_{p_N}\left(\frac{\partial\chi}{\partial\mu_{p_N}}\right)_{0,0}$$
$$+\frac{1}{2}\sum_{p=1}^{3}\sum_{q=1}^{3}B_pB_q\left(\frac{\partial_2\chi}{\partial B_p\partial B_q}\right)_{0,0}$$
$$+\frac{1}{2}\sum_{N=1}^{N_A}\sum_{p=1}^{3}\sum_{q=1}^{3}\mu_{p_N}\mu_{q_N}\left(\frac{\partial_2\chi}{\partial\mu_{p_N}\partial\mu_{q_N}}\right)_{0,0} \tag{54}$$
$$+\sum_{N=1}^{N_A}\sum_{p=1}^{3}\sum_{q=1}^{3}B_p\mu_{q_N}\left(\frac{\partial_2\chi}{\partial B_p\partial\mu_{q_N}}\right)_{0,0}+\cdots$$

$$\chi=\chi^0+\chi^1+\chi^2+\cdots\sum_{p=0}^{n}\chi^p \tag{55}$$

Identification of both expansions and evaluation of the derivatives, gives, for NDDO:

$$\chi^1=\sum_{p=1}^{3}B_p\chi^{1B_p}+\sum_{N=1}^{N_A}\sum_{p=1}^{3}\mu_{p_N}\chi^{1\mu_p} \tag{56}$$

$$\chi^2=\sum_{p=1}^{3}\sum_{q=1}^{3}B_pB_q\chi^{2B_{pq}}+\sum_{N=1}^{N_A}\sum_{p=1}^{3}\sum_{q=1}^{3}\mu_{p_N}\mu_{q_N}\chi^{2\mu_{pq}}+\sum_{N=1}^{N_A}\sum_{p=1}^{3}\sum_{q=1}^{3}B_p\mu_{q_N}\chi^{2B\mu_{pq}} \tag{57}$$

$$h_{rs}^{1B_p}=\left(\frac{\partial h_{rs}^1}{\partial B_p}\right)_{0,0}=\frac{ie}{2\hbar c}S_{rs}^p h_{rs}^0+\frac{e}{4mc}I_{rs}^p \tag{58}$$

$$h_{rs}^{1\mu_p}=\left(\frac{\partial h_{rs}^1}{\partial\mu_{p\kappa}}\right)_{0,0}=\frac{e}{2mc}K_{rs}^p \tag{59}$$

$$h_{rs}^{2B_{pq}}=\frac{1}{2}\left(\frac{\partial_2 h_{rs}^2}{\partial B_p\partial B_q}\right)_{0,0}=-\frac{e^2}{8\hbar c^2}s_{rs}^p s_{rs}^q h_{rs}^0+\frac{e^2}{16mc^2}J_{rs}^{pq}-\frac{ie^2}{16\hbar mc^2}\left[s_{rs}^p I_{rs}^q+s_{rs}^q I_{rs}^p\right] \tag{60}$$

$$h_{rs}^{2\mu_{pq}}=\frac{1}{2}\left(\frac{\partial_2 h_{rs}^2}{\partial\mu_{p\kappa}\partial\mu_{q\kappa}}\right)_{0,0}=\frac{e^2}{2mc^2}L_{rs}^{pq} \tag{61}$$

$$h_{rs}^{2B\mu_{pq}}=\left(\frac{\partial_2 h_{rs}^2}{\partial B_p\partial\mu_{q\kappa}}\right)_{0,0}=\frac{ie^2}{4\hbar mc^2}s_{rs}^p K_{rs}^q+\frac{e^2}{4mc^2}M_{rs}^{pq} \tag{62}$$

$$G_{rs}^{1w_p} = -\sum_{t(S)}\sum_{u(R)}\frac{1}{2}(1-\delta_{RS})P_{tu}^{1w_p}\gamma_{ruts} \quad w = B, \mu \tag{63}$$

$$\gamma_{ruts} = \langle \varphi_r\varphi_u | \varphi_t\varphi_s \rangle \tag{64}$$

$$P_{rs}^{1w_p} = \sum_{i=1}^{m}\sum_{j=m+1}^{n}\frac{2}{e_j^0 - e_i^0}\left[F_{ij}^{1w_p}c_{ri}^0c_{sj}^0 + F_{ji}^{1w_p}c_{rj}^0c_{si}^0\right] \tag{65}$$

$$F_{ij}^{1w_p} = \left\langle \Psi_i^0 | F^{1w_p} | \Psi_j^0 \right\rangle = \sum_{t=1}^{n}\sum_{u=1}^{n} c_{ti}^{0*}c_{uj}^0 F_{tu}^{1w_p} \tag{66}$$

$$F_{tu}^{1w_p} = h_{tu}^{1w_p} + G_{tu}^{1wp}\left(P^{1w_p}\right) \quad w = B, \mu \tag{67}$$

The above formalism shows P^1 depends on F^1, in turn a function of P^1 via $G^1(P^1)$. To determine P^1, an iterative process is initialized by $G^1(P^1) = 0 \Rightarrow F^1 = h^1$. Giving P^1 used in $F^1 = h^1 + G^1$. At convergence P^1 for INDO is determined $G^{1wp}\hat{a}$ using the following:

$$G_{r(R)s(R)}^{1w_p} = \sum_{t(R)}\sum_{u(R)} P_{tu}^{1w_p}\left(\gamma_{rstu} - \frac{1}{2}\gamma_{ruts}\right) \tag{68}$$

$$G_{r(R)s(S)}^{1w_p} = \frac{1}{2}P_{rs}^{1w_p}\gamma_{rrss} \tag{69}$$

For E^2 this formalism leads to the relation:

$$\begin{aligned} 2E^2 = &\sum_{p=1}^{3}\sum_{q=1}^{3} B_pB_q tr\left\{E^{2B_{pq}}\right\} + \sum_{\kappa=1}^{N_A}\sum_{p=1}^{3}\sum_{q=1}^{3}\mu_{p\kappa}\mu_{q\kappa} tr\left\{E^{2\mu_{pq}}\right\} \\ &+ \sum_{\kappa=1}^{N_A}\sum_{p=1}^{3}\sum_{q=1}^{3} B_p\mu_{q\kappa} tr\left\{E^{2B\mu_{pq}}\right\} \end{aligned} \tag{70}$$

$$E^{2w_{pq}} = \left(P^{1w_p}\right)^T h^{1w_q} + 2P^{0T}h^{2w_{pq}}, \quad w = B, \mu \tag{71}$$

$$E^{2B\mu_{pq}} = \left(P^{1B_p}\right)^T h^{1w_q} + \left(P^{1\mu_q}\right)^T h^{1B_p} + 2P^{0T}h^{2B\mu_{pq}} \tag{72}$$

7. DOUBLE PERTURBED HF

Coupled HF can be formulated for a double perturbation.[27–31] Setting $\chi_h = F - e_h$, the HF equation to order one using the DPCHF method is written

$$\left[\chi_h^{10}\Psi_h^{00} + \chi_h^{00}\Psi_h^{10}\right] + \left[\chi_h^{01}\Psi_h^{00} + \chi^{00_h}\Psi_h^{01}\right] = 0 \tag{73}$$

$$\langle\Psi_j^{10}|\Psi_h^{00}\rangle + \langle\Psi_j^{00}|\Psi_h^{10}\rangle + \left[\langle\Psi_j^{01}|\Psi_h^{00}\rangle + \langle\Psi_j^{00}|\Psi_h^{01}\rangle\right] = 0 \tag{74}$$

Terms have two indices successively to express order of the two perturbations. From (73) and (74) shows a sum of terms.

$$\left(F^{10} - e_h^{10}\right)\Psi_h^{00} + \left(F^{00} - e_h^{00}\right)\Psi_h^{10} = 0 \tag{75}$$

$$\langle\Psi_j^{10}|\Psi_h^{00}\rangle + \langle\Psi_j^{00}|\Psi_h^{10}\rangle = 0 \tag{76}$$

$$\left(F^{01} - e_h^{01}\right)\Psi_h^{00} + \left(F^{00} - e_h^{00}\right)\Psi_h^{01} = 0 \tag{77}$$

$$\langle\Psi_j^{01}|\Psi_h^{00}\rangle + \langle\Psi_j^{00}|\Psi_h^{01}\rangle = 0 \tag{78}$$

In the limit of a complete basis the solution of Eqs. (75–78) is

$$e_h^{10} = F_{hh}^{10}; \quad e_h^{01} = F_{hh}^{01} \tag{79}$$

$$a_{jh}^{10} = \frac{F_{jh}^{10}}{e_h^{00} - e_j^{00}}; \quad a_{jh}^{01} = \frac{F_{jh}^{01}}{e_h^{00} - e_j^{00}} \tag{80}$$

$$a_{jh}^{*10} + a_{hj}^{10} = 0; \quad a_{jh}^{*01} + a_{hj}^{01} = 0 \tag{81}$$

$$a_{hh}^{10} = 0; \quad a_{hh}^{01} = 0 \tag{82}$$

8. EXCHANGE OF PERTURBATION IN HF

Multiply Eq. (75) by Ψ_h^{01} and Eq. (77) by Ψ_h^{10} and integrate. The following identity is obtained

$$\langle\Psi_h^{01}|F^{10}|\Psi_h^{00}\rangle = \langle\Psi_h^{10}|F^{01}|\Psi_h^{00}\rangle^* \quad h \in [1, n] \tag{83}$$

Thus identity permits the exchange of perturbations.[32–34]

9. HFR EXCHANGE OF PERTURBATIONS

Adding Eq. (83) and its conjugate. Adding the first m equations corresponding to the m occupied orbitals:

$$\sum_{i=1}^{m}\left[\left\langle\Psi_h^{00}|F^{10}|\Psi_h^{01}\right\rangle + \left\langle\Psi_h^{01}|F^{10}|\Psi_h^{00}\right\rangle\right] = \sum_{i=1}^{n}\left[\left\langle\Psi_h^{00}|F^{01}|\Psi_h^{10}\right\rangle + \left\langle\Psi_h^{10}|F^{01}|\Psi_h^{00}\right\rangle\right] \tag{84}$$

The left-hand side of Eq. (84) is called T_1, takes the form:

$$T_1 = \sum_{h=1}^{m}\sum_{j=1}^{m} t_{jh} + \sum_{h=1}^{m}\sum_{j=m+1}^{n} t_{jh} \tag{85}$$

$$t_{jh} = a_{jh}^{01} F_{hj}^{10} + a_{jh}^{*01} F_{jh}^{10} \tag{86}$$

From Eqs. (81) and (82), we show the first term of Eq. (85) is zero. expanding F_{hj}^{10} and F_{jh}^{10} in an unperturbed LCAO, from the right of Eq. (84), gives the HFR exchange theorem:

$$tr\left\{\left(P^{01}\right)^{T} F^{10}\right\} = tr\left\{\left(P^{10}\right)^{T} F^{01}\right\} \tag{87}$$

P^{01} is the first order correction to the density matrix due to the second perturbation, μ_κ, e.g.,:

$$P_{rs}^{01} = \sum_{h=1}^{m}\sum_{j=m+1}^{n} \frac{2}{e_h^{00} - e_j^{00}}\left[F_{hj}^{01} c_{rh}^{0} c_{sj}^{0*} + F_{jh}^{01} c_{rj}^{0} c_{sh}^{0*}\right] \tag{88}$$

P^{10} is defined like F^{01} by F^{10}.

10. PERTURBATION EXCHANGE FOR MOLECULAR DIAMAGNETISM

The matrix $\chi^1 = h^1, G^1, F^1, P^1$ is obtained from Eq. (56). However, χ^1 is written from DPCHF as:

$$\chi^1 = \chi^{10} + \chi^{01} \equiv \chi^{1B} + \chi^{1\mu} \tag{89}$$

Comparing Eqs. (56) and (89) shows

$$\chi^{1B} = \sum_{p=1}^{3} B_p \chi^{1B_p}; \quad \chi^{1\mu} = \sum_{\kappa=1}^{N_A}\sum_{p=1}^{3} \mu_{p\kappa} \chi^{1\mu_p} \tag{90}$$

Replacing Eq. (90); $\chi = P, F$; in Eq. (87):

$$\begin{aligned} &\sum_{\kappa=1}^{N_A}\sum_{p=1}^{3}\sum_{q=1}^{3} B_p \mu_{q\kappa} tr\left\{\left(P^{1\mu_q}\right)^{T} F^{1B_p}\right\} \\ &\sum_{\kappa=1}^{N_A}\sum_{p=1}^{3}\sum_{q=1}^{3} B_p \mu_{q\kappa} tr\left\{\left(P^{1B_p}\right)^{T} F^{1\mu_q}\right\} \end{aligned} \tag{91}$$

Deriving both sides of Eq. (91) with respect to B_h and μ_{eN}, the exchange theorem takes the form

$$tr\left\{\left(P^{1\mu_e}\right)^T F^{1B_h}\right\} = tr\left\{\left(P^{1B_h}\right)^T F^{1\mu_e}\right\} \tag{92}$$

Within NDDO and INDO we can show

$$tr\left\{\left(P^{1\mu_e}\right)^T G^{1B_h}\right\} = tr\left\{\left(P^{1B_h}\right)^T G^{1\mu_e}\right\} \tag{93}$$

From Eqs. (30) and (66) with $w = B$, the left-hand side of Eq. (93) becomes

$$tr\left\{\left(P^{1\mu_e}\right)^T G^{1B_h}\right\} = \sum_{R=1}^{N_A}\sum_{S=1}^{N_A}\sum_{r(R)}\sum_{s(S)}\sum_{t(S)}\sum_{u(R)} -\frac{1}{2}(1-\delta_{RS})P_{tu}^{1B_h}P_{rs}^{1\mu_e}\gamma_{ruts} \tag{94}$$

The indices r and u, s and t are equivalent and can be permuted. Hermiticity of P^1 and the identity $\gamma_{urts} \equiv \gamma_{ruts}$, lead to:

$$tr\left\{\left(P^{1\mu_e}\right)^T G^{1B_h}\right\} = \sum_{R=1}^{N_A}\sum_{S=1}^{N_A}\sum_{r(R)}\sum_{s(S)} P_{rs}^{1B_h}\sum_{t(S)}\sum_{u(R)} -\frac{1}{2}(1-\delta_{RS})P_{tu}^{1\mu_e}\gamma_{ruts} \tag{95}$$

Again substituting Eq. (66) with $w = \mu$ in Eq. (95) using Eq. (30), we establish the theorem Eq. (93) within NDDO and INDO using Eq. (30) we write

$$tr\left\{\left(P^{1\mu_e}\right)^T G^{1B_h}\right\} = \sum_{R=1}^{N_A}\sum_{r(R)}\sum_{s(R)} P_{rs}^{1\mu_e}G_{rs}^{1B_h} + \sum_{R=1}^{NA}\sum_{S\neq R}\sum_{r(R)}\sum_{s(S)} P_{rs}^{1\mu_e}G_{rs}^{1B_h} \tag{96}$$

Replacing the single and two-center elements of $G_{rs}^{1B_h}$ using Eqs. (68) and (69) with $w = B$; equivalent indices and indistinguishable electrons, using Eqs. (68) and (69) with $w = \mu$, we obtain Eq. (93). Given that $F^{1w_v} = h^{1w_v} + G^{1w_v}$; $w = B$, μ and $v = h$, e, combining Eqs. (92) and (93) leads to the theorem:

$$tr\left\{\left(P^{1\mu_e}\right)^T h^{1B_h}\right\} = tr\left\{\left(P^{1B_h}\right)^T h^{1\mu_e}\right\} \tag{97}$$

11. THEOREM OF EXCHANGE OF PERTURBATION DIRECTIONS

From Eq. (70) shows that indices p and q are equivalent. For E^2 permuting p and q shows this equivalence. The difference is zero:

$$\sum_{p=1}^{3}\sum_{q=1}^{3}B_pB_q tr\{E^{2B_{pq}}-E^{2B_{qp}}\}+\sum_{\kappa=1}^{N_A}\sum_{p=1}^{3}\sum_{q=1}^{3}\mu_{p\kappa}\mu_{q\kappa}$$
$$tr\{E^{2\mu_{pq}}-E^{2\mu_{qp}}\}+\sum_{\kappa=1}^{N_A}\sum_{p=1}^{3}\sum_{q=1}^{3}B_p\mu_{q\kappa}tr\{E^{2B\mu_{qp}}\} \tag{98}$$
$$-\sum_{\kappa=1}^{N_A}\sum_{p=1}^{3}\sum_{q=1}^{3}B_q\mu_{p\kappa}tr\{E^{2B\mu_{qp}}\}\equiv 0$$

The last two terms of Eq. (98) give zero from the equivalence of p and q. This relation holds for any field and moment: $h^{2w_{qp}}\equiv h^{2w_{qp}}$, $w=B,\mu$, with Eq. (71) and Eq. (72), give the following:

$$tr\left\{\left(P^{1w_q}\right)^T h^{1w_p}\right\}=tr\left\{\left(P^{1w_p}\right)^T h^{1w_q}\right\};\quad w=B,\mu \tag{99}$$

12. THE TENSORS χ_{he}, $\Phi_{\kappa,he}$, AND $\sigma_{\kappa,he}$

The components he of the tensors χ, Φ, and σ are defined by Eq. (3). Given $E\left(\vec{B},\vec{\mu}\right)=E^0+E^1+\triangle E$, deriving Eq. (70) leads to:

$$\chi_{he}=-\frac{1}{2}tr\left\{\left(P^{1B_h}\right)^T h^{1B_e}+\left(P^{1B_e}\right)^T h^{1B_h}+4P^{0T}h^{2B_{he}}\right\} \tag{100}$$

$$\Phi_{N,he}=-\frac{1}{2}tr\left\{\left(P^{1\mu_h}\right)^T h^{1\mu_e}+\left(P^{1\mu_e}\right)^T h^{1\mu_h}+4P^{0T}h^{2\mu_{he}}\right\} \tag{101}$$

$$\sigma_{N,he}=\frac{1}{2}tr\left\{\left(P^{1B_h}\right)^T h^{1\mu_e}+\left(P^{1\mu_e}\right)^T h^{1B_h}+2P^{0T}h^{2B\mu_{he}}\right\} \tag{102}$$

These relations for the component he require two distinct matrices; P^1 different, the theorems for exchange of perturbations Eq. (97) and their directions Eq. (99), lead to considerable simplification:

$$\chi_{he}=-tr\left\{\left(P^{1B_h}\right)^T h^{1B_e}+2P^{0T}h^{2B_{he}}\right\} \tag{103}$$

$$\Phi_{\kappa,he}=-tr\left\{\left(P^{1\mu_h}\right)^T h^{1\mu_e}+2P^{0T}h^{2\mu_{he}}\right\} \tag{104}$$

$$\sigma_{N,he}=tr\left\{\left(P^{1B_h}\right)^T h^{1\mu_e}+P^{0T}h^{2B\mu_{he}}\right\} \tag{105}$$

Only the density matrix P^1 with a single iterative process is then required to evaluate the hl of the tensors χ, Φ_N, and σ_N. For χ and σ_N, a single matrix P^{1B_h} used to determine the susceptibility tensor is subsequently used for the nuclear screening tensor.

13. DISCUSSION AND CONCLUSION

This ab initio CPHF work was implemented in the Slater-type orbital package (from 2003), based on the work of Joudieh. The integrals were programmed by Hoggan and students.[35–39]

An extensive numerical integration DFT package also offers accurate NMR observables from calculation. Its grid integration takes us beyond the scope of this analytically based work, however, to place this work in context and illustrate the comparisons made in Ref. 1 please see the relevant work behind the NMR applications of the Amsterdam density functional (ADF) code.[40–42]

This present work has focused on underlying formalism for CPHF local magnetic tensors. Some NDDO approximations were shown in Ref. 1 to provide fast and accurate chemical shifts and improved methods to obtain them are also presented here.

REFERENCES

1. Vieille, L.; Berlu, L.; Combourieu, B.; Hoggan, P. E. A Quantum Chemistry GIAO Molecular Site Approach of NMR Chemical Shifts Generalized to the Whole Periodic Table. *J. Theor. Comput. Chem.* **2002**, *1*, 295–308.
2. Stewart, J. J. P. MOPAC: A Semi-Empirical Molecular Orbital Program. *QCPE* **1983**, *11*, 455.
3. Stewart, J. J. P. Optimization of Parameters for Semiempirical Methods I. *Method. J. Comput. Chem.* **1989**, *10*, 209.
4. Dewar, M. J. S.; Zoebisch, E. G.; Healy, E. F. Stewart, J. J. P. Development and Use of Quantum Mechanical Molecular Models. 76. AM1: A New General Purpose Quantum Mechanical Molecular Model. *J. Am. Chem. Soc.* **1985**, *107*, 3902.
5. Frisch, M. J.; et al. *Gaussian*. Gaussian, Inc.: Pittsburg, PA, 2003.
6. Joudieh, N. Theorie SCF-GIAO-NDDO du Diamagnetisme Moleculaire et Modelisation des Susceptibilités Diamagnétiques Moyennes et des Deplacements Chimiques de RMN dans le Cadre de MOPAC. PhD Thesis, Université de Rouen, France 1998.
7. Pople, J. A. The Theory of Chemical Shifts in Nuclear Magnetic Resonance. I. Induced Current Densities. *Proc. R. Soc. London* **1957**, *A239*, 541.
8. Pople, J. A. Molecular Orbital Theory of Diamagnetism. I. An Approximate LCAO Scheme. *J. Chem. Phys.* **1962**, *37*, 53.
9. Ditchfield, R. Molecular Orbital Theory of Magnetic Shielding and Magnetic Susceptibility. *J. Chem. Phys.* **1972**, *56*, 5688.
10. Ditchfield, R. Self-Consistent Perturbation Theory of Diamagnetism. *Mol. Phys.* **1974**, *27*, 789.
11. Roothaan, C. C. J. New Developments in Molecular Orbital Theory. *Rev. Mod. Phys.* **1951**, *23*, 69.
12. Hall, G. G. The Molecular Orbital Theory of Chemical Valency. VIII. A Method of Calculating Ionization Potentials. *Proc. Roy. Soc.* **1951**, *A205*, 541.
13. Stevens, R. M.; Pitzer, P. M.; Lipscomb, W. N. Perturbed Hartree-Fock Calculations. I. Magnetic Susceptibility and Shielding in the LiH Molecule. *J. Chem. Phys.* **1963**, *38*, 550.

14. Stevens, R. M.; Lipscomb, W. N. Perturbed Hartree-Fock Calculations. II. Further Results for Diatomic Lithium Hydride. *J. Chem. Phys.* **1964**, *40*, 2238.
15. Stevens, R. M.; Lipscomb, W. N. Perturbed Hartree-Fock Calculations. III. Polarizability and Magnetic Properties of the HF Molecule. *J. Chem. Phys.* **1964**, *41*, 184.
16. Stevens, R. M.; Lipscomb, W. N. Perturbed Hartree-Fock Calculations. IV. Second-Order Properties of the Fluorine Molecule. *J. Chem. Phys.* **1964**, *41*, 3710.
17. Stevens, R. M.; Lipscomb, W. N. Perturbed Hartree-Fock Calculations. V. Magnetic Properties of the BH Molecule. *J. Chem. Phys.* **1965**, *42*, 3666.
18. Stevens, R. M.; Lipscomb, W. N. Magnetic Properties of the Li_2 Molecule. *J. Chem. Phys.* **1965**, *42*, 4302.
19. McWeeny, R. Perturbation Theory for the Fock-Dirac Density Matrix. *Phys. Rev.* **1962**, *126*, 1028.
20. McWeeny, R. Self-Consistent Perturbation Theory. *Chem. Phys. Lett.* **1968**, *1*, 567.
21. Diercksen, G.; McWeeny, R. Self-Consistent Perturbation Theory. I. General Formulation and Some Applications. *J. Chem. Phys* **1966**, *44*, 3554.
22. Dewar, M. J. S.; Theil, W. Ground States of Molecules. 38. The MNDO Method. Approximations and Parameters. *J. Am. Chem. Soc.* **1978**, *99*, 4899.
23. Dewar, M. J. S.; Theil, W. A Semiempirical Model for the Two-Center Repulsion Integrals in the NDDO Approximation. *Theor. Chim. Acta* **1977**, *46*, 89.
24. Dewar, M. J. S.; Rzepa, H. S. Ground States of Molecules. MNDO Results for Molecules Containing Fluorine. *J. Am. Chem. Soc.* **1978**, *100*, 58.
25. Dewar, M. J. S.; McKee, M. L. Ground States of Molecules. 41. MNDO Results for Molecules Containing Boron. *J. Am. Chem. Soc.* **1977**, *99*, 5231.
26. London, F. Theorie Quantique des Courants Interatomiques dans les Combinaisons aromatiques. *J. Phys. Radium* **1937**, *8*, 397.
27. Aihara, J. I. Theoretical Relationship Between Diamagnetic Susceptibility Exaltation and Aromatic Stabilization. *J. Am. Chem. Soc.* **1979**, *101*, 558.
28. Morales, R. G. E.; Parrini, F.; Vargas, V. Intermolecular Proton Transfer Process in Sulfur Tautomeric Heterocycles. I. A Quantum Chemical Approach. *Phosphorus Sulfur Silicon* **1998**, *133*, 1.
29. Arrighini, G. P.; Maestro, M.; Moccia, R. Magnetic Properties of Polyatomic Molecules. I. Magnetic Susceptibility of H_2O, NH_3, CH_4, H_2O_2. *J. Chem. Phys.* **1968**, *49*, 882.
30. Arrighini, G. P.; Guidotti, C.; Maestro, M.; Moccia, R.; Salvetti, O. SCF-MOs and Molecular Properties of Methane. *J. Chem. Phys.* **1968**, *49*, 2224.
31. Arrighini, G. P.; Giovanni, P.; Guidotti, C.; Salvetti, O. SCF MOs and Molecular Properties of H_2O. *J. Chem. Phys.* **1970**, *52*, 1037.
32. Smith, C. M.; Amos, R. D.; Handy, N. C. Theoretical Calculations of the Nuclear Magnetic Shielding Tensors for the Ethylenic Carbon Atoms in Cyclopropenes. *Mol. Phys.* **1992**, 77, 381.
33. Hameka, H. F. On the Nuclear Magnetic Shielding in the Hydrogen Molecule. *Mol. Phys.* **1958**, *1*, 203.
34. Hameka, H. F. Theory of Magnetic Properties of Molecules With Particular Emphasis on the Hydrogen Molecule. *Rev. Mod. Phys.* **1962**, *34*, 87.
35. Barnett, M. P.; Coulson, C. A. The Evaluation of Integrals Occurring in the Theory of Molecular Structure. Parts I & II. *Philos. Trans. Roy. Soc.* **1951**, *243*, 221.
36. Bouferguene, A.; Rinaldi, D. L. The Evaluation of Integrals Occurring in the Theory of Molecular Structure. Parts I & II. *Int. J. Quantum. Chem.* **1994**, *50*, 21.
37. Bouferguene, A.; Fares, M.; Hoggan, P. E. STOP: A Slater-Type Orbital Package for Molecular Electronic Structure Determination. *Int. J. Quantum. Chem.* **1996**, *57*, 801.
38. Bouferguene, A.; Hoggan, P. E. STOP: A Slater Type Orbital Package. *QCPE* **1996**, *667*, 16.

39. Hoggan, P. E. Note on Hydrogenic Atomic Orbitals to Evaluate Sensitive Properties of Molecules: An Example of NMR Chemical Shifts. *J. Theor. Comput. Chem.* **2004**, *3*, 163–168.
40. Schreckenbach, G.; Ziegler, T. Calculation of NMR Shielding Tensors Using Gauge-Including Atomic Orbitals and Modern Density Functional Theory. *J. Phys. Chem.* **1995**, *99*(2), 606–661.
41. Autschbach, J. NMR Calculations for Paramagnetic Molecules and Metal Complexes. Dixon, D.A. (Ed.), *Annual Reports in Computational Chemistry*; Vol. 11. Elsevier: Amsterdam, 2015; pp 3–36.
42. Falls, Z.; Zurek, E.; Autschbach, J. Computational Prediction and Analysis of the ^{2}7Al Solid-State NMR Spectrum of Methylaluminoxane (MAO) at Variable Temperatures and Field Strengths. *Phys. Chem. Chem. Phys.* **2016**, *18*, 24106–24118.

SECTION II

Electron Correlation in Atomic Systems

CHAPTER TEN

Atomic Electronic Structure Computations With Hylleraas-CI Wave Functions

Frank E. Harris[1]
University of Utah, Salt Lake City, UT, United States
Quantum Theory Project, University of Florida, Gainesville, FL, United States
[1]Corresponding author: e-mail address: harris@qtp.ufl.edu

Contents

1. Introduction 188
2. Wave Functions and Potential-Energy Matrix Elements 189
   2.1 One-Electron Integrals 191
   2.2 Hylleraas Two-Electron Integrals 191
   2.3 Exponentially Correlated Two-Electron Integrals 194
3. General Methods for Exponential Correlation 196
4. Hylleraas Integrals 197
5. Unlinked Three- and Four-Electron Integrals 198
6. The Auxiliary Two-Body Integrals g_i 201
7. Check of $\mathcal{K}_2$ 203
8. Kinetic Energy 204
9. Kinetic Energy in Hylleraas-CI 206
Appendix. Angular Momentum Formulas 207
References 208

Abstract

The Hylleraas-CI (Hy-CI) method is conventionally defined as based on superposition-of-configurations (also called configuration interaction) wave functions in which each term (configuration) is built from an orbital product to which is appended at most one linear factor r_{ij}, where r_{ij} is the distance between particles i and j. The functions comprising an orbital product are usually chosen to be Slater-type orbitals that include spherical-harmonic angular dependence. We consider here both the conventional definition and its generalization (called *extended Hy-CI* or E-Hy-CI) in which the correlation factor r_{ij} is replaced by a more general function $f(r_{ij})$. The present communication reviews the mathematical methods presently available for the fully analytical treatment of the integrals arising when these types of explicitly correlated wave functions are used within the framework of both the usual and extended Hylleraas-CI to study atomic and more general single-center systems.

Advances in Quantum Chemistry, Volume 76
ISSN 0065-3276
https://doi.org/10.1016/bs.aiq.2017.04.001

The analysis includes novel elements that may improve the efficiency of computations; the chapter also calls attention to new formulas for treating kinetic-energy integrals in Hy-CI methods.

1. INTRODUCTION

Ever since the early days of quantum mechanics investigators have sought methods for describing the electronic structures of atoms and molecules that are more rapidly convergent than superposition-of-configurations (also called configuration interaction) expansions of the electronic wave function in orbital products. Probably the earliest endeavor of this type was that of Hylleraas, whose study of the He atom[1] used a wave function that included as a multiplicative factor the explicit appearance of the interelectron distance r_{12}. In the half-century following this work by Hylleraas, a relatively small number of papers confirmed the effectiveness of explicitly correlated wave functions; significant contributions during that period included work on the hydrogen molecule by James and Coolidge,[2] a nearly quantitative study of that system by Kolos and Roothaan,[3] an examination of three-body systems by Delves and Kalotas,[4] and a detailed study of the lithium atom by Larsson.[5]

More recent developments in the technology of Hylleraas-type methods have included analytically based work by King[6] and by Büsse et al.,[7] as well as studies focused on calculations for molecules, based on numerical methods of types suggested by Kutzelnigg.[8] These studies (often referred to as $\mathbf{r}_{12}$ methods) are related to but outside the scope of the present contribution and have been the subject of two recent reviews.[9,10]

Despite the advances made possible by the work referenced above, attempts to apply Hylleraas methods to larger systems revealed that the occurrence of a wide variety of combinations of r_{ij} factors (and higher powers thereof) led to exceedingly complicated computations, and as early as 1971 it was proposed by Sims and Hagstrom,[11] and independently by Woźnicki[12] to consider configurations (wave function terms) that contained at most a single, linear r_{ij} factor. Methods based on wave functions of this type, now usually referred to as *Hylleraas-CI*, were over time more fully developed and applied to a variety of atomic problems. Representative work in that area is in Refs. 13–16.

An alternative to the Hylleraas-CI development might be the use of exponentially correlated wave functions [using factors $\exp(-\beta r_{ij})$].

This type of wave function was proposed in the mid-1960s by Bonham,[17,18] but its use on even as small a system as the Li atom then seemed impractical. However, the introduction of Gaussian-type correlation factors (in conjunction with Gaussian-type orbitals) was indeed practical, as was shown in work by Rychlewski et al.[19] A potential drawback of this approach is that the individual wave function terms do not have the correct limiting behavior at either short or long distance; this deficiency is usually overcome either by building orbitals that are linear combinations of Gaussians (*contracted* functions) or by carrying out a relatively extended superposition of configurations. Completely analytical studies of exponential correlation on four-body systems (e.g., the Li atom) only became practical after publication of an extraordinary paper by Fromm and Hill.[20]

Recently the Nakatsuji group proposed a method they call *extended Hy-CI* (E-Hy-CI), containing in any wave function configuration at most one correlation factor that may contain both a power of r_{ij} and an exponential $e^{\beta_{ij} r_{ij}}$, and provided analytical formulas for the unlinked integrals that then arise.[21]

The present contribution reviews some aspects of atomic electronic structure computations by these Hylleraas-inspired methods, focusing on problems that can at present be treated analytically and that are relevant to conventional or generalized Hylleraas-CI computations. The exposition includes a preliminary discussion of analytical techniques that appear not to have previously been applied to Hylleraas-type four-electron integrals.

For exponential correlation and traditional Hylleraas calculations, our focus restricts consideration to systems of at most four particles, i.e., to a complexity corresponding to that of the Li isoelectronic series. No such restriction applies to either traditional or extended Hylleraas-CI methods.

2. WAVE FUNCTIONS AND POTENTIAL-ENERGY MATRIX ELEMENTS

We define potential-energy matrix elements as those not containing an explicit occurrence of the differential operator ∇_i for any particle i (this definition therefore includes integrals of overlap type). Using language that implies applications to atoms, we consider here potential-energy matrix elements that explicitly contain the interelectron distances r_{ij} as factors or linearly in exponents. These integrals can be identified with diagrams that are formed by

(1) Introducing a numbered vertex corresponding to each particle not at the origin of the coordinate system (i.e., not the nucleus);

(2) For each distinct r_{ij} occurring anywhere in the integral, drawing a line in the diagram that connects vertices i and j.

Any matrix element that is represented by a diagram whose vertices are not all connected by lines can be factored into integrals each of which is represented by a connected subdiagram. A connected (sub)diagram containing only one or two vertices corresponds to a one- or two-electron integral that can be evaluated analytically by well-known methods (briefly discussed below). The difficulty of evaluating connected (sub)diagrams with three or more vertices depends upon whether the (sub)diagram is linked (a diagram is defined as *linked* if it contains one or more closed loops). Any part of a diagram not included in a closed loop (its *unlinked part*) corresponds to a conceptually straightforward three-dimensional integration over the coordinates describing the unlinked $\mathbf{r}_{ij}$. But if there are several unlinked r_{ij} the integration becomes quite tedious.

Linked (sub)diagrams pose difficulties because the functional dependence on the internal coordinates cannot be separated. A completely analytical approach is known only for the cases of two and three vertices (three- and four-body problems). Three-body matrix elements with exponential correlation were treated by Thakkar and Smith[22] using the interparticle separations as coordinates. While that approach is also in principle possible for the four-body problem, it proved to be extremely difficult, resisting analytical solution until the contribution of Fromm and Hill;[20] a more compact solution restricted to Hylleraas functions was, however, found by Remiddi.[23]

In view of the above discussion, completely analytical treatment of exponentially correlated or general Hylleraas calculations are at present (and in this author's opinion for the foreseeable future) only available for systems with a maximum of three electrons. But because no matrix element in a Hylleraas-CI computation contains more than one factor of type r_{ij} from each of its two wave functions (plus perhaps one more from the interelectron potential energy), no Hylleraas-CI diagram will contain more than three lines. The limitation to three lines means that a Hylleraas-CI connected (sub)diagram, if linked, will contain three vertices, and if unlinked, no more than four.

We therefore proceed to further discussion of integrals for a maximum of four electrons. To keep the present contribution reasonable in size, we omit some implementation details and place emphasis on the underlying principles.

2.1 One-Electron Integrals

These integrals are elementary but we examine them in part to introduce our notations. Defining the unnormalized orbital a as

$$a \equiv a_l^m(1) = g_a Y_{l_a}^{m_a}(\Omega_1) \text{ with } g_a = r_1^{n_a-1} e^{-\alpha_a r_1}, \tag{1}$$

and with other orbitals b, … defined analogously, we consider integrals of the form

$$I_1 = \int a^*(1)\, r_1^p\, Y_L^M(\Omega_1)\, b(1)\, d\tau_1, \tag{2}$$

where $d\tau_1 = r_1^2 dr_1 \sin\theta_1 d\theta_1\, d\phi_1 = r_1^2 dr_1 d\Omega_1$ and $(r_1, \theta_1, \phi_1) = (r_1, \Omega_1)$ are the spherical polar coordinates of Electron 1 in a coordinate system with origin at the nucleus, and $Y_l^m(\Omega)$ is a spherical harmonic at Condon–Shortley phase (see Appendix. Angular Momentum Formulas). We consider here values of p that are integral and in a range such that I_1 converges. If we set $p = L = M = 0$, I_1 becomes proportional to the overlap integral $\langle a|b\rangle$:

$$I_1(p = L = M = 0) = \left\langle a(1) \left| \frac{1}{\sqrt{4\pi}} \right| b(1) \right\rangle = \frac{1}{\sqrt{4\pi}} \langle a|b\rangle. \tag{3}$$

To evaluate I_1 for general parameters, we note that the angular integration reduces to a Gaunt coefficient, represented by an array in square brackets—see Appendix. Angular Momentum Formulas, while the radial integration can be identified as a gamma function (here a factorial). The result is

$$I_1 = \langle a | r^p\, Y_L^M(\Omega) | b \rangle = (-1)^{m_a} \begin{bmatrix} l_a & L & l_b \\ -m_a & M & m_b \end{bmatrix} \frac{(n_a + n_b + p)!}{(\alpha_a + \alpha_b)^{n_a + n_b + p + 1}}. \tag{4}$$

2.2 Hylleraas Two-Electron Integrals

We consider integrals of the form

$$I_2 = \langle a(1) b(2) | r_1^{p_1} r_2^{p_2} r_{12}^{p} | c(1) d(2) \rangle, \tag{5}$$

where $r_{12} = |\mathbf{r}_1 - \mathbf{r}_2|$, and with integer exponents p_1, p_2, p for which I_2 converges, with $p \geq -1$.

The evaluation of I_2 proceeds by different routes depending upon whether p is even or odd. For even p we discuss here only the case $p = 2$ (larger even p, which are not needed for the usual Hylleraas-CI, can be reached by analyses similar to that now illustrated). Writing $r_{12}^2 = r_1^2 + r_2^2 - 2\mathbf{r}_1 \cdot \mathbf{r}_2$, we note that I_2 can be written as a sum of one-electron integrals plus an integral containing a dot product of two one-electron forms. One can write $\mathbf{r}_1 \cdot \mathbf{r}_2$ as $r_1 r_2 \cos\chi$, where χ is the angle between $\mathbf{r}_1$ and $\mathbf{r}_2$ and use the fact that $\cos\chi$ has a well-known form, but for pedagogical purposes we prefer to introduce vector spherical-harmonic concepts[24] and write the unit vector $\mathbf{r}_i$ as

$$\hat{\mathbf{r}}_i = \sqrt{\frac{4\pi}{3}} \sum_{\mu=-1}^{1} (-1)^{\mu} Y_1^{-\mu}(\Omega_i)\hat{\mathbf{e}}_{\mu}, \tag{6}$$

where $\hat{\mathbf{e}}_{\mu}$ is a unit vector from the set

$$\hat{\mathbf{e}}_1 = -\frac{\hat{\mathbf{x}} + i\hat{\mathbf{y}}}{\sqrt{2}}, \quad \hat{\mathbf{e}}_{-1} = \frac{\hat{\mathbf{x}} - i\hat{\mathbf{y}}}{\sqrt{2}}, \quad \hat{\mathbf{e}}_0 = \hat{\mathbf{z}}, \tag{7}$$

with orthonormality relation

$$\hat{\mathbf{e}}_{\mu} \cdot \hat{\mathbf{e}}_{-\nu} = (-1)^{\nu} \delta_{\mu\nu}. \tag{8}$$

Using Eqs. (6) and (8) and writing $\mathbf{r}_i = r_i \hat{\mathbf{r}}_i$, we reach

$$\mathbf{r}_1 \cdot \mathbf{r}_2 = \frac{4\pi r_1 r_2}{3} \sum_{\mu=-1}^{1} (-1)^{\mu} Y_1^{\mu}(\Omega_1) Y_1^{-\mu}(\Omega_2). \tag{9}$$

Eq. (9) is equivalent to the usual formula for $r_1 r_2 \cos\chi$.

Identifying

$$r_{12}^2 = r_1^2 + r_2^2 - \frac{8\pi r_1 r_2}{3} \sum_{\mu=-1}^{1} (-1)^{\mu} Y_1^{\mu}(\Omega_1) Y_1^{-\mu}(\Omega_2), \tag{10}$$

substituting the explicit forms of orbitals a, b, c, d, recognizing the angular integrals as Gaunt coefficients and the radial integrals as factorials, making the definitions $n_1 = n_a + n_c + p_1 - 1$, $n_2 = n_b + n_d + p_2 - 1$, $\alpha_1 = \alpha_a + \alpha_c$, $\alpha_2 = \alpha_b + \alpha_d$, and defining $(n)_p$ as the Pochhammer symbol $[(n)_0 = 1, (n)_1 = n, (n)_2 = n(n+1)$, etc.], we reach

$$I_2 = \langle a(1)b(2)|r_1^{p_1} r_2^{p_2} r_{12}^2|c(1)d(2)\rangle$$
$$= \frac{(-1)^{m_a+m_b} n_1!\, n_2!}{\alpha_1^{n_1+1}\alpha_2^{n_2+1}} \left\{ \left[\frac{(n_1+1)_2}{\alpha_1^2} + \frac{(n_2+1)_2}{\alpha_2^2}\right] \delta_{l_a l_c}\delta_{m_a m_c}\delta_{l_b l_d}\delta_{m_b m_d} - \frac{8\pi}{3}\frac{(n_1+1)(n_2+1)}{\alpha_1\alpha_2} \sum_{\mu=-1}^{1} (-1)^\mu \begin{bmatrix} l_a & 1 & l_c \\ -m_a & \mu & m_c \end{bmatrix} \begin{bmatrix} l_b & 1 & l_d \\ -m_b & -\mu & m_d \end{bmatrix} \right\}. \tag{11}$$

The second term within the braces of Eq. (11) will vanish unless both $l_a - l_c = \pm 1$ and $l_b - l_d = \pm 1$ and will even then yield a nonzero contribution only if both $\mu = m_a - m_c$ and $\mu = m_d - m_b$.

For p odd, we discuss here only the cases arising for the usual Hylleraas-CI, namely $p = \pm 1$. For $p = -1$ we expand $1/r_{12}$ using the Laplace expansion of $1/r_{12}$, eq. (16.66) of Ref. 25, but written using $(-1)^M Y_L^{-M}(\Omega_1)$ to avoid introducing a complex conjugate:

$$\frac{1}{r_{12}} = \sum_{L=0}^{\infty} \frac{4\pi}{2L+1} \frac{r_<^L}{r_>^{L+1}} \sum_{M=-L}^{L} (-1)^M Y_L^{-M}(\Omega_1) Y_L^M(\Omega_2). \tag{12}$$

Here $r_>$ and $r_<$ refer to the larger and smaller of r_1 and r_2. Inserting this expansion,

$$I_2 = \left\langle a(1)b(2)\left|\frac{r_1^{p_1} r_2^{p_2}}{r_{12}}\right|c(1)d(2)\right\rangle$$
$$= \sum_{L=0}^{\infty} \frac{4\pi}{2L+1} F_L \sum_{M=-L}^{L} (-1)^{m_a+m_b+M} \begin{bmatrix} l_a & L & l_c \\ -m_a & -M & m_c \end{bmatrix} \begin{bmatrix} l_b & L & l_d \\ -m_b & M & m_d \end{bmatrix}, \tag{13}$$

where F_L is a two-dimensional radial integral

$$F_L = \int_0^\infty dr_1 \int_0^\infty dr_2 \frac{r_1^{n_1} r_2^{n_2} r_<^L}{r_>^{L+1}} e^{-\alpha_1 r_1 - \alpha_2 r_2} \tag{14}$$

that is usually evaluated by dividing into subranges $r_1 < r_2$ and $r_1 > r_2$. Properties of the Gaunt coefficients cause the extent of the L and M summations in Eq. (13) to be severely limited, with the M summation containing at most one nonzero term.

For $p = +1$, the formula is similar to that for $p = -1$, except that we use the Laplace-type expansion of r_{12}, probably first written by Jen:[26]

$$r_{12}=\sum_{L=0}^{\infty}\frac{4\pi r_>}{2L+1}\left[\frac{(r_</r_>)^{L+2}}{2L+3}-\frac{(r_</r_>)^{L}}{2L-1}\right]\sum_{M=-L}^{L}(-1)^M Y_L^{-M}(\Omega_1)Y_L^M(\Omega_2). \tag{15}$$

We find

$$I_{2a}=\langle a(1)b(2)|r_1^{p_1}r_2^{p_2}r_{12}|c(1)d(2)\rangle$$
$$=\sum_{L=0}^{\infty}\frac{4\pi}{2L+1}G_L\sum_{M=-L}^{L}(-1)^{m_a+m_b+M}\begin{bmatrix} l_a & L & l_c \\ -m_a & -M & m_c\end{bmatrix}\begin{bmatrix} l_b & L & l_d \\ -m_b & M & m_d\end{bmatrix}, \tag{16}$$

with

$$G_L=\int_0^{\infty}dr_1\int_0^{\infty}dr_2\left[\frac{r_1^{n_1}r_2^{n_2}r_<^{L+2}}{(2L+3)r_>^{L+1}}-\frac{r_1^{n_1}r_2^{n_2}r_<^{L}}{(2L-1)r_>^{L-1}}\right]e^{-\alpha_1 r_1-\alpha_2 r_2}. \tag{17}$$

G_L has usually been evaluated by dividing the integration into subranges.

2.3 Exponentially Correlated Two-Electron Integrals

These integrals have integrands that include the exponential form $\exp(-\alpha_1 r_1-\alpha_2 r_2-\beta r_{12})$ and may also contain powers of r_1, r_2, or r_{12}. There may also be spherical harmonics whose arguments are the angular coordinates Ω_1 and Ω_2. Even with exponential correlation these integrals can be evaluated; following the approach of Thakkar and Smith,[22] we use r_1, r_2, and r_{12} as coordinates. Two of the three remaining internal coordinates can be the angles θ_1, ϕ_1 that specify the direction of $\mathbf{r}_1$; the final coordinate (which we call χ) describes the rotation of the triangle formed by r_1, r_2, and r_{12} about the axis $\mathbf{r}_1$. The coordinate system $(r_1, r_2, r_{12}, \theta_1, \phi_1, \chi)$ has volume element $d\tau=r_1r_2r_{12}\sin\theta_1\, dr_1\, dr_2\, dr_{12}\, d\theta_1\, d\phi\, d\chi=r_1r_2r_{12}\, dr_1\, dr_2\, dr_{12}\, d\Omega_1\, d\chi$.

After some initial rearrangement to reach a standard form, these integrals can be written

$$I_{2x}=\langle a(1)b(2)|r_1^{p_1}r_2^{p_2}r_{12}^{p}e^{-\beta r_{12}}|c(1)d(2)\rangle$$
$$=\int r_1^{p_1}r_2^{p_2}r_{12}^{p}e^{-\alpha_1 r_1-\alpha_2 r_2-\beta r_{12}}Y_{l_a}^{m_a}(\Omega_1)^*Y_{l_c}^{m_c}(\Omega_1)Y_{l_b}^{m_b}(\Omega_2)^*Y_{l_d}^{m_d}(\Omega_2)d\tau. \tag{18}$$

We start by combining the spherical harmonics of argument Ω_2 according to

$$Y_{l_b}^{m_b}(\Omega_2)^* Y_{l_d}^{m_d}(\Omega_2) = \sum_{L_2} (-1)^{m_d} \begin{bmatrix} l_b & l_d & L_2 \\ -m_b & m_d & -M_2 \end{bmatrix} Y_{L_2}^{M_2}(\Omega_2), \tag{19}$$

where $M_2 = m_d - m_b$. We next rewrite $Y_{L_2}^{M_2}$ in the coordinate system to be used here; we do so by expressing it as a linear combination of harmonics whose axial direction is that of $\mathbf{r}_1$, for which the polar angle is θ_{12} (the angle between $\mathbf{r}_1$ and $\mathbf{r}_2$), and for which the azimuthal coordinate is χ. While θ_{12} is not one of our new coordinates, it is conveniently expressed in them, with

$$\cos\theta_{12} = \frac{r_1^2 + r_2^2 - r_{12}^2}{2r_1 r_2}. \tag{20}$$

The new coordinates are reached from the original ones by a rotation through Euler angles (in the notation of Ref. 25, pp. 140ff) $\alpha = \phi_1$, $\beta = \theta_1$. The Euler angle γ is arbitrary, as its only effect is to determine the direction in which $\chi = 0$, which we will soon find to be irrelevant. The coordinate rotation is described by

$$Y_{L_2}^{M_2}(\Omega_2) = \sum_M D_{L_2}^{M_2 M}(\phi_1, \theta_1, \gamma) Y_{L_2}^{M}(\theta_{12}, \chi). \tag{21}$$

The quantities D are called Wigner rotation coefficients; they depend upon the Euler angles describing the coordinate-system rotation.

After the rotational transformation, we integrate over χ; this integration yields a zero result except for $M = 0$, for which (irrespective of the direction of $\chi = 0$) we get

$$\int_0^{2\pi} Y_{L_2}^{M_2}(\Omega_2) d\chi = 2\pi D_{L_2}^{M_2 0} Y_{L_2}^{0}(\theta_{12}, \gamma) = 2\pi D_{L_2}^{M_2 0} \sqrt{\frac{2L_2+1}{4\pi}} P_{L_2}(\cos\theta_{12}). \tag{22}$$

The expression of $Y_{L_2}^0$ as a Legendre polynomial corresponds to Eq. (A.4).

For $M = 0$ the Wigner coefficient assumes a simple form[27] and does not depend upon Euler angle γ: For the rotation to align with $\mathbf{r}_1$, it is[28]

$$D_{L_2}^{M_2 0} = \sqrt{\frac{4\pi}{2L_2+1}} Y_{L_2}^{M_2}(\Omega_1). \tag{23}$$

The result after integrating over $\boldsymbol{\Omega}_1$ (yielding a Gaunt coefficient) is

$$I_{2x} = 2\pi \sum_{L_2} (-1)^{m_a + m_d} \begin{bmatrix} l_b & l_d & L_2 \\ -m_b & m_d & -M_2 \end{bmatrix} \begin{bmatrix} l_a & l_c & L_2 \\ -m_a & m_c & M_2 \end{bmatrix} I_{2r}. \tag{24}$$

In Eq. (24) we must set $M_2 = m_d - m_b$; I_{2x} will vanish unless also $M_2 = m_a - m_c$. The quantity I_{2r} is an integral over the radial variables:

$$I_{2r} = \int dr_1\, dr_2\, dr_{12} r_1^{p_1} r_2^{p_2} r_{12}^{p}\, e^{-\alpha_1 r_1 - \alpha_2 r_2 - \beta r_{12}} P_{L_2}(\cos\theta_{12}), \qquad (25)$$

where the ranges of r_1, r_2, and r_{12} are over all nonnegative values such that these lengths can form a triangle. Note that before carrying out the integrations in Eq. (25) $\cos\theta_{12}$ is to be replaced by its representation in the current coordinate system, as given in Eq. (20).

The denominator $r_1 r_2$ of Eq. (20) does not produce negative powers of any r_i in the integrand of Eq. (25) because the maximum power of $\cos\theta_{12}$ is L_2, which is bounded by the smaller of $l_a + l_c$ and $l_b + l_d$, and is therefore smaller than all occurring values of p_1 and p_2.

After insertion of the formula for $\cos\theta_{12}$, the integrand of Eq. (25) will consist of a polynomial in the interparticle distances multiplying an exponential. The integration is elementary.

We defer a discussion of three- and four-electron integrals until after a review of general methods for the treatment of exponential correlation in small atoms.

3. GENERAL METHODS FOR EXPONENTIAL CORRELATION

For three-body (two-electron) systems exponentially correlated wave functions do not present major problems because the three interparticle distances r_1, r_2, and r_{12} can be chosen as internal coordinates, with the position and orientation of the three-particle triangle described by specifying the center of mass (or the position of one particle) and the Euler angles of the triangle orientation. In those coordinates, the internal "volume element" is proportional to $r_1 r_2 r_{12}$, and the main computational issue is to deal with the ranges of these distances.

For four-body (three-electron) systems it is still possible to use all the r_i and r_{ij} as explicit coordinates, but the internal volume element (contrary to a claim in Ref. 29) is extremely complicated, leading to integrals long thought to be intractable. Evaluation of the four-body integrals needed for exponentially correlated electronic structure computations was the problem solved by Fromm and Hill in 1987.[20] Fromm and Hill's solution, however, suffered from the difficulty that its implementation required laborious tracking of the branches of the dilogarithm functions it contained. Avoidance of this inconvenience and various other technical improvements were described in a paper by the present author in 1997.[30] That 1997 paper also resolved another issue: Remiddi[23] had published completely analytical formulas for certain integrals arising in Hylleraas calculations, and they agreed numerically with the corresponding special cases of the Fromm/Hill formula. But an analytical demonstration to that effect was missing. A formula presented in Ref. 30 was shown to lead to the desired analytical result.

As indicated above, the integrals needed for three-electron Hylleraas wave functions were special cases of those needed for general exponentially correlated wave functions, so in principle all such integrals could now be computed analytically. However, the formulas involved were rather unwieldy, and the computational process was greatly simplified by the development of recurrence formulas enabling all the needed integrals to be obtained from a single starting value. This objective was achieved, first for traditional Hylleraas integrals only, by Pachucki and Puchalski,[31] and later, for general exponential correlation, by the present author.[32]

4. HYLLERAAS INTEGRALS

While the general development for exponentially correlated wave functions in principle provided closed analytic formulas for many Hylleraas integrals, those formulas were often more laborious to evaluate than the expansions that were already in use. Those expansions were based on the Laplace expansion of $1/r_{ij}$[25] and its generalization to the related quantity r_{ij}.[26]

For the usual Hylleraas-CI, the limitations in the occurrence of r_{ij} factors caused the expansions of most of the integrals to reduce to finite sums; the exception was a linked integral containing a product of the form $r_{ij}r_{ik}/r_{jk}$; such "triangle" integrals were first discussed by Szasz[33] and an analytic formula for a basic triangle integral was included in the work of Remiddi.[23] This integral and its analog that includes exponential correlation are in principle evaluable analytically. However, a number of workers have found it more convenient to use numerical methods of the convergence-acceleration type[14] for evaluating these linked integrals.

For integrals other than the triangle integral, an alternative to the Laplace-type expansion is to carry out the evaluation in a coordinate system in which the r_{ij} are coordinates. That approach has been followed by Ruiz in Refs. 34,35, and other papers.

5. UNLINKED THREE- AND FOUR-ELECTRON INTEGRALS

The unlinked three-electron integrals needed for Hy-CI (including E-Hy-CI) are of the form

$$\mathcal{J} = \langle a(1)b(2)c(3)|r_{12}^{p_{12}} r_{13}^{p_{13}} e^{-\beta_{12}r_{12}-\beta_{13}r_{13}}|d(1)e(2)f(3)\rangle. \tag{26}$$

For the usual Hy-CI, both β are zero, $p_{12} = 1$, and $p_{13} = \pm 1$. There are three distinct four-electron integral types:

$$\mathcal{K}_1 = \left\langle a(1)b(2)c(3)d(4) \left| \frac{r_{13}^{p_{13}} r_{34}^{p_{34}} e^{-\beta_{13}r_{13}-\beta_{34}r_{34}}}{r_{12}} \right| e(1)f(2)g(3)h(4) \right\rangle, \tag{27}$$

$$\mathcal{K}_2 = \left\langle a(1)b(2)c(3)d(4) \left| \frac{r_{12}^{p_{12}} r_{34}^{p_{34}} e^{-\beta_{12}r_{12}-\beta_{34}r_{34}}}{r_{13}} \right| e(1)f(2)g(3)h(4) \right\rangle, \tag{28}$$

$$\mathcal{L} = \left\langle a(1)b(2)c(3)d(4) \left| \frac{r_{13}^{p_{13}} r_{14}^{p_{14}} e^{-\beta_{13}r_{13}-\beta_{14}r_{14}}}{r_{12}} \right| e(1)f(2)g(3)h(4) \right\rangle. \tag{29}$$

For the usual Hy-CI, all the β are zero and all $p = 1$.

It is not appropriate here to develop detailed formulas for all the above integral types, but it is instructive to examine techniques that can be used for their evaluation. We use $\mathcal{K}_2$ as an example illustrating the use of evaluation methods that appear not to have been used previously. A more detailed discussion will be published elsewhere.[36]

We use as coordinates r_1, r_2, r_3, r_4, $\mathbf{\Omega}_1 \equiv (\theta_1, \phi_1)$, r_{12}, χ_2, $\mathbf{\Omega}_3 \equiv (\theta_3, \phi_3)$, r_{34}, χ_4, where $\mathbf{\Omega}_1$ and $\mathbf{\Omega}_3$ are spherical polar angular coordinates, while χ_2 and χ_4 are the azimuthal angles for the rotation of $\mathbf{r}_2$ about the $\mathbf{r}_1$ axis and that of $\mathbf{r}_4$ about the $\mathbf{r}_3$ axis. The volume element for this coordinate set is

$$d\tau = r_1 r_2 r_{12} r_3 r_4 r_{34} dr_1 dr_2 dr_{12} dr_3 dr_4 dr_{34} d\mathbf{\Omega}_1 d\mathbf{\Omega}_3 d\chi_2 d\chi_4, \tag{30}$$

where $d\Omega_i = \sin\theta_i d\theta_i d\phi_i$. The ranges of the r_i are $(0, \infty)$, those of the r_{ij} are $(|r_i - r_j|, r_i + r_j)$, and those of the χ_i are $(0, 2\pi)$.

Using the Laplace expansion for $1/r_{13}$, Eq. (12), we initially reduce $\mathcal{K}_2$ to

$$\begin{aligned}\mathcal{K}_2 = &\sum_{L=0}^{\infty} \frac{4\pi}{2L+1} \sum_{\mu=-L}^{L} \int_0^{\infty} r_1 dr_1 \int d\mathbf{\Omega}_1 a^*(1)e(1) Y_L^{\mu}(\mathbf{\Omega}_1)^* \\ &\times \int_0^{\infty} r_3 dr_3 \int d\mathbf{\Omega}_3\, c^*(3)g(3) Y_L^{\mu}(\mathbf{\Omega}_3) \left[\frac{\min(r_1, r_3)^L}{\max(r_1, r_3)^{L+1}} \right] I_{12}(r_1) I_{34}(r_3),\end{aligned} \tag{31}$$

where

$$I_{12}(r_1) = \int_{\mathcal{R}_{12}} dr_2\, dr_{12} r_2 r_{12} \int_0^{2\pi} d\chi_2\, b^*(2) f(2) r_{12}^{p_{12}} e^{-\beta_{12} r_{12}}, \tag{32}$$

$$I_{34}(r_3) = \int_{\mathcal{R}_{34}} dr_4\, dr_{34} r_4 r_{34} \int_0^{2\pi} d\chi_4\, d^*(4) h(4) r_{34}^{p_{34}} e^{-\beta_{34} r_{34}}. \tag{33}$$

Here $\mathcal{R}_{ij}$ denotes a region of integration for fixed r_i and for all r_j and r_{ij} consistent therewith (i.e., for all r_j and r_{ij} that with r_i can form a triangle).

Next, we combine the orbitals for each electron, as illustrated here for Electron 1:

$$r_1 a^*(1) e(1) = \sum_{L_1} \begin{bmatrix} l_a & l_e & L_1 \\ -m_a & m_e & -M_1 \end{bmatrix} r_1^{n_1} e^{-\alpha_1 r_1} Y_{L_1}^{M_1}(\Omega_1), \tag{34}$$

where $n_1 = n_a + n_e - 1$, $\alpha_1 = \alpha_a + \alpha_e$, and $M_1 = m_e - m_a$, the array in brackets is a Gaunt coefficient (see Appendix. Angular Momentum Formulas), and the summation is over all L_1 with nonvanishing contributions. Formulas for other electrons have corresponding indices n_i, α_i, and M_i.

We continue by expressing $Y_{L_2}^{M_2}(\Omega_2)$ as a sum of spherical harmonics $Y_{L_2}^{\mu}$ in a coordinate system whose axis is aligned with $\mathbf{r}_1$, an axial direction defined by the angular coordinates (θ_1, ϕ_1). This is the same transformation that was described in Eqs. (21)–(24), with the result that

$$I_{12} = 2\pi \sum_{L_2} (-1)^{m_f} \begin{bmatrix} l_b & l_f & L_2 \\ -m_b & m_f & -M_2 \end{bmatrix} Y_{L_2}^{M_2}(\Omega_1) F_{12}(r_1), \tag{35}$$

with

$$F_{12}(r_1) = \int_{\mathcal{R}_{12}} dr_2\, dr_{12}\, r_2^{n_2} r_{12}^{p_{12}+1} e^{-\alpha_2 r_2 - \beta_{12} r_{12}} P_{L_2}(\cos\theta_{12}). \tag{36}$$

Corresponding treatment of the integrations of I_{34} yields

$$I_{34} = 2\pi \sum_{L_4} (-1)^{m_h} \begin{bmatrix} l_d & l_h & L_4 \\ -m_d & m_h & -M_4 \end{bmatrix} Y_{L_4}^{M_4}(\Omega_3) F_{34}(r_3), \tag{37}$$

with

$$F_{34}(r_3) = \int_{\mathcal{R}_{34}} dr_4\, dr_{34}\, r_4^{n_4} r_{34}^{p_{34}+1} e^{-\alpha_4 r_4 - \beta_{34} r_{34}} P_{L_4}(\cos\theta_{34}). \tag{38}$$

Remember that

$$\cos\theta_{12} = \frac{r_1^2 + r_2^2 - r_{12}^2}{2r_1 r_2}, \qquad \cos\theta_{34} = \frac{r_3^2 + r_4^2 - r_{34}^2}{2r_3 r_4}. \tag{39}$$

Incorporating the above elements into the formula for $\mathcal{K}_2$, we reach

$$\mathcal{K}_2 = 16\pi^3 \sum_{\text{all } L} \sum_{\mu=-L}^{L} (-1)^M \, C_{12} C_{34} \, G_L \times \int d\Omega_1 \, Y_{L_1}^{M_1}(\Omega_1) Y_{L_2}^{M_2}(\Omega_1) Y_L^{\mu}(\Omega_1)^* \int d\Omega_3 \, Y_{L_3}^{M_3}(\Omega_3) Y_{L_4}^{M_4}(\Omega_3) Y_L^{\mu}(\Omega_3), \tag{40}$$

with

$$G_L = \frac{1}{2L+1} \int_0^\infty dr_1 r_1^{n_1} e^{-\alpha_1 r_1} F_{12}(r_1) \times \int_0^\infty dr_3 r_3^{n_3} e^{-\alpha_3 r_3} F_{34}(r_3) \left[\frac{\min(r_1, r_3)^L}{\max(r_1, r_3)^{L+1}} \right] \tag{41}$$

and

$$C_{12} = \begin{bmatrix} l_a & l_e & L_1 \\ -m_a & m_e & -M_1 \end{bmatrix} \begin{bmatrix} l_b & l_f & L_2 \\ -m_b & m_f & -M_2 \end{bmatrix}, \tag{42}$$

$$C_{34} = \begin{bmatrix} l_c & l_g & L_3 \\ -m_c & m_g & -M_3 \end{bmatrix} \begin{bmatrix} l_d & l_h & L_4 \\ -m_d & m_h & -M_4 \end{bmatrix}. \tag{43}$$

The angular integrations in Eq. (40) can be written as Gaunt coefficients, and at least one of these Gaunt coefficients will be zero unless $\mu = M_{12} = -M_{34}$. Assigning $\mu = M_{12}$ and deleting the μ summation, our expression for $\mathcal{K}_2$ becomes

$$\mathcal{K}_2 = 16\pi^3 \sum_{\text{all } L} (-1)^{M+M_{12}} C_{12} C_{34} C_{13} \, G_L, \tag{44}$$

where

$$C_{13} = \begin{bmatrix} L_1 & L_2 & L \\ M_1 & M_2 & -M_{12} \end{bmatrix} \begin{bmatrix} L_3 & L_4 & L \\ M_3 & M_4 & M_{12} \end{bmatrix} \tag{45}$$

and $M = m_e + m_f + m_g + m_h$. The angular dependence of $\mathcal{K}_2$ has now been encapsulated in the sign factor and the C_{ij} product.

It is useful to consider a rearrangement (new to the present work) of the radial integrals G_L, to the form

$$G_L = \int_0^\infty \frac{d\xi}{\xi^{2L+2}} g_1(\xi) g_3(\xi), \tag{46}$$

with

$$g_1(\xi) = \int_0^\xi dr_1 \, r_1^{n_1+L} e^{-\alpha_1 r_1} F_{12}(r_1), \tag{47}$$

$$g_3(\xi) = \int_0^\xi dr_3 \, r_3^{n_3+L} e^{-\alpha_3 r_3} F_{34}(r_3). \tag{48}$$

Eq. (46) can be proved by applying to it an integration by parts on the range $\xi = (0, \infty)$, where $\int u\, dv = uv - \int v\, du$, with $u = g_1(\xi) g_3(\xi)$ and $v = -\xi^{-2L-1}/(2L+1)$. The uv term of this transformation vanishes at $\xi = 0$ because both g_1 and g_3 each approach zero as $\xi \to 0$ at least as rapidly as ξ^{L+1}, and uv also vanishes as $\xi \to \infty$ due to the factor ξ^{-2L-1} and the fact that the integrals g_1 and g_3 are convergent for all ξ.

The computationally most demanding part of the evaluation of $\mathcal{K}_2$ is the integral G_L, and the major effort in its evaluation is in the computation of the quantities $g_i(\xi)$, each of which (as we shall see) reduces to a polynomial in ξ, with each term multiplied by an exponential. But it is of great practical importance to note the advantages provided by the novel transformation represented by Eq. (46):

(1) Any given g_i depends only on the parameters of a product of two orbitals,

(2) The same g_i occurs in a large number of different overall integrals $\mathcal{K}_2$ and can be computed once and then reused,

(3) The use of the Laplace expansion and fixed-point arithmetic (*vide infra*) cause no divergences to be encountered in the sample computations for $\mathcal{K}_2$ presented in this paper.

6. THE AUXILIARY TWO-BODY INTEGRALS g_i

To complete the specification of a practical method for the evaluation of $\mathcal{K}_2$, we need now to discuss in more detail the integrals g_i, defined in Eqs. (47) and (48).

Discussing here g_1 (the treatment of g_3 is analogous), for $L_2 = 0$ the integral g_1 has the generic form

$$g(p_1,p_2,p_{12};\gamma_1,\gamma_2,\gamma_{12};\xi) = \int_0^{\xi} ds_1 \int_0^{\infty} ds_2 \int_{|s_1-s_2|}^{s_1+s_2} ds_{12}\, s_1^{p_1} s_2^{p_2} s_{12}^{p_{12}} e^{-\gamma_1 s_1-\gamma_2 s_2-\gamma_{12} s_{12}}, \tag{49}$$

and can be written as derivatives of $g(0, 0, 0; \gamma_1, \gamma_2, \gamma_{12}; \xi)$:

$$g(p_1,p_2,p_{12};\gamma_1,\gamma_2,\gamma_{12};\xi) = \left(-\frac{\partial}{\partial\gamma_1}\right)^{p_1}\left(-\frac{\partial}{\partial\gamma_2}\right)^{p_2}\left(-\frac{\partial}{\partial\gamma_{12}}\right)^{p_{12}} g_0, \tag{50}$$

where

$$g_0 = \int_0^{\xi} ds_1 \int_0^{\infty} ds_2 \int_{|s_1-s_2|}^{s_1+s_2} ds_{12}\, e^{-\gamma_1 s_1-\gamma_2 s_2-\gamma_{12} s_{12}}. \tag{51}$$

The integration of g_0 over s_2 and s_{12} is of a type investigated long ago by Perkins[37] and rederived recently by Padhy.[38] That result is

$$\int_0^{\infty} ds_2 \int_{|s_1-s_2|}^{s_1+s_2} ds_{12}\, e^{-\gamma_2 s_2-\gamma_{12} s_{12}} = 2\,\frac{e^{-\gamma_{12}s_1}-e^{-\gamma_2 s_1}}{\gamma_2^2-\gamma_{12}^2}. \tag{52}$$

Inserting the formula of Eq. (52) into Eq. (51) and integrating over s_1, we find

$$g_0 = \frac{2}{(s_1+s_2)(s_1+s_{12})(s_2+s_{12})} + \frac{2}{s_2^2-s_{12}^2}\left[\frac{e^{-(s_1+s_2)\xi}}{s_1+s_2} - \frac{e^{-(s_1+s_{12})\xi}}{s_1+s_{12}}\right]. \tag{53}$$

For nonzero L_2, g_1 becomes a linear combination of the functions g of Eq. (50):

$$g_p(p_1,p_2,p_{12},\gamma_1,\gamma_2,\gamma_{12},L_2,\xi) = \sum_{\mathbf{g}} C_{L_2}(\mathbf{q}) g(q_1,q_2,q_{12},\gamma_1,\gamma_2,\gamma_{12},\xi), \tag{54}$$

where $\mathbf{q}$ stands for the index set q_1, q_2, q_{12} and the coefficients C_L can be computed using Eq. (A.5) and the multinomial expansion of $\cos\theta_{12}$, Eq. (20), in powers of r_1, r_2, and r_{12}. Details will be given elsewhere.[36]

7. CHECK OF $\mathcal{K}_2$

One can use Maple to check the formula for $\mathcal{K}_2$ against literature values. To do so,

(1) Obtain formulas for g_1 and g_3 (as functions of ξ) for a sufficient range of derivatives of g_0, evaluating them for parameters s_1, s_2, and s_{12} that are given as fractions (not as floating-point numbers, because the latter occur in combinations that will not cancel to exact zeros).

(2) If L_2 or L_4 is nonzero, compute the algebraic forms corresponding to the relevant quantities g_p,

(3) Combine the g_0 and/or g_p and compute G_L according to Eq. (46). The exact cancelation of the fractional numbers avoids a spurious divergence in the integral of Eq. (46) at $\xi = 0$.

(4) Insert G_L into Eq. (44), also evaluating the coefficients C_{ij}.

We carried out two sets of checks, the first of which was to verify the resolution of the formula into angular and radial parts and the integration-by-parts formula, by comparing with results reported for the usual Hy-CI formulation by Ruiz in table 3 of Ref. 35. For all the integrals we checked, the orbitals for Electrons 1 and 2 had $\alpha = 3.685$, corresponding to $\alpha_1 = \alpha_2 = 7.37$, while the orbitals for Electrons 3 and 4 had $\alpha = 0.956$, corresponding to $\alpha_3 = \alpha_4 = 1.912$. We proceeded by applying our MAPLE procedure `Make_G(G,alpha,alpha',beta,xi)`, which makes an array `G[p1,p2,p12]` containing the algebraic form $g(p_1, p_2, p_{12}, \alpha, \alpha', \beta, \xi)$ as defined in Eq. (50) for a range of p_1, p_2, and p_{12}. This program was executed for arguments `G1,737/100,737/100,0,xi` and `G3,1912/1000,1912/1000,0,xi` to make arrays of g_1 in `G1[p1,p2,p12]` and arrays of g_3 in `G3[p3,p4,p34]`. We also used our program `Make_GP(GP,G)` to form the array `GP[q1,q2,q12,L]` containing in algebraic form the quantities g_p defined in Eq. (54) for the parameter values used to make `G`.

The quantity G_L was then given as an integral on the range $\xi = (0, \infty)$ with an integrand containing polynomials multiplying negative exponentials; these expressions were integrated using Maple (to yield a numeric fraction, which we then converted into decimal form). Details of the coding are given in Table 1. For the three cases we examined (lines 1, 2, and 3 of Ruiz' data), the only set of L values that contributed were all zeros, and the product $C_{12}C_{34}C_{13}$ evaluated to

Table 1 Maple Coding for Checking the Integral $\mathcal{K}_2$

			Integrand, Eq. (46)
Ruiz	`Make_G(G1,737/100,`	Line 1	$\xi^{-2}G1[1, 1, 2]\ G3[3, 3, 2]$
Table 3	`737/100,0,xi)`	Line 2	$\xi^{-2}G1[3, 3, 2]\ G3[3, 3, 2]$
	`Make_G(G3,1912/1000,`	Line 3	$\xi^{-2}G1[1, 3, 2]\ G3[3, 3, 2]$
	`1912/1000,0,xi)`		
Wang et al.	`Make_G(G1,6,8,-1/2,xi)`	Line 1	$\xi^{-2}G1[1, 1, 1]\ G3[1, 1, 1]$
Table IV	`Make_G(G3,4,2,-1/2,xi)`	Line 2	$\xi^{-2}G1[2, 2, 1]\ G3[2, 2, 1]$
	`Make_GP(GP1,G1)`	Line 4	$\xi^{-4}GP1[3, 2, 1, 1]\ G3[3, 2, 1]$
	`Make_GP(GP3,G3)`	Line 5	$\xi^{-4}GP1[3, 2, 1, 1]\ GP3[3, 2, 1, 1]$
		Line 6	$\xi^{-4}G1[3, 3, 1]\ G3[3, 3, 1]$
		Line 9	$\xi^{-6}GP1[5, 3, 1, 2]\ GP3[5, 3, 1, 2]$
		Line 10	$\xi^{-6}GP1[5, 3, 2, 2]\ GP3[5, 3, 2, 2]$

$1/64\pi^3$, so the value of $\mathcal{K}_2$ was in all these cases $G_0/4$. All three test cases yielded agreement with the data of Ruiz to all the significant figures are listed.

A second set of checks was against data of the Nakatsuji group, from table IV of Ref. 21. These checks were intended to confirm the analysis when the exponential parameters β_{ij} are nonzero. Because the Nakatsuji data reported only the values of radial integrals, the check was between those data and our values of $G_0/4$. Details of the coding for these tests are in the bottom section of Table 1. Because the Nakatsuji data were only reliable to the extent it was printed in bold characters, we made calculations both at the precision that group used (34 significant digits) and at higher precision (to determine the extent of a loss in precision of calculation). Our calculations are compared with those of the Nakatsuji group in Table 2. Our conclusions were that for all the seven cases we studied we agreed completely with the Nakatsuji data to the extent of its reliability, and in cases where the Nakatsuji calculations were relatively inaccurate (their lines 9 and 10) our calculations improved the accuracy by about five significant figures.

8. KINETIC ENERGY

Unexpectedly, a bottleneck in Hylleraas-CI computations has been the computation of kinetic-energy matrix elements. In completely orbital formulations, the derivatives describing the kinetic energy in quantum

Table 2 Comparison of Values of the Integral $\mathcal{K}_2$: Wang, 34-Digit Arithmetic[21] vs This Research (34- and 60-Digit Arithmetic)

Line 1	Wang	**1.465 423 281 172 008 340 718 675 056 269** 763 × 10^{-6}
	This research (34S)	1.465 423 281 172 008 340 718 675 056 269 759
	(60S)	1.465 423 281 172 008 340 718 675 056 269 759
Line 2	Wang	**3.455 948 258 119 907 308 258 045 381 210 14**0 × 10^{-7}
	This research (34S)	3.455 948 258 119 907 308 258 045 381 210 142
	(60S)	3.455 948 258 119 907 308 258 045 381 210 142
Line 4	Wang	−**4.479 303 653 926 613 814 855 805 199 71**6 874 × 10^{-9}
	This research (34S)	−4.479 303 653 926 613 814 855 805 199 712 225
	(60S)	−4.479 303 653 926 613 814 855 805 199 712 565
Line 5	Wang	**5.623 732 368 534 378 098 402 718 867** 978 85 × 10^{-10}
	This research (34S)	5.623 732 368 534 378 098 402 718 867 890 250
	(60S)	5.623 732 368 534 378 098 402 718 867 890 717
Line 6	Wang	**2.799 519 792 819 307 124 798 406 613 728 066** × 10^{-7}
	This research (34S)	2.799 519 792 819 307 124 798 406 613 728 070
	(60S)	2.799 519 792 819 307 124 798 406 613 728 070
Line 9	Wang	−**5.381 773 582 248 076 958 003** 411 278 978 8× 10^{-13}
	This research (34S)	−5.381 773 582 248 076 958 003 393 325 45
	(60S)	−5.381 773 582 248 076 958 003 393 322 387 867
Line 10	Wang	−**1.438 964 823 159 370 296 803 3**16 838 538 9× 10^{-12}
	This research (34S)	−1.438 964 823 159 370 296 803 388 684 872 5
	(60S)	−1.438 964 823 159 370 296 803 388 684 608 452

Line numbers refer to table IV of Wang et al.

mechanics simply change the powers of r_i in a matrix element, making it simple to identify the kinetic energy in terms of potential-energy integrals. However, the presence of factors r_{ij}, either as powers or in exponents, generates new complications.

When the wave function contains no explicit angular factors, the kinetic-energy operator $\hat{T}$ can be reduced to the form[39]

$$\hat{T} = -\frac{1}{2}\sum_{i<j}\left(\frac{1}{m_i}+\frac{1}{m_j}\right)\left(\frac{\partial^2}{\partial r_{ij}^2}+\frac{2}{r_{ij}}\frac{\partial}{r_{ij}}\right) - \sum_i \frac{1}{m_i}\sum_{\substack{j<k \\ j,k\neq i}} \cos\theta_{ijk}\,\frac{\partial^2}{\partial r_{ij}\partial r_{ik}}. \tag{55}$$

The indices in Eq. (55) run over all the particles (including any nuclei, whether or not they are assumed to be of infinite mass), m_i is the mass of particle i, and $\cos\theta_{ijk}$ is the cosine of the angle between $\mathbf{r}_{ij}$ and $\mathbf{r}_{ik}$. It can be evaluated as

$$\cos\theta_{ijk} = \frac{r_{ij}^2 + r_{ik}^2 - r_{jk}^2}{2r_{ij}r_{ik}}. \tag{56}$$

Because Eq. (55) is general, its use yields the kinetic energy even when the particles are all of finite mass, thereby removing the need for an estimate of the nonphysical quantity called "mass polarization."

For wave functions containing explicit angular factors (e.g., spherical harmonics), the kinetic-energy operator requires additional terms. This topic is discussed in Refs. 40–42.

Evaluation of the kinetic energy for exponentially correlated wave functions was examined in 1993 by Rebane,[43] who showed how the kinetic-energy matrix elements could be written in terms of suitable potential-energy contributions. Rebane's derivation was later simplified by the author's research group.[44] Unfortunately Rebane's formulation involves all the exponents of the exponentially correlated wave function and does not apply to the usual Hylleraas functions.

9. KINETIC ENERGY IN HYLLERAAS-CI

In the absence of formulas relating Hylleraas kinetic-energy matrix elements to those of potential energy, expressions based on the Laplace expansion can be differentiated, leading after some complication to the needed formulas. The situation is even more cumbersome if, following Ruiz, one uses the interparticle coordinates directly. In any case, relations involving kinetic-energy integrals would be useful in reducing computational effort and/or in providing additional checks for the numerical work. We have analyzed this problem, showing how the use of concepts associated with vector spherical harmonics[24] can be applied to derive the missing kinetic-energy formulas. Details have been reported elsewhere.[45]

APPENDIX. ANGULAR MOMENTUM FORMULAS

The spherical harmonics $Y_l^m(\theta,\phi)$, alternatively written $Y_l^m(\Omega)$, can be defined with the sign convention chosen by Condon and Shortley[46] (*Condon–Shortley phase*) by the Rodrigues formula (see Ref. 25, pp. 753 and 757)

$$Y_l^m(\Omega) = N_{lm}\frac{(-1)^m}{2^l l!}(1-u^2)^{m/2}\frac{d^{l+m}}{du^{l+m}}(u^2-1)^l e^{im\phi}, \tag{A.1}$$

where $u = \cos\theta$ (θ is the polar coordinate in Ω) and N_{lm} is the factor

$$N_{lm} = \sqrt{\frac{(2l+1)(l-m)!}{4\pi(l+m)!}} \tag{A.2}$$

that makes the Y_l^m orthonormal. With these definitions,

$$Y_l^m(\Omega)^* = (-1)^m Y_l^{-m}(\Omega), \tag{A.3}$$

$$Y_l^0(\Omega) = \sqrt{\frac{2l+1}{4\pi}}\,P_l(\cos\theta), \tag{A.4}$$

where P_l is a Legendre polynomial.

The Legendre polynomial $P_l(x)$ has the following power-series expansion (see Ref. 25, p. 718), needed for the evaluation of Eqs. (47) and (48):

$$P_l(x) = \sum_{k=0}^{[l/2]}(-1)^k\frac{(2l-2k)!}{2^l k!(l-k)!(l-2k)!}x^{l-2k}. \tag{A.5}$$

Here $[l/2]$ stands for the largest integer $\leq l/2$.

A product of spherical harmonics of the same argument Ω can be expanded into a sum of harmonics of that argument. The coefficients in that expansion are known as *Gaunt coefficients*. Unfortunately there is no unanimity as to the definition of the Gaunt coefficient. Choosing the definition of Pinchon and Hoggan[47], we introduce a bracket notation that we hope will become adopted:

$$\begin{bmatrix} l_1 & l_2 & l_3 \\ m_1 & m_2 & m_3 \end{bmatrix} = \int Y_{l_1}^{m_1}(\Omega) Y_{l_2}^{m_2}(\Omega) Y_{l_3}^{m_3}(\Omega)\, d\Omega. \tag{A.6}$$

Expansion of the spherical harmonic product $Y_{l_1}^{m_1} Y_{l_2}^{m_2}$ in the orthonormal set Y_L^M, carried out by taking scalar products with $(Y_L^M)^*$, leads after use of Eqs. (A.3) and (A.6) to

$$Y_{l_1}^{m_1}(\Omega) Y_{l_2}^{m_2}(\Omega) = \sum_{LM} (-1)^M \begin{bmatrix} l_1 & l_2 & L \\ m_1 & m_2 & -M \end{bmatrix} Y_L^M(\Omega). \tag{A.7}$$

Because harmonics with upper indices m_1 and m_2 form a product all of whose nonvanishing terms have the same value of M, Eq. (A.7) can be simplified by dropping the M summation, setting $M = m_1 + m_2$.

The Gaunt coefficients can be written in terms of Wigner 3-j symbols[28]. Using the standard notation for that symbol (an array of l and m values in ordinary parentheses), the Gaunt coefficients as defined here can be written

$$\begin{bmatrix} l_1 & l_2 & l_3 \\ m_1 & m_2 & m_3 \end{bmatrix} = \sqrt{\frac{(2l_1+1)(2l_2+1)(2l_3+1)}{4\pi}} \begin{pmatrix} l_1 & l_2 & l_3 \\ m_1 & m_2 & m_3 \end{pmatrix} \begin{pmatrix} l_1 & l_2 & l_3 \\ 0 & 0 & 0 \end{pmatrix}. \tag{A.8}$$

Well-documented computer programs exist for the evaluation of the 3-j symbols, making it straightforward to evaluate expressions involving Gaunt coefficients.

REFERENCES

1. Hylleraas, E. Neue Berechnung der Energie des Heliums im Grundzustande, sowie des Tiefsten Terms von Ortho-Helium. *Z. Phys.* **1929**, *54*, 347–366.
2. James, H. M.; Coolidge, A. S. The Ground State of the Hydrogen Molecule. *J. Chem. Phys.* **1933**, *1*, 825–835.
3. Kolos, W.; Roothaan, C. C. J. Correlated Orbitals for the Ground State of the Hydrogen Molecule. *Rev. Mod. Phys.* **1960**, *32*, 205–210.
4. Delves, L. M.; Kalotas, T. Variational Calculations for Three-Body S-States With Coulomb Interactions. *Aust. J. Phys.* **1968**, *21*, 431–438.
5. Larsson, S. Calculations on the ^{2}S Ground State of the Lithium Atom Using Wave Functions of Hylleraas Type. *Phys. Rev.* **1968**, *169*, 49–54.
6. King, F. W. Analysis of Some Integrals Arising in the Atomic Four-Electron Problem. *J. Chem. Phys.* **1993**, *99*, 3622–3628.
7. Büsse, G.; Kleindienst, H.; Lüchow, A. Nonrelativistic Energies for the Be Atom: Double-Linked Hylleraas-CI Calculation. *Int. J. Quantum Chem.* **1998**, *66*, 241–247.

8. Kutzelnigg, W. R_{12}-Dependent Terms in the Wave Function as Closed Sums of Partial Wave Amplitudes for Large *l*. *Theor. Chim. Acta* **1985**, *68*, 445–469.
9. Klopper, W.; Manby, F. R.; Ten-No, S.; Valeev, E. F. R_{12} Methods in Explicitly Correlated Molecular Electronic Structure Theory. *Int. Rev. Phys. Chem.* **2006**, *25*(03), 427–468.
10. Kong, L.; Bischoff, F. A.; Valeev, E. F. Explicitly Correlated R12/F12 Methods for Electronic Structure. *Chem. Rev.* **2012**, *112*, 75–107.
11. Sims, J. S.; Hagstrom, S. A. One-Center r_{ij} Integrals Over Slater-Type Orbitals. *J. Chem. Phys.* **1971**, *55*, 4699–4710.
12. Woźnicki, W. On the Method of Constructing the Variational Wave Function for Many-Electron Systems. In: *Theory of Electronic Shell in Atoms and Molecules,* Jucys, A. Ed.; Mintis: Vilnius, 1971; p. 103.
13. Pestka, G.; Woźnicki, W. Hylleraas-Type Calculations for Lithium. *Chem. Phys. Lett.* **1996**, *255*, 281–286.
14. Sims, J. S.; Hagstrom, S. A. Mathematical and Computational Science Issues in High Precision Hylleraas-Configuration Interaction Variational Calculations: I. Three-Electron Integrals. *J. Phys. B* **2004**, *37*, 1519–1540.
15. Sims, J. S.; Hagstrom, S. A. Mathematical and Computational Science Issues in High Precision Hylleraas-Configuration Interaction Variational Calculations: II. Kinetic Energy and Electron-Nucleus Interaction Integrals. *J. Phys. B* **2007**, *40*, 1575–1587.
16. Sims, J. S.; Hagstrom, S. A. Mathematical and Computational Science Issues in High Precision Hylleraas-Configuration Interaction Variational Calculations: III. Four-Electron Integrals. *J. Phys. B* **2015**, *48*, 175003.
17. Bonham, R. A. On Certain Atomic Integrals Involving Products of r_{ij}. *J. Mol. Spectrosc.* **1965**, *15*, 112–116.
18. Bonham, R. A. On Certain Atomic Integrals Involving Products of r_{ij}: Part II. The Case of a Square Diagram With One Diagonal. *J. Mol. Spectrosc* **1966**, *20*, 197–199.
19. Rychlewski, J.; Cencek, W.; Komasa, J. The Equivalence of Explicitly Correlated Slater and Gaussian Functions in Variational Quantum Chemistry Computations: The Ground State of H_2. *Chem. Phys. Lett.* **1994**, *229*, 657–660.
20. Fromm, D. M.; Hill, R. N. Evaluation of Three-Electron Integrals. *Phys. Rev. A* **1987**, *36*, 1013–1044.
21. Wang, C.; Mei, P.; Kurokawa, Y.; Nakashima, H.; Nakatsuji, H. Analytical Evaluations of Exponentially Correlated Unlinked One-Center, Three- and Four-Electron Integrals. *Phys. Rev. A* **2012**, *85*, 042512.
22. Thakkar, A. Smith, V. H., Jr. Compact and Accurate Integral-Transform Wave Functions. I. The 1^1S State of the Helium-Like Ions From H^- Through Mg^{10+}. *Phys. Rev. A* **1977**, *15*, 1–15.
23. Remiddi, E. Analytic Value of the Atomic Three-Electron Correlation Integral With Slater Wave Functions. *Phys. Rev. A* **1991**, *44*, 5492–5502.
24. Morse, P. M.; Feshbach, H. *Methods of Theoretical Physics. Part II.* McGraw-Hill: New York, 1953;1898–1901.
25. Arfken, G. B.; Weber, H. J.; Harris, F. E. *Mathematical Methods for Physicists.* Academic Press: New York, 2013.
26. Jen, C. K. The Continuous Electron Affinity Spectrum of Hydrogen. *Phys. Rev.* **1933**, *43*, 540–547.
27. The formula for the Wigner coefficient is given incorrectly in a number of recent papers[34,35,48,49] but because the formula was actually used in those papers only for a zero upper index (for which it is correct) the overall conclusions of those papers remain valid.
28. Edmonds, A. R. *Angular Momentum in Quantum Mechanics.* Princeton Univ. Press: Princeton, NJ, 1960.

29. Frolov, A. M. Smith, V. H. J. Generalization of the Exponential Variational Ansatz in Relative Coordinates for Bound State Calculations in Four-Body Systems. *J. Chem. Phys.* **2001**, *115*, 1187–1196.
30. Harris, F. E. Analytic Evaluation of Three-Electron Integrals With Slater Wave Functions. *Phys. Rev. A* **1997**, *55*, 1820–1831.
31. Pachucki, K.; Puchalski, M.; Remiddi, E. Recursion Relations for the Generic Hylleraas Three-Electron Integral. *Phys. Rev. A* **2004**, *70*(03), 032502.
32. Harris, F. E. Recurrence Formulas for Fully Exponentially Correlated Four-Body Wavefunctions. *Phys. Rev. A* **2009**, *79*(03), 032517.
33. Szasz, L. Atomic Many-Body Problem. I. General Theory of Correlated Wave Functions. *Phys. Rev.* **1962**, *126*, 169–181.
34. Ruiz, M. B. Evaluation of Hylleraas-CI Atomic Integrals by Integration Over the Coordinates of One Electron. I. Three-Electron Integrals. *J. Math. Chem.* **2009**, *46*, 24–64.
35. Ruiz, M. B. Evaluation of Hylleraas-CI Atomic Integrals by Integration over the Coordinates of One Electron. II. Four-Electron Integrals. *J. Math. Chem.* **2009**, *46*, 1322–1355.
36. Harris, F. E. To be published.
37. Perkins, J. F. Atomic Integrals Containing r_{ij}^{p} Correlation Factors With Unlinked Indices. *J Chem. Phys.* **1969**, *50*, 2819–2823.
38. Padhy, B. Analytic Evaluation of Some Three- and Four-Electron Atomic Integrals Involving s STO's and Exponential Correlation with Unlinked. **2017**, ArXiv: 1701.05701v1 [physics.atom-ph].
39. Frost, A. A. The Use of Interparticle Coordinates in Electronic Energy Calculations for Atoms and Molecules. *Theor. Chim. Acta* **1962**, *1*, 36–41.
40. Harris, F. E. Integrals for Exponentially Correlated Four-Body Systems of General Angular Symmetry. In: *Fundamental World of Quantum Chemistry: A Tribute to Per-Olov Löwdin*; Brändas, E. J., Kryachko, E. S., Eds.; Vol. 3. Kluwer: Dordrecht, 2004; pp. 115–128.
41. Harris, F. E. Current Methods for Coulomb Few-Body Problems. *Adv. Quantum Chem.* **2004**, *47*, 129–155.
42. Harris, F. E. Angular Symmetry and Hylleraas Coordinates in Four-Body Problems. *Adv. Quantum Chem.* **2005**, *50*, 61–75.
43. Rebane, T. K. To the Energy Calculation of Coulomb System. *Opt. Spektrosk.* **1993**, *75*, 945–953 (*Opt. Spectrosc.* **1993**, *75*, 557).
44. Harris, F. E.; Frolov, A. M.; Smith, V. H. Exponential Variational Expansion in Relative Coordinates for Highly Accurate Bound State Calculations in Four-Body Systems. *J. Chem. Phys.* **2003**, *119*, 8833–8841.
45. Harris, F. E. Kinetic-Energy Matrix Elements for Atomic Hylleraas-CI Wave Functions. *J. Chem. Phys.* **2016**, *144*, 204110 (Erratum 2016, 145, 129901).
46. Condon, E. U.; Shortley, G. *The Theory of Atomic Spectra.* Cambridge University Press: Cambridge, 1951.
47. Pinchon, D.; Hoggan, P. E. New Index Functions for Storing Gaunt coefficients. *Int. J. Quantum Chem.* **2007**, *107*, 2186–2196.
48. Ruiz, M. B. Evaluation of Hylleraas-CI Atomic Integrals. III. Two-Electron Kinetic Energy Integrals. *J. Math. Chem.* **2011**, *49*, 2457–2485.
49. Ruiz, M. B. Evaluation of Hylleraas-CI Atomic Integrals by Integration Over the Coordinates of One Electron. IV. An Improved Algorithm for Three-Electron Kinetic Energy Integrals. *J. Math. Chem.* **2016**, *54*, 1083–1109.

CHAPTER ELEVEN

One-Particle Effective Potential for Helium Atom

Daniel Gebremedhin[1], Charles Weatherford[1]
Florida A&M University, Tallahassee, FL, United States
[1]Corresponding authors: e-mail address: daniel1.gebremedhin@famu.edu; charles.weatherford@famu.edu

Contents

1. Introduction 211
2. Preliminary 212
3. Construction of Basis Set 214
4. Eigen Equation and the Iterative Scheme 216
5. Results and Discussion 217
6. Conclusion 221
Acknowledgments 221
References 222

Abstract

A single-particle pseudo-potential that splits the effect of the electron–electron repulsive potential of Helium (He) atom into two noninteracting identical particle potentials is numerically computed. This is done by minimizing the expectation value of the difference between the approximate and exact Hamiltonians over the Hilbert space of He atom. The one-particle potential is expanded in a spatial basis set which leads to an overdetermined system of linear equation that was solved using a least square approximation. The method involves a self-consistent iterative scheme where a converged solution valid for any state of the atom can be calculated. The total ground state energy for these two noninteracting particles under the calculated potential is found to be − 2.861 68, which is the Hartree–Fock limit for the He atom.

1. INTRODUCTION

It is well known that the feasibility of highly accurate numerical calculations for the multielectron Schrödinger equation (SE) is restricted to only a small number of electrons. The starting point of most such calculations is the Hartree–Fock (HF) method which results in an effective single-particle equation with an averaged (including only the static correlation)

Advances in Quantum Chemistry, Volume 76
ISSN 0065-3276
https://doi.org/10.1016/bs.aiq.2017.05.006

potential field. HF theory is a well-defined theory but it fails to include the dynamic correlation potential which is very significant in most applications.

Density functional theory (DFT), on the other hand, is in principle exact in that it establishes the existence of an exchange–correlation functional and leads to an effective one-body potential which includes all the static and dynamic correlation effects in the SE. However, a systematic method of calculating this exchange–correlation functional and the resulting one-body potentials has yet to be developed. The main difficulty associated with any theory that reduces to an effective single-particle potential stems from the necessity of including the effects of the two-electron coulomb potential.

The main purpose of this chapter is to have a fresh look at correlation from a purely numerical perspective. We split the electron–electron repulsive potential of the He atom into two one-particle potentials. These potentials are then calculated and the resulting Hamiltonian should have the same eigenvalue as the exact Hamiltonian. This will give us a chance to explore the possibility that this approximation can be systematically improved by means of numerical techniques. The helium (He) atom is the simplest system that can be used as a test bed for such an attempt. We will use the ground state of He atom, for which highly accurate numerical results are available,[1] to make a comparison between our result and the exact answer.

In this chapter, we will use bold upper case letters C, $\vec{C}$ to represent matrices and vectors, respectively. A spherical coordinate system will be used throughout, where for the three-dimensional position vector $\mathbf{r}$, $r = |\mathbf{r}|$ will represent its magnitude and $\hat{r} = \mathbf{r}/r$ a unit vector giving direction. We shall also use $r_{12} = |\mathbf{r}_1 - \mathbf{r}_2|$ for the distance between two electrons. The familiar *bra–ket* notation will be used wherever appropriate.

2. PRELIMINARY

With the following definition of a core-Hamiltonian for a nucleus of atomic number Z,

$$h(Z;\mathbf{r}) = -\frac{1}{2}\nabla_{\mathbf{r}}^2 - \frac{Z}{r} \tag{1}$$

the nonrelativistic Hamiltonian of a Helium atom is given by

$$H(Z;\mathbf{r}_1,\mathbf{r}_2) = h(\mathbf{r}_1) + h(\mathbf{r}_2) + \frac{1}{r_{12}}. \tag{2}$$

The corresponding time independent SE is

$$H|\Psi(\mathbf{r}_1, \mathbf{r}_2)\rangle = E|\Psi(\mathbf{r}_1, \mathbf{r}_2)\rangle. \tag{3}$$

The main objective of this chapter is to come up with a single-particle potential $v(r)$ such that the above Hamiltonian is replaced by an approximate Hamiltonian $\mathcal{H}$ of two noninteracting particles given below.

$$\mathcal{H}(Z; \mathbf{r}_1, \mathbf{r}_2) = h(\mathbf{r}_1) + v(\mathbf{r}_1) + h(\mathbf{r}_2) + v(\mathbf{r}_2) \tag{4}$$

The approximate potential v is obtained by insisting that the expectation value of $(H - \mathcal{H})$, in the domain spanned by the Hilbert space of the He atom, vanishes.

$$\left\langle \Psi(\mathbf{r}_1, \mathbf{r}_2) \left| \frac{1}{r_{12}} - v(\mathbf{r}_1) - v(\mathbf{r}_2) \right| \Psi(\mathbf{r}_1, \mathbf{r}_2) \right\rangle_{1,2} = 0 \tag{5}$$

At this point, we need to define a basis with which to expand the wave function Ψ. For a reason which will be clear, we chose the eigenfunctions of $\mathcal{H}$, which are a complete set. Namely, we expand

$$\Psi(\mathbf{r}_1, \mathbf{r}_2) = \sum_{\mu} B_\mu |\Phi_\mu(\mathbf{r}_1, \mathbf{r}_2)\rangle \tag{6}$$

where

$$\mathcal{H}|\Phi_\mu(\mathbf{r}_1, \mathbf{r}_2)\rangle = |\Phi_\mu(\mathbf{r}_1, \mathbf{r}_2)\rangle \mathcal{E}_\mu. \tag{7}$$

Upon substitution of the expansion in Eq. (6) into Eq. (5), we get

$$\sum_{\mu'\mu} B^*_{\mu'} \left\langle \Phi_{\mu'}(\mathbf{r}_1, \mathbf{r}_2) \left| \frac{1}{r_{12}} - v(\mathbf{r}_1) - v(\mathbf{r}_2) \right| \Phi_\mu(\mathbf{r}_1, \mathbf{r}_2) \right\rangle_{1,2} B_\mu = 0. \tag{8}$$

Since the coefficients B are linearly independent, the above equation must be valid element-wise. Moreover, there is no distinction between the two electrons and thus, the following is consistent with Eq. (8).

$$\langle \Phi_{\mu'}(\mathbf{r}_1, \mathbf{r}_2) | v(\mathbf{r}_1) | \Phi_\mu(\mathbf{r}_1, \mathbf{r}_2) \rangle_{1,2} = \frac{1}{2} \left\langle \Phi_{\mu'}(\mathbf{r}_1, \mathbf{r}_2) \left| \frac{1}{r_{12}} \right| \Phi_\mu(\mathbf{r}_1, \mathbf{r}_2) \right\rangle_{1,2} \tag{9}$$

We now expand v in some basis set g, which will be specified later on.

$$v(\mathbf{r}) = \sum_{q=0}^{Q-1} g_q(\mathbf{r}) D_q \tag{10}$$

Substituting the above expansion into Eq. (8) gives the following set of simultaneous equations for each state μ

$$\sum_q \Omega^{\mu}_{\mu' q} D_q = \Gamma^{\mu}_{\mu'} \tag{11}$$

where,

$$\Omega^{\mu}_{\mu' q} = \langle \Phi_{\mu'}(\mathbf{r}_1, \mathbf{r}_2) | g_q(\mathbf{r}_1) | \Phi_{\mu}(\mathbf{r}_1, \mathbf{r}_2) \rangle_{1,2} \tag{12}$$

$$\Gamma^{\mu}_{\mu'} = \frac{1}{2} \left\langle \Phi_{\mu'}(\mathbf{r}_1, \mathbf{r}_2) \left| \frac{1}{r_{12}} \right| \Phi_{\mu}(\mathbf{r}_1, \mathbf{r}_2) \right\rangle_{1,2}. \tag{13}$$

Eq. (11) is simultaneous equations $\mathbf{\Omega} \in \mathbb{R}^{N \times Q}$ and is usually an overdetermined system (with $N > Q$). We multiplied from the left by its transpose ($\mathbf{\Omega}^T$) before solving the resulting square linear system of size Q. This results in the least square approximation for the solution coefficients $\vec{D}$.[2] A good reference for more on least square problems can be found in Ref. 3.

The unknown potential v and its corresponding eigenfunction Φ are mutually dependent, and hence, the solution can only proceed by iteration using a self-consistent method. Once the iteration is converged, it will allow us to represent the He atom by its equivalent single-particle SE shown below.

$$\left[-\frac{1}{2}\nabla^2 - \frac{Z}{r} + \sum_q g_q(\mathbf{r}) D_q - \frac{\mathcal{E}}{2} \right] \psi(\mathbf{r}) = 0 \tag{14}$$

Notice that the choice of the eigenfunctions Φ as a basis set in Eq. (6) was crucial to linking $\mathcal{H}$ and H with the added price that the solution can only proceed iteratively.

3. CONSTRUCTION OF BASIS SET

The radial part of the basis set used will be the Coulomb Sturmian (CS) functions given below[4,5]

$$R_{nl}(\alpha, r) = \sqrt{\frac{\alpha}{n} \frac{(n-l-1)!}{(n+l)!}} (2\alpha) e^{-\alpha r} (2\alpha r)^l L^{2l+1}_{n-l-1}(2\alpha r) \tag{15}$$

where L is the associated Laguerre polynomial, α a positive real number. Also l and n are interrelated angular and principal quantum numbers, respectively, with ranges $l = 0, 1, \ldots, l_{\max}$ and $n = l + 1, l + 2, \ldots, l + n_{\max}$. The quantity $(n_{\max} - 1)$ is the order of the polynomial L.

$$\Phi_\mu(\mathbf{r}_1, \mathbf{r}_2) = \sum_{\nu=0}^{N-1} \phi_\nu(\mathbf{r}_1, \mathbf{r}_2) C_{\nu\mu} \tag{16}$$

where μ, ν label an ordered set of six quantum numbers. We also introduce a label σ, another ordered set, for two screening parameters shown below.

$$\mu = (n_1, l_1, n_2, l_2, L, M), \qquad \sigma = (\alpha_1, \alpha_2) \tag{17}$$

This allows us to define the following spatial two-particle configuration state function

$$F_\nu(\sigma, \mathbf{r}_1, \mathbf{r}_2) = R_{n_1 l_1}(\alpha_1, r_1) R_{n_2 l_2}(\alpha_2, r_2) X_{l_1 l_2}^{LM}(\hat{r}_1, \hat{r}_2) \tag{18}$$

where X is the spherical bipolar harmonics.[6] The basis functions ϕ must take into account for the exchange of the two electrons ($\mathbf{r}_1 \leftrightarrow \mathbf{r}_2$) according to Pauli's exclusion principle. For that purpose, ϕ in Eq. (16) will be

$$\phi_\nu(\sigma, \mathbf{r}_1, \mathbf{r}_2) = F_\nu(\sigma, \mathbf{r}_1, \mathbf{r}_2) + aF_\nu(\sigma, \mathbf{r}_2, \mathbf{r}_1) \tag{19}$$

where $a = +1$ and $a = -1$ for singlet and triplet states, respectively. In practice, it would be convenient to reverse the order of the coordinates in the last term of Eq. (19). For this purpose, given μ and σ as specified in Eq. (17), one can define the corresponding $\overline{\mu}$ and $\overline{\sigma}$

$$\overline{\mu} = (n_2, l_2, n_1, l_1, L, M), \qquad \overline{\sigma} = (\alpha_2, \alpha_1) \tag{20}$$

and rewrite Eq. (19) as:

$$\phi_\nu(\sigma, \mathbf{r}_1, \mathbf{r}_2) = F_\nu(\sigma, \mathbf{r}_1, \mathbf{r}_2) + aF_{\overline{\nu}}(\overline{\sigma}, \mathbf{r}_1, \mathbf{r}_2). \tag{21}$$

The bipolar harmonics would have been an exception to the above arrangement because they obey the following identity.[7]

$$X_{l_2 l_1}^{LM}(\hat{r}_2, \hat{r}_1) = (-1)^{l_1 + l_2 - L} X_{l_1 l_2}^{LM}(\hat{r}_1, \hat{r}_2) \tag{22}$$

However, the angular momentum quantum numbers of the two electrons must conserve parity which is expressed by $(-1)^{l_1 + l_2} = (-1)^L$.[8,9] Hence, Eq. (21) correctly specifies the properly symmetrized basis functions and thus is implemented in this work.

In addition to this, we have the usual requirement for the angular quantum numbers in bipolar harmonics that (l_1, l_2, L) be triangular and $|M| \leq L$. In order to avoid numerical problems due to redundancy in Eq. (21), $n_1 \leq n_2$ is used whenever $l_1 = l_2$. The ground state of the He atom has a spherical symmetry, so in all the calculation done in this chapter, the total angular momentum and its projection will be set to $L = M = 0$. Hence, total size of the expansion N given in Eq. (16) will be determined completely by the quantities $n_{\max}$ and $l_{\max}$.

4. EIGEN EQUATION AND THE ITERATIVE SCHEME

Substitution of the expansion given in Eq. (16) into Eq. (7) and evaluating the necessary projection integrals leads to the following equations defining a generalized eigenvalue equation of size N

$$\sum_{\nu=0}^{N-1} \mathcal{H}_{\nu'\nu} C_{\nu\mu} = \sum_{\nu=0}^{N-1} O_{\nu'\nu} C_{\nu\mu} \mathcal{E}_{\mu} \tag{23}$$

where the matrix elements are numerical values of the integrals.

$$\mathcal{H}_{\nu'\nu} = \langle \phi_{\nu'}(\mathbf{r}_1, \mathbf{r}_2) | \mathcal{H} | \phi_{\nu}(\mathbf{r}_1, \mathbf{r}_2) \rangle_{1,2} \tag{24}$$

$$O_{\nu'\nu} = \langle \phi_{\nu'}(\mathbf{r}_1, \mathbf{r}_2) | \phi_{\nu}(\mathbf{r}_1, \mathbf{r}_2) \rangle_{1,2} \tag{25}$$

Evaluation of the integral in Eq. (24) requires knowledge of the potential v (or the array $\vec{D}$) indicated in Eq. (10), which must be supplied from the previous iterate. In this work, the first iteration has been seeded by ignoring v altogether and solving equation (23) for the resulting Hamiltonian of two noninteracting Hydrogen-like atoms. This Hamiltonian matrix of Hydrogen-like atoms and the overlap matrix $\mathbf{O}$ may be stored since they remain the same throughout the iteration. Once the first eigenpair $\left\{\mathbf{C}, \vec{\mathcal{E}}\right\}$ is obtained this way and the state of interest μ is specified (in this work, only the ground state $\mu = 0$ is discussed), the procedure for rest of the iteration is sketched below.

1. By employing the definition of $\mathbf{\Phi}$ given in Eq. (16), carry out all the integrals shown in Eqs. (12) and (13) and construct the system of linear equation (11).
2. Solve Eq. (11). Generate the least square solution of $\vec{D}$ as described above, if $N \neq Q$.
3. With v as specified in Eq. (10), evaluate the desired integrals in Eqs. (24) and (25). Only the terms involving $v(\mathbf{r}_2)$ and $v(\mathbf{r}_2)$ in Eq. (24) are to be changed during each iteration.

4. Solve the generalized eigensystem in Eq. (23) for a new eigenpair $\left\{\mathbf{c}, \vec{\mathcal{E}}\right\}$.
5. Check the eigenvalue of the specified state $\mathcal{E}_\mu$ for convergence. Go back to step (1) and repeat until the convergence of $\mathcal{E}_\mu$ is satisfactory.

5. RESULTS AND DISCUSSION

In order to test the efficacy of the basis functions Φ and the quality of our numerical implementation, we chose to do bound state calculations for the ground state energy of He atom and compare with published results before proceeding. For this purpose, the total wave function Ψ of the He atom was expanded directly in the basis set ϕ according to Eq. (16) and was used to diagonalize Eq. (3). This leads to an eigensystem, which is identical in form to Eq. (23), for the full Hamiltonian H in Eq. (2). Notice that all the resulting integrals are needed later when dealing with the approximate Hamiltonian $\mathcal{H}$.

The calculated total ground state energies E_0 for $l_{\max} = 0$–6 with fixed $n_{\max} = 15$ are shown in Table 1. The values of $n_{\max}$ and $l_{\max}$ along with the corresponding total number of terms N are adapted from Ref. 8 for comparison purposes. Ref. 9 used expansions of a larger size with $n_{\max} = 20$. The screening parameters must be identical due to symmetry of the ground state. We used the value of σ with $\alpha_1 = \alpha_2$ as a variational parameter to do a one-dimensional search for the minimum E_0. The accuracy of our calculated E_0, with respect to the known exact value,[1] generally sits in between these

Table 1 For $n_{\max} = 15$ and $l_{\max} = 0$–6, the Matrix Size N and the Calculated Ground State Energy for the Helium Atom E_0 are Shown

$l_{\max}$	N	Ref. 8	Present Work	Ref. 9
0	120	−2.879 027 923	−2.879 027 803 148 13	−2.879 027 97
1	240	−2.900 515 058	−2.900 513 890 717 26	−2.900 513 86
2	360	−2.902 772 510	−2.902 762 736 989 56	−2.902 762 09
3	480	−2.903 327 013	−2.903 314 945 150 55	−2.903 313 21
4	600	−2.903 524 584	−2.903 510 283 999 86	−2.903 506 82
5	720	−2.903 613 398	−2.903 595 032 451 61	−2.903 589 25
6	840	−2.903 660 341	−2.903 636 673 568 45	−2.903 628 16

Columns 3 and 4 are E_0 from the indicated references. Exact value is $E_0 = -2.903\ 724\ 377$.

two previous works. Ref. 8 implemented improved basis functions, which are generalized Sturmians[5] for the Yukawa potential, generated by expanding in the CS (or R) functions. The basis set in Ref. 9 is essentially identical to ours but implemented with fixed $\alpha_1 = \alpha_2 = 2.0$, which may mainly explain our superior results gained with smaller N.

Let us now return to calculating the approximate potential v and its corresponding ground state energy $\mathcal{E}_0$, which is the main intent of this chapter. The basis set used for the potential v in Eq. (10) is still based on the CS functions,

$$g_q(\gamma, \mathbf{r}) = R_{nl}(\gamma, r)\ Y_l^m(\hat{r}) \tag{26}$$

where q is now a composite of the three quantum numbers (n, l, m) and Y is the spherical harmonic function. For the ground state, which has spherically symmetric total angular momentum of $L = M = 0$, it can be shown that the angular integrals in Eqs. (12) and (13) identically vanish unless $l = m = 0$ in q. So Eq. (10) is actually implemented here only in its purely radial form according to

$$v(\gamma, r) = \sum_{n=1}^{Q} R_{n0}(\gamma, r) D_n \tag{27}$$

with fixed parameters $\gamma = 1.0$ and $Q = n_{\max}$. Our numerical tests revealed that, although relaxing the value of γ is of little consequence, allowing $Q > n_{\max}$, on the other hand, caused the convergence of the iteration to proportionally deteriorate. The two screening parameters on the two-electron basis set ϕ were, however, still kept equal $\alpha_1 = \alpha_2$, but varied in search for the minimum value of $\mathcal{E}_0$.

In Table 2 the numerical values for the calculated total ground state energy $\mathcal{E}_0$ of the approximate Hamiltonian $\mathcal{H}$ are shown. For such a minimal size of the expansion, $\mathcal{E}_0$ quickly approaches the Hartree–Fock limit[10] of the He atom ground state energy and saturates. Although not shown on the table, further increase in $l_{\max}$ and/or $n_{\max}$ did not push the value of $\mathcal{E}_0$ any lower. This is the main finding reported in this chapter. We have also indicated E_0 of H, in a similar fashion to Table 1, under the same parameters regarding expansion size. E_0 in Table 2 can be viewed as the value $\mathcal{E}_0$ is expected approach as $\mathcal{H}$ approximates H more closely. The case in the last row of Table 2 is picked to display further details of a particular calculation.

Fig. 1 shows a plot of the calculated one-particle potential $v(r)$. This plot is identical in shape to the familiar effective potentials available in Hartree–Fock theory of atoms, which is consistent with its corresponding HF limit value of $\mathcal{E}_0 = -2.861\,68$. However, unlike those typical HF

Table 2 The Total Ground State Energy $\mathcal{E}_0$ and E_0 for the Approximate and Full Hamiltonians $\mathcal{H}$ & H, Respectively

l_{max}	n_{max}	N	$\mathcal{E}_0$	E_0
0	1	1	−2.847 656 250 000 00	−2.847 656 250 000 00
0	2	3	−2.847 656 250 000 00	−2.860 487 304 441 45
0	3	6	−2.861 590 054 663 49	−2.878 305 210 250 11
0	4	10	−2.861 590 054 663 48	−2.878 636 224 485 79
0	5	15	−2.861 679 675 254 11	−2.878 916 961 719 77
0	6	21	−2.861 679 675 254 14	−2.878 987 818 139 47
1	1	2	−2.847 656 250 000 00	−2.862 486 194 058 53
1	2	6	−2.847 656 250 000 00	−2.881 304 993 112 48
1	3	12	−2.861 590 054 663 49	−2.899 222 283 077 92
1	4	20	−2.861 590 054 663 48	−2.899 842 247 185 35
1	5	30	−2.861 679 675 254 12	−2.900 254 354 558 96
1	6	42	−2.861 679 675 254 15	−2.900 377 272 952 80

HF limit for the He atom is −2.861 679 995 6.

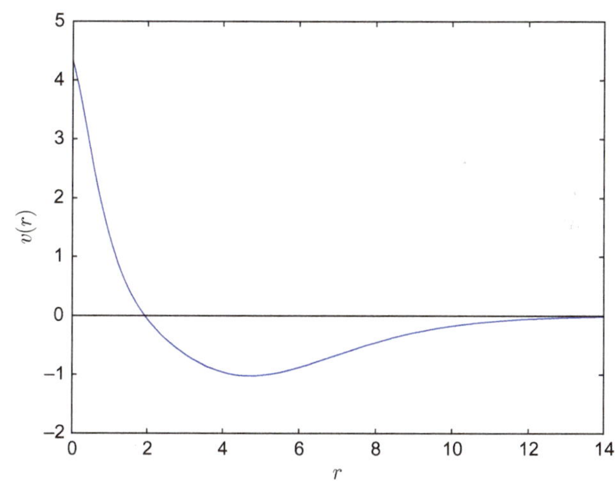

Fig. 1 Plot of the one-particle potential v of He atom for $l_{max} = 1$ and $n_{max} = 6$.

effective potentials which are singular at the origin, the plot intercepts the vertical-axis at a finite value due to the nature of the basis functions R. Table 3 shows the numerical values of the iterates of $\mathcal{E}_0$ starting from that of two noninteracting particles ($\mathcal{E}_0 \approx -4.0$) and which progress toward

Table 3 Convergence of the Iteration on $\mathcal{E}_0$ for $l_{max}=1$ and $n_{max}=6$

Iteration	$\mathcal{E}_0$
0	−4.000 000 000 000 00
1	−2.891 378 411 879 48
2	−2.864 437 570 753 50
3	−2.861 911 308 721 43
4	−2.861 699 821 422 73
5	−2.861 681 410 389 82
6	−2.861 679 825 133 26
7	−2.861 679 688 189 49
8	−2.861 679 676 370 81
9	−2.861 679 675 350 53
10	−2.861 679 675 262 46
11	−2.861 679 675 254 86
12	−2.861 679 675 254 20
13	−2.861 679 675 254 14
14	−2.861 679 675 254 15

The optimum value for the screening parameters is $\alpha_1=\alpha_2=1.966\,462\,285\,145\,22$.

Table 4 Coefficients D_n of the One-Particle Potential as Defined in Eq. (27)

n	D_n
1	1.189 281 599 296 86
2	3.698 243 308 230 93
3	−5.277 833 556 221 71
4	4.639 513 856 308 60
5	−2.192 164 800 202 36
6	0.1 127 360 491 781 20

These correspond to $\mathcal{E}_0=-2.861\,679\,675\,254\,15$ and $\gamma=1.0$.

the HF limit as the iteration proceeds. All iterations for all of the cases shown in Table 2 have similarly converged to values very close to the working precision. Finally, the expansion coefficients $\vec{D}$ for the calculated one-particle potential given in Eq. (27) are shown in Table 4 for demonstration purposes.

One can directly plug this particular potential v into the one-particle Hamiltonian shown in Eq. (14), diagonalize using their favorite method, and verify that the ground state eigenvalue is the one shown in the table.

6. CONCLUSION

The aim of this work has been to force the approximate Hamiltonian $\mathcal{H}$ to look like the He atom Hamiltonian H as closely as possible by solving simultaneous equations and in the process, to capture some of the dynamic correlation effects. Our results clearly indicate that our method can get to the HF limit of the ground state energy, but no more. There may be several reasons for why the method managed to produce the static correlation and yet would not go beyond it. We believe the explanation is not merely numerical such as the quality of basis set, or the size of the expansion or other similar effects of optimization. The main trait that our one-particle potential has in common with its HF analog is that it averages the two-electron correlation effect over all space and does not include the effect of the "Coulomb hole." But unlike HF theory which uses a single determinate state function, the present method is a many-determinant theory. Both the HF theory and the present theory are variational.

The effective potential in both HF and density functional theories is explicitly a functional of the electronic density ρ, i.e., $v(\mathbf{r}) \rightarrow v[\rho](\mathbf{r})$. In fact, DFT asserts not only does such a potential exist, but it also is exact.

There are several avenues to explore suggested by the results of the present work. Instead of expanding the one-particle potential in a spatial wave function as is done here, we could make it a functional of electronic density and see how it behaves. Alternatively, the potential ansatz for the model Hamiltonian $\mathcal{H}$ can be modified so as to include a function that would include some aspect of the Coulomb hole. In this sense, this work can be considered as providing some insight into a systematic way of computing the elusive dynamic correlation potential, and possibly in a way to be described in future work, the correlation functional in DFT.

ACKNOWLEDGMENTS

D.H.G. and C.A.W. were partially supported by the Department of Energy, National Nuclear Security Administration, under Award Number(s) DE-NA0002630. C.A.W. was also supported in part by the Defense Threat Reduction Agency.

REFERENCES

1. Nakashima, H.; Nakatsuji, H. Solving the Schrödinger Equation for Helium Atom and Its Isoelectronic Ions With the Free Iterative Complement Interaction (ICI) Method. *J. Chem. Phys.* **2007**, *127*(22), 224104, https://doi.org/10.1063/1.2801981.
2. Zhou, X.; Lin, C. D. Linear-Least-Squares Fitting Method for the Solution of the Time-Dependent Schrödinger Equation: Applications to Atoms in Intense Laser Fields. *Phys. Rev. A* **2000**, *61*, 053411, https://doi.org/10.1103/PhysRevA. 61.053411.
3. Lawson, C.; Hanson, R. *Solving Least Squares Problems.* Society for Industrial and Applied Mathematics: Philadelphia, PA, 1995; https://doi.org/10.1137/1.9781611971217.
4. Calderini, D.; Cavalli, S.; Coletti, C.; Grossi, G.; Aquilanti, V. Hydrogenoid Orbitals Revisited: From Slater Orbitals to Coulomb Sturmians. *J. Chem. Sci.* **2012**, *124*(1), 187–192, https://doi.org/10.1007/s12039-012-0215-7.
5. Avery, J.; Avery, J. *Generalized Sturmians and Atomic Spectra.* World Scientific: Singapore, 2012;1–8, https://doi.org/10.1142/9789812773593_0001.
6. Varshalovich, D. A.; Moskalev, A. N.; Khersonskii, V. K. *Quantum Theory of Angular Momentum* (Chapter 5). World Scientific: Singapore, 2013; pp. 130–169, https://doi.org/10.1142/9789814415491_0006.
7. Gebremedhin, D.; Weatherford, C. Canonical Two-Range Addition Theorem for Slater-Type Orbitals. *Int. J. Quantum Chem.* **2013**, *113*(1), 71–75, https://doi.org/10.1002/qua.24319.
8. Frapiccini, A. L.; Randazzo, J. M.; Gasaneo, G.; Colavecchia, F. D. Sturmian Expansions for Two-Electron Atomic Systems: Singly and Doubly Excited States. *Phys. Rev. A* **2010**, *82*, 042503, https://doi.org/10.1103/PhysRevA. 82.042503.
9. Foumouo, E.; Kamta, G. L.; Edah, G.; Piraux, B. Theory of Multiphoton Single and Double Ionization of Two-Electron Atomic Systems Driven by Short-Wavelength Electric Fields: An Ab Initio Treatment. *Phys. Rev. A* **2006**, *74*, 063409, https://doi.org/10.1103/PhysRevA.74.063409.
10. Davis, C. L.; Jensen, H. A.; Monkhorst, H. J. Accurate Hartree? Fock Wave Functions Without Exponent Optimization. *J. Chem. Phys.* **1984**, *80*(2), 840–855, https://doi.org/10.1063/1.446738.

CHAPTER TWELVE

Configuration Interaction Study of the ^{3}P Ground State of the Carbon Atom

María Belén Ruiz[1], Robert Tröger
Friedrich-Alexander-University Erlangen-Nürnberg, Erlangen, Germany
[1]Corresponding author: e-mail address: maria.belen.ruiz@fau.de

Contents

1. Introduction 223
2. Configuration Interaction 224
3. Selection of Configurations 227
4. Calculations 229
5. Conclusions and Perspectives 234
Acknowledgments 236
References 236
Further Reading 238

Abstract

Configuration Interaction (CI) calculations on the ground state of the C-atom are carried out using a small basis set of Slater orbitals [7s6p5d4f3g]. The configurations are selected according to their contribution to the total energy. One set of exponents is optimized for the whole expansion. Using some computational techniques to increase efficiency, our computer program is able to perform partially parallelized runs of 1000 configuration term functions within a few minutes. With the optimized computer program we were able to test a large number of configuration types and chose the most important ones. The energy of the ^{3}P ground state of carbon atom with a wave function of angular momentum $L=1$ and $M_L=0$ and spin eigenfunction with $S=1$ and $M_S=0$ leads to -37.83526523 h, which is millihartree accurate. We discuss the state of the art in the determination of the ground state of the carbon atom and give an outlook about the complex spectra of this atom and its low-lying states.

1. INTRODUCTION

The spectrum of the isolated carbon atom is the most complex one among the light atoms. The ground state of carbon atom is a triplet ^{3}P state and its low-lying excited states are singlet ^{1}D, ^{1}S, and ^{1}P states, more stable

Advances in Quantum Chemistry, Volume 76
ISSN 0065-3276
https://doi.org/10.1016/bs.aiq.2017.07.001

than the corresponding triplet excited ones ^3D and ^3S against the Hund's rule of maximal multiplicity. While a quintet spin ^5S state is the third excited state, no other quintet is stable as a bound state of the carbon atom. In this work, we study the ^3P ground state of carbon atom.

The first calculations on the ^3P ground state of carbon atom were carried out by Boys,[1] Clementi,[2] and Bagus and Moser[3] using small basis sets and by Schaefer et al.[4] with a larger basis. The short superposition of configurations used by Weiss[5] with only 40 terms and a basis set of $n=5$ or [5s4p3d2f] basis serves as a guide to start the calculations on the ground state. The first accurate Configuration Interaction (CI) calculations on the ground state of carbon atom were performed by A. Bunge[6] and posterior works of A. Bunge and C.F. Bunge in the early 1970s.[7,8] A few years later Sasaki and Yoshimine[9] carried out very sophisticated CI calculations with the ALCHEMY computer program of the theoretical chemistry group at IBM in San Jose. They used natural orbitals of s, p, d, f, g, h, and i symmetry formed with large numbers of atomic orbitals (in total 35) and large numbers of configurations. They obtained the best CI energy to date. This energy has only been improved recently by the use of the method of Exponential Correlated Gaussians (ECGs).[10] Other accurate approaches used in atoms are Quantum Monte Carlo[11] and Multiconfigurational explicit correlated wave function.[12] A larger number of low-lying excited states of the carbon atom have been calculated using Variational and Diffusion Monte Carlo approaches.[13]

In this work, we show a simple methodology to perform CI calculations obtaining a good total energy using very little computer time. We show how to construct the configurations and select them and how to find the most important ones with respect to the energy. Section 4 below contains the wave function expansion written in detail. Finally, we compare the obtained energy for the ground state with the results obtained by other authors and methods.

2. CONFIGURATION INTERACTION

The method of CI proposed by Löwdin[14] in 1955, the so-called Father of Quantum Chemistry, is a variational systematic method to calculate straightforwardly the electron correlation energy in atoms and molecules. In a recent volume celebrating the 100th anniversary of the birth of Per-Olov Löwdin, one of his students, Carlos F. Bunge, dedicated a chapter to him.[15] Another chapter about the state of the art in highly accurate CI calculations on atoms and molecules[16] is also recommended.

The CI method endeavors to recover rapidly a great part of the correlation energy, but the method converges extremely slowly for obtaining highly accurate energy values. This is due to the shortcomings of the CI wave function which form does not fulfill the electronic cusp condition. The CI wave function contains excitations like p(1)p(2), d(1)d(2), which are proportional to $r_{12}^2, r_{12}^4, \ldots, r_{12}^{2n}$ terms, but not to a linear r_{12} term, which is required by the cusp. Physically, the interelectronic coordinate r_{ij} represents the situation when the distance between two electrons gets close to zero. Therefore, the CI wave function needs a huge number of Slater determinants to converge. Conversely, methods including r_{ij} are the explicitly correlated methods, introduced by the pioneering work of Hylleraas in 1929[17]; they converge faster to the exact solution of the Schrödinger equation. Nevertheless, in this work we employ the CI method as a first step for further Hylleraas-Configuration Interaction (Hy-CI) calculations on the ground state of carbon atom. For more information on the Hy-CI method, see Refs. 18,19.

The CI wave function is constructed as a linear combination of N_{conf} configurations, where the coefficients C_{p} are determined variationally by solving the eigenvalue problem which follows from the Schrödinger equation:

$$\Psi = \sum_{p=1}^{N_{\text{conf}}} C_{\text{p}} \Phi_{\text{p}} \tag{1}$$

The configurations Φ_{p} are symmetry adapted since they are eigenfunctions of the square of the angular momentum operator $\hat{L}^2$. Every configuration is then a linear combination of function terms and these ones are a linear combination of Slater determinants:

$$\Phi_{\text{p}} = \hat{O}\left(\hat{L}^2\right)\hat{\mathcal{A}}\psi_{\text{p}}\chi \tag{2}$$

In this work, the symmetry-adapted configurations are constructed "a priori" so that they are eigenfunctions of $\hat{L}^2$. Another possibility would be the posterior projection of the configurations over the proper space, as indicated in Eq. (2) by the projection operator $\hat{O}\left(\hat{L}^2\right)$. The Slater determinants are constructed with the help of the antisymmetrization operator $\hat{\mathcal{A}}$, which acts not only over the spatial orbitals but also over the spin function part.

The configurations are also eigenfunctions of the square of the spin operator $\hat{S}^2$. χ is a spin eigenfunction. In the case of the ^{3}P ground state of carbon atom, a triplet state with $S=1$, we have chosen for convenience a spin eigenfunction with $M_S=0$:

$$\chi = (\alpha\beta - \beta\alpha)(\alpha\beta - \beta\alpha)(\alpha\beta + \beta\alpha). \tag{3}$$

This spin function differs only by a sign from the singlet one $S=1$ for six electrons, and in this way the computer program can be used for singlet or triplet states only by changing one sign. There are more spin eigenfunctions for $S=1$, namely, the ones with $M_S=1$ and $M_S=-1$, but they are all degenerate with respect to the energy. This means, it is indifferent which one we use and since they are orthogonal, it is sufficient to use only one of them.

As discussed in the case of the Li atom in Ref. 20 and calculations of the Be atom,[21] it is also sufficient to consider only one spin function. The spatial part of the basis functions consists of Hartree products of Slater-Type Orbitals (STOs):

$$\psi_p = \prod_{k=1}^{n_e} \phi(r_k, \theta_k, \varphi_k) \tag{4}$$

Every Slater orbital ϕ is represented by only one orbital s, p, d, f, g, or i; therefore, this is named the minimal basis set. The unnormalized STOs are defined:

$$\phi(r, \theta, \varphi) = r^{n-1} e^{-\alpha r} Y_l^m(\theta, \varphi) \tag{5}$$

where Y_l^m are the spherical harmonics.[20]

The Schrödinger equation to be solved is:

$$\hat{H}\Psi = E\Psi \tag{6}$$

The atomic Hamiltonian for a fixed nucleus written in Hylleraas coordinates, see Ref. 22, for a CI wave function reduces effectively to:

$$\hat{H} = -\frac{1}{2}\sum_{i=1}^{n_e} \frac{\partial^2}{\partial r_i^2} - \sum_{i=1}^{n_e} \frac{1}{r_i}\frac{\partial}{\partial r_i} - \sum_{i=1}^{n_e} \frac{Z}{r_i} + \sum_{i<j}^{n_e} \frac{1}{r_{ij}} + \sum_{i=1}^{n_e} \frac{1}{r_i^2} l_i(l_i+1) \tag{7}$$

where l_i are the angular quantum numbers of the orbitals or spherical harmonics. From the variational principle, we have to solve the following matrix eigenvalue problem:

$$(\mathbf{H}-E\mathbf{S})\mathbf{C}=0 \tag{8}$$

with the matrix elements of the Hamiltonian **H** and overlap **S** matrices:

$$\begin{aligned}\mathbf{H}_{kl} &= \int \Phi_k \hat{H} \Phi_l \partial\tau \\ \mathbf{S}_{kl} &= \int \Phi_k \Phi_l \partial\tau\end{aligned} \tag{9}$$

The matrix elements are sums of one- and two-electron integrals. Their expressions and calculation have been described in previous papers (see Refs. 21,23).

3. SELECTION OF CONFIGURATIONS

In this work, we use unnormalized nonorthogonal orbitals STOs. Our basis set can be called minimal basis, since for every electron a single STO is employed. This basis is [7s6p5d4f3g]. This is a difference with respect to the natural orbitals employed in the works of Bunge and Bunge[7] and Sasaki and Yoshimine,[9] where an atomic orbital is a linear combination of basis orbitals and in addition they are optimized. In this work, we have the restriction of same orbital exponent per pair of electrons. Later this issue will be further discussed.

The ground-state leading configuration of the C-atom is $1s^2\,2s^2\,2p^2$, in our nomenclature sssspp (i.e., s(1)s(2)s(3)s(4)p(5)p(6)). The other configurations with large contributions to the energy for the P-symmetry ($L=1$) six-electron state are, ordered by decreasing energy contribution, sssppd, sspppp, ssssdd, ssppdd, pppppp, sssddd, ssssff, ssspdf, ssssgg, ssspfg, ssffpp, and ssggpp. The permutations of these configurations were also included. For instance, the configuration type sssspp, with the restriction of same exponent per electron pair, includes the configurations ssspsp, spspss, spsssp, ppssss, and ssppss. Note that configurations like ppsspp or ddsspp do have a great energy contribution, because the inner excitations resemble implicitly the term r_{12}. The same permutations have been considered for the other configurations.

Other configurations than these have been tried and sorted out because of their very low energy contribution. As pointed out by Bunge,[7,9] we also encountered two sets of configurations with the correct symmetry $L=1$, those that interact with the ground-state main configuration or HF configuration and some others which do not interact at all with the first configuration. These noninteracting configurations are: ssssssp, ssspd, ssssdf, ssppp, among other possible ones and their permutations. After trying these configurations for low n, they were sorted out of the expansion.

The quantum number $M_L = 0$ was chosen, because for this case the same spin primitives except for a sign than in the singlet state are required. We systematically selected the CI configurations according to their energy contribution. This was done by calculations on blocks constructed for all possible configurations. In these blocks, all excitations are included from single, double, etc. up to the maximal number of excitations, in this case sextuple, whereas seldom sextuple-excited configurations showed even a moderate contribution. The eigenvalue equation was diagonalized upon each addition of a configuration. In this manner, the contribution of every single configuration and of each block of a given type to the total energy was evaluated. If the energy difference was less than the threshold $|E_{i-1} - E_i| < 1 \times 10^{-6}$ hartree (h), the new configuration was discarded. In this manner, all configurations were checked, leading to a relatively compact CI wave function. The procedure of selection of the configurations is similar to that described in our previous works (see Refs. 21,22).

We have noticed that usually the larger the contribution of a configuration, the smaller the sum of the l quantum numbers of the employed orbitals $l_1 + l_2 + l_3 + l_4 + l_5 + l_6$ is; i.e., the contribution of the configuration sssspp > sspppp for a P-state. Among the many possibilities to construct configurations of these symmetries, energetically important configurations were proven to be those with an inner S-shell and the outer pair of electrons with the P symmetry, for example, ppsspp and ddsspp. Other important configurations for the ground state of carbon atom are of the form: ppd, pdf, and pfg and all their permutations.

Note that there are more possible "degenerate L-eigenfunction" solutions with a larger number of Slater determinants. Specifically, these are degenerate with respect to the quantum numbers L and M, but with possible different energy contribution, i.e., nondegenerate with respect to the energy.[7] Although the inclusion of various degenerate configurations has been shown to improve the energy of the state, such a contribution is very small. This becomes important for very accurate CI calculations, as reported by Bunge et al.[16] In our work, we have constructed the configurations as linear combinations of all different terms occurring in the different degenerate configuration (see Table 1).

In Table 1, a list of two- and three-electron configurations of ^{1}S and ^{3}P symmetry is shown. The total configuration is constructed as the product of these configurations. It is clear that the product of $S \times P = P$ leads to

Table 1 Configuration Terms Needed for the Construction of Two- and Three-Electron Configurations With $M_L = 0$

Configuration	Configuration Terms
(^{1}S pp)	$p_0 p_0$, $p_1 p_{-1}$, $p_{-1} p_1$
(^{1}S dd)	$d_0 d_0$, $d_1 d_{-1}$, $d_{-1} d_1$, $d_2 d_{-2}$, $d_{-2} d_2$
(^{1}S ff)	$f_0 f_0$, $f_1 f_{-1}$, $f_{-1} f_1$, $f_2 f_{-2}$, $f_{-2} f_2$, $f_3 f_{-3}$, $f_{-3} f_3$
(^{1}S gg)	$g_0 g_0$, $g_1 g_{-1}$, $g_{-1} g_1$, $g_2 g_{-2}$, $g_{-2} g_2$, $g_3 g_{-3}$, $g_{-3} g_3$, $g_4 g_{-4}$, $g_{-4} g_4$
(^{3}P pp)	$p_1 p_{-1}$, $p_{-1} p_1$
(^{3}P dd)	$d_1 d_{-1}$, $d_{-1} d_1$, $d_2 d_{-2}$, $d_{-2} d_2$
(^{3}P ff)	$f_1 f_{-1}$, $f_{-1} f_1$, $f_2 f_{-2}$, $f_{-2} f_2$, $f_3 f_{-3}$, $f_{-3} f_3$
(^{3}P ff)	$g_1 g_{-1}$, $g_{-1} g_1$, $g_2 g_{-2}$, $g_{-2} g_2$, $g_3 g_{-3}$, $g_{-3} g_3$, $g_4 g_{-4}$, $g_{-4} g_4$
(^{3}P ppd)	$p_1 p_1 d_{-2}$, $p_{-1} p_{-1} d_2$, $p_1 p_{-1} d_0$, $p_{-1} p_1 d_0$, $p_0 p_1 d_{-1}$, $p_0 p_{-1} d_1$, $p_1 p_0 d_{-1}$, $p_{-1} p_0 d_1$
(^{3}P pdf)	$p_1 d_2 f_{-3}$, $p_{-1} d_{-2} f_3$, $p_0 d_2 f_{-2}$, $p_0 d_{-2} f_2$, $p_1 d_1 f_{-2}$, $p_{-1} d_{-1} f_2$, $p_1 d_{-2} f_1$, $p_{-1} d_2 f_{-1}$, $p_1 d_{-1} f_0$, $p_{-1} d_1 f_0$, $p_0 d_1 f_{-1}$, $p_0 d_{-1} f_1$, $p_1 d_0 f_{-1}$, $p_{-1} d_0 f_1$
(^{3}P pfg)	$p_1 f_3 g_{-4}$, $p_{-1} f_{-3} g_4$, $p_1 f_{-3} g_2$, $p_{-1} f_3 g_{-2}$, $p_1 f_2 g_{-3}$, $p_{-1} f_{-2} g_3$, $p_0 f_3 g_{-3}$, $p_0 f_{-3} g_3$, $p_0 f_2 g_{-2}$, $p_0 f_{-2} g_2$, $p_1 f_1 g_{-2}$, $p_{-1} f_{-1} g_2$, $p_1 f_{-2} g_1$, $p_{-1} f_2 g_{-1}$, $p_1 f_{-1} g_0$, $p_{-1} f_1 g_0$, $p_0 f_1 g_{-1}$, $p_0 f_{-1} g_1$, $p_1 f_0 g_{-1}$, $p_{-1} f_0 g_1$
(^{3}P ddd)	$d_2 d_{-2} d_0$, $d_{-2} d_2 d_0$, $d_0 d_2 d_{-2}$, $d_0 d_{-2} d_2$, $d_2 d_0 d_{-2}$, $d_{-2} d_0 d_2$, $d_2 d_{-1} d_{-1}$, $d_{-2} d_1 d_1$, $d_1 d_1 d_{-2}$, $d_{-1} d_{-1} d_2$, $d_1 d_{-2} d_1$, $d_{-1} d_2 d_{-1}$, $d_1 d_{-1} d_0$, $d_{-1} d_1 d_0$, $d_0 d_1 d_{-1}$, $d_0 d_{-1} d_1$, $d_1 d_0 d_{-1}$, $d_{-1} d_0 d_1$

P symmetry. In this list, sums, subtractions, and coefficients are missing, because they all represent a configuration term and the corresponding coefficient is determined variationally during the calculation. Only the constituents of the configurations are given. For example, the configuration ppsspp is the product of (^{1}S pp) (^{1}S ss) (^{3}P pp). Other configurations not appearing in Table 1 have been already given in previous papers and are here omitted (see Refs. 20,22).

4. CALCULATIONS

We have written a CI computer program for six-electron atomic systems in Fortran 95. Numerical calculations have been conducted with double precision arithmetic. The program is an extension of our four- and five-electron CI programs. The four-electron computer program has been thoroughly checked by comparing the results of our numerical calculations with the results by Sims and Hagstrom for the Be atom.[24] In these calculations, we obtained a complete agreement, and the five-electron CI program has been tested by comparing the calculations with the ones of Bunge and Froese Fischer as reported in Ref. 23.

In this work, we start with a full-CI wave function for a basis $n=7$ (or [7s6p5d4f3g]) testing all possible configurations and retaining all which contribute more than 1×10^{-6} h, which at the beginning of the calculation are almost all of them. First for a relatively short expansion, the orbital exponents are optimized. We use the same technique than in previous calculations of the Be and Li atoms.[20–22] A set of three exponents is used (one for the K-shell, other for the electrons of the inner L-shell, and other for the pair of outer electrons of the L-shell) and kept equal for all configurations. This technique accelerates computations, while still producing sufficiently accurate wave functions to determine the bound-state properties. During the calculations and optimizations, we use the virial theorem to control the quality of the wave function and guide the numerical optimization:

$$\sigma = -\frac{\langle V\rangle}{\langle T\rangle} \tag{10}$$

After this, the orbital exponents are optimized once more and kept equal for the rest of the calculations. The wave function is very compact; it contains many different configurations and linear dependences produced by very similar configurations are avoided. This means that it is computationally favorable to vary the configurations within a quantum number than to make a calculation with one configuration for all quantum numbers. We have added in this work configurations up to $n=7$. At the end, it is difficult to find configurations which achieve to add something to the energy. It is then clear that the contribution of a configuration depends on the order where it has been added. Nevertheless, this simple method is very stable and helps to retain the most important configurations.

In Table 2, the truncated CI wave function expansion is shown. The configurations with larger contribution are the ground-state configuration sssspp; the open-shell configuration ssspsp; the ssppd and sssdpp ^{3}P configurations; followed by the configuration with inner ^{1}S excitation ppsspp;

Table 2 Selected CI Wave Function Expansion for the ^3P Ground State of Carbon Atom. N_{conf} is the Number of Symmetry Adapted Configurations and n the Basis Set, i.e. n=3 Means [3s2p1d]

Conf.	n	N_{conf}	$N_{conf,tot}$	Energy (h)	−Diff. (μh)	Virial
sssspp	6	181	181	−37.72179917	0	2.00006
ssspsp	6	100	281	−37.72756352	5764	2.00003
ssppss	3	1	282	−37.72756615	2	2.00003
spsssp	5	44	326	−37.72856844	1002	1.99999
spspss	4	6	332	−37.72859600	27	1.99999
sspppp	6	50	382	−37.75059877	22002	1.99999
spsppp	5	56	438	−37.75377264	3173	1.99999
spppsp	3	1	439	−37.75377312	0	1.99999
ppsspp	7	78	517	−37.77733780	23564	2.00002
ppspsp	5	14	531	−37.77749030	152	2.00001
pppppp	5	6	537	−37.77797403	483	2.00002
sssppd	5	61	598	−37.80802592	30051	2.00000
sssdpp	5	41	639	−37.81502320	6997	2.00000
sspspd	5	20	659	−37.81515432	131	2.00002
sspdps	5	14	673	−37.81523446	80	2.00003
ssdspp	5	2	675	−37.81523913	4	2.00003
spsspd	6	18	693	−37.81759021	2351	2.00002
spsdps	7	36	729	−37.81823607	645	2.00002
sppdss	3	4	733	−37.81827887	42	2.00001
sdsspp	4	9	742	−37.81848443	205	2.00001
sdspsp	5	36	778	−37.81856553	81	2.00001
sdppss	3	2	780	−37.81856602	0	2.00001
pdsssp	5	31	811	−37.81881301	246	2.00002
pdspss	5	19	830	−37.81885580	42	2.00002
ssppdd	3	2	832	−37.81910329	247	2.00001
sspdpd	5	21	853	−37.82326956	4166	2.00006
ssddpp	7	14	867	−37.82360092	331	2.00003

Continued

Table 2 Selected CI Wave Function Expansion for the ^3P Ground State of Carbon Atom. N_{conf} is the Number of Symmetry Adapted Configurations and n the Basis Set, i.e. n=3 Means [3s2p1d]—cont'd

Conf.	n	N_{conf}	$N_{conf,tot}$	Energy (h)	−Diff. (μh)	Virial
spsdpd	3	2	869	−37.82361411	13	2.00003
sppdsd	4	2	871	−37.82363361	19	2.00003
spdpsd	3	1	872	−37.82363620	2	2.00003
sdsdpp	6	6	878	−37.82392325	287	2.00002
ppssdd	3	3	881	−37.82392335	0	2.00002
ddsspp	5	26	907	−37.82630085	2377	2.00009
ssssdd	4	13	920	−37.82913117	2830	2.00011
sssdsd	5	21	941	−37.82951963	388	2.00009
ppddpp	3	1	942	−37.82953897	19	2.00009
sssddd	4	3	945	−37.82966403	125	2.00010
ssddsd	4	6	951	−37.82971989	55	2.00009
ssspdf	5	7	958	−37.83022935	509	2.00010
sssdpf	7	16	974	−37.83205398	1824	2.00012
sssfpd	6	11	985	−37.83303456	980	2.00013
sssfdp	7	4	989	−37.83307198	37	2.00013
sspdsf	4	8	997	−37.83309783	25	2.00013
sspfsd	5	10	1007	−37.83316362	65	2.00013
ssdfsp	6	20	1027	−37.83350195	338	2.00010
dfsssp	4	4	1031	−37.83365871	156	2.00010
sspppf	5	4	1035	−37.83385580	197	2.00009
ssssff	5	5	1040	−37.83411703	261	2.00010
sssfsf	5	9	1049	−37.83416895	51	2.00009
ssffpp	5	4	1053	−37.83463042	461	2.00010
ffsspp	5	4	1057	−37.83474788	117	2.00011
ssspfg	5	1	1058	−37.83475897	11	2.00011
sssfpg	7	9	1067	−37.83500422	245	2.00012
sssgpf	6	5	1072	−37.83512518	120	2.00012

Table 2 Selected CI Wave Function Expansion for the ^{3}P Ground State of Carbon Atom. N_{conf} is the Number of Symmetry Adapted Configurations and n the Basis Set, i.e. n=3 Means [3s2p1d]—cont'd

Conf.	n	N_{conf}	$N_{conf,tot}$	Energy (h)	−Diff. (μh)	Virial
ssfgsp	5	1	1073	−37.83513520	10	2.00012
fgsssp	5	1	1074	−37.83516054	25	2.00013
ssggpp	6	3	1077	−37.83523933	78	2.00013
ssssgg	5	1	1078	−37.83524671	7	2.00013
sssgsg	6	2	1080	−37.83526523	18	2.00013

The orbitals exponents are $\alpha=\alpha'=6.39437109$, $\beta=\beta'=1.78356693$, and $\gamma=1.68267232$ for all the expansion. The unit $1\,\mu h=1\times10^{-6}\,a.u.$

Table 3 Angular Energy Limits for the ^{3}P Ground State of Carbon Atom

n	Basis Set	N_{conf}	$N_{conf,tot}$	Energy (a.u.)	−Diff. (μh)	Virial
$n=2$	[2s1p]	66	66	−37.67686631	7981595	1.98191
$n=3$	[3s2p1d]	332	398	−37.81179154	134925	1.99911
$n=4$	[4s3p2d1f]	358	756	−37.83030344	18511	2.00021
$n=5$	[5s4p3d2f1g]	288	1044	−37.83458668	4283	2.00019
$n=6$	[6s5p4d3f2g]	27	1071	−37.83510448	517	2.00015
$n=7$	[7s6p5d4f3g]	9	1080	−37.83526520	160	−2.00013

n is the basis set indicated by the highest principal quantum number.

afterward we find the sspppp, which shows the quasi-degeneration of the s and p orbitals of the L-shell; the configuration ssssdd has contributed less as would be expected by the low l_i quantum numbers; and higher excitations sspdpd, ddsspp, and ssffpp follow among others.

If we order the energy for the basis set n, we encounter the angular energy limits, assuming the wave function is saturated for these limits. The contribution to the total energy decreases with increasing n (see Table 3). That the contribution of $n=7$ is still not close to zero indicates that the wave function has not fully converged. We would need to add more terms with $n=8$ and higher. We estimate that the contribution to the energy of the missing g, as well h,i, and higher orbital configurations is about 4200 μh (microhartree).

Our energy result is very close to that of Bunge using similar basis set. In that work, Bunge used about 5000 Slater determinants. In our work, the number of configurations is larger than in other works, because we use a basis of one orbital for one electron. The calculation of Sasaki and Yoshimine is highly elaborated using up to i-orbitals and 900 configurations. We could estimate that they used more than 5000 of our type configurations. They have managed to recover a great amount of correlation energy by using the CI method. Their calculation can be considered as the CI benchmark for the carbon atom. We can observe that the calculation using the ECG method, which includes all interelectronic coordinates, is close in value to Sasaki and Yoshimine calculation (Table 4).

5. CONCLUSIONS AND PERSPECTIVES

We have calculated the ground state of carbon atom with millihartree accuracy and modest computational effort. We have determined the important configurations of the ground states. The relevant configurations together with the leading one sssspp are ssppd and followed by sspppp and ssssdd, showing that the excitation ppd is more important than the quasi-degeneration of the s and p levels as it was the case in the Be atom. The weight of the configurations ppsspp, ddsspp, ssffpp can be interpreted as the description of the cusp r_{12} with inner singlet ^1S excitations pp and dd. Finally, the configuration ssspsp adds the orbital splitting of the orbitals, which was missing in sssspp due to the imposed double occupancy.

In the same way, the low-lying states of carbon atom can be determined. The spectrum of the low-lying excited states of carbon atom looks very complex. Maldonado et al. calculated a large number of low-lying excited states employing excitation energies obtained with Variational and Diffusion Monte Carlo approaches.[13] Their spectrum serves as a guide for future CI calculations. It agrees very well with our estimations calculated adding to the estimated exact energy of the ground state −37.8450 h the excitation energies (in hartree) of the levels, which are collected in the Atomic Data Base of NIST.[26] These estimated energies of bound states of the carbon atom are given in Table 5 and compared with the energy values obtained from Monte Carlo excitation energies. The results show that carbon atom represents a vast area of research.

Table 4 Comparison of Variational Upper Bonds to the Energy of the ^3P Ground State of Carbon Atom Calculated by Different Methods

Method	Orb.	Authors	Year	Ref.	Basis Set	N_{confs}	Energy (h)
CI	STO	Weiss	1967	5	[4s3p2d2f]	40	−37.77888
VMC		Sarsa et al.	2016	12			−37.81537
FCI	STO	Bunge	1970	8	[7s6p4d3f]	234	−37.83378
CI	STO	This work	2017		[7s6p5d4f3g]	1080	−37.835265
DMC		Maldonado et al.	2011	13			−37.83544(9)
CI(SDTQ)	STO	Sasaki and Yoshimine	1974	9	[10s9p8d8f6g4h2i]	993	−37.8393
ECG	GTO	Sharkey et al.	2010	10		500	−37.84012879
MC		Seth et al.	2011	11			−37.84446
Estimated exact		Chakravorty et al.	1993	25			−37.8450

Table 5 Estimated Total Energies of the Low-Lying States of C-Atom

Configuration	State	Energy (cm^{-1})[26]	Estimated Exp. (h)	DMC[13]
$2s^2\ 2p^2$	^3P	0.0	−37.8450	−37.83544
$2s^2\ 2p^2$	^1D	10,193	−37.7985573	−37.78966(6)
$2s^2\ 2p^2$	^1S	21,648	−37.7463645	−37.73830(5)
$2s\ 2p^3$	^5S	33,735	−37.691292	−37.69026(3)
$2s^2$ 2p3s	2^3P	60,333	−37.5701026	−37.56103(10)
$2s^2$ 2p3s	^1P	61,981	−37.5625938	−37.55352(6)
$2s\ 2p^3$	^3D	64,091	−37.5529799	−37.54374(8)
$2s^2$ 2p3p	2^1P	68,856	−37.5312690	−37.52335(7)
$2s^2$ 2p3p	2^3D	69,689	−37.5274735	−37.51974(13)
$2s^2$ 2p3p	^3S	70,743	−37.5226712	−37.51523(5)
$2s^2$ 2p3p	3^3P	71,352	−37.5198964	−37.49697(5)[a]
$2s^2$ 2p3p	2^1D	72,611	−37.5141599	−37.49096(3)[a]
$2s^2$ 2p3p	2^1S	73,976	−37.5079405	−37.47429(3)[a]
$2s\ 2p^3$	4^3P	75,254	−37.5021175	−37.46760(10)[a]
$2s^2$ 2p3d	3^1D	77,680	−37.4910639	−37.48274(8)
$2s^2$ 2p4s	5^3P	78,148	−37.4889315	−37.46084(3)[a]
$2s^2$ 2p3d	^3F	78,199	−37.4886991	−37.48097(13)
Ioniz.		90,820	−37.4311936	

[a]Energies calculated with the Variational Monte Carlo approach.[13]
For the energy conversion, we have used 1 h = 219,474.6305 cm^{-1}. The ground-state energy is taken as −37.8450 h.

ACKNOWLEDGMENTS

One of us (M.B.R.) would like to thank Carlos Bunge for very interesting discussions on the construction of configurations in the CI method and to James S. Sims for providing integral values of two-electron integrals to test our codes and fruitful discussions about the general treatment of angular momentum in the wave functions. Both of us are very thankful to the Editor of this volume Philip E. Hoggan for the kind invitation to participate in it and for reading the manuscript.

REFERENCES

1. Boys, S. F. Electronic Wave Functions. IV. Some General Theorems for the Calculation of Schrodinger Integrals Between Complicated Vector-Coupled Functions for Many-Electron Atoms. *Proc. R. Soc.* **1953**, *A217*, 235.
2. Clementi, E. Correlation Energy for Atomic Systems. *J. Chem. Phys.* **1963**, *38*, 2248.

3. Bagus, P. S.; Moser, C. M. Calculation of Energy Levels Which Arise From the p^2 Configuration of the Ground State of Carbon. Multiconfiguration Hartree-Fock Calculations. *Phys. Rev.* **1968**, *167*, 13.
4. Schaefer, H. F.; Klemm, R. A.; Harris, F. E. Atomic Hyperfine Structure. II. First-Order Wave Functions for the Ground States of B,C,N,O, and F. *Phys. Rev.* **1969**, *181*, 137.
5. Weiss, A. W. Superposition of Configurations and Atomic Oscillator Strengths—Carbon I and II. *Phys. Rev.* **1967**, *162*, 71.
6. Bunge A. PhD Thesis; University of Florida: Gainesville, FL, 1968, Which May be Obtained as Thesis 69-10899 From University Microfilms, Inc., Ann Arbor, MI.
7. Bunge, A.; Bunge, C. F. Correlation Energy of the Ground State of C. *Phys. Rev. A* **1970**, *1*, 1599.
8. Bunge, A. Electronic Wavefunctions for Atoms. III. Partition of Degenerate Spaces and Ground State of C. *J. Chem. Phys.* **1970**, *53*, 20.
9. Sasaki, F.; Yoshimine, M. Configuration-Interaction Study of Atoms. I. Correlation Energies of B, C, N, O, F, and Ne. *Phys. Rev. A* **1974**, *9*, 17.
10. Sharkey, K. L.; Bubin, S.; Adamowicz, L. Analytical Energy Gradient in Variational Calculations of the Two Lowest ^{3}P States of the Carbon Atom With Explicitly Correlated Gaussian Basis Functions. *J. Chem. Phys.* **2010**, *132*, 184106.
11. Seth, P.; Rios, P. L.; Needs, R. J. Quantum Monte Carlo Study of the First-Row of Atoms and Ions. *J. Chem. Phys.* **2011**, *134*, 084105.
12. Sarsa, A.; Buendía, E.; Gálvez, F. J. Multi-configurational Explicitly Correlated Wave Functions for the Study of Confined Many Electron Atoms. *J. Phys. B At. Mol. Opt. Phys.* **2016**, *49*, 145003.
13. Maldonado, P.; Sarsa, A.; Buendía, E.; Gálvez, F. J. Jastrow Correlated and Quantum Monte Carlo Calculations for the Low-Lying States of the Carbon Atom. *J. Chem. Phys.* **2011**, *134*, 134102.
14. Löwdin, P.-O. Quantum Theory of Many-Particle Systems. I. Physical Interpretations by Means of Density Matrices, Natural Spin-Orbitals, and Convergence Problems in the Method of Configurational Interaction. *Phys. Rev.* **1955**, *97*, 1474.
15. Bunge, C. F. Specifics on the Scientific Legacy of Per-Olov Löwdin. *Adv. Quantum Chem.* **2017**, *74*, 53.
16. Almora-Díaz, C. X.; Rivera-Arrieta, H. I.; Bunge, C. F. Recent Progress in the Variational Orbital Approach to Atomic and Molecular Electronic Structure. *Adv. Quantum Chem.* **2016**, *72*, 129.
17. Hylleraas, E. A. Neue Berechnung der Energie des Heliums im Grundzustande, sowie des tiefsten Terms von Ortho-Helium. *Z. Phys.* **1929**, *54*, 347.
18. Sims, J. S.; Hagstrom, S. A. Combined Configuration-Interaction-Hylleraas-Type Wave-Function Study of the Ground State of the Beryllium Atom. *Phys. Rev. A* **1971**, *4*, 908.
19. Sims, J. S.; Hagstrom, S. A. One-Center r_{ij} Integrals Over Slater-Type Orbitals. *J. Chem. Phys.* **1971**, *55*, 4699.
20. Ruiz, M. B.; Margraf, J. T.; Frolov, A. M. Hylleraas-Configuration-Interaction Analysis of the Low-Lying States in the Three-Electron Li Atom and Be^+ Ion. *Phys. Rev. A* **2013**, *88*, 012505.
21. Frolov, A. M.; Ruiz, M. B. Bound State Spectrum of the Triplet States of Be Atom. *Chem. Phys. Lett.* **2014**, *595–596*, 197.
22. Ruiz, M. B.; Latorre, F.; Frolov, A. M. Singlet and Triplet Bound State Spectra in the Four-Electron Be-Like Atomic Systems. *Adv. Quantum Chem.* **2016**, *73*, 119.
23. M.B. Ruiz, M. Rojas, G. Chicón, and P. Otto. Configuration Interaction Calculations on the ^{2}P Ground State of Boron Atom and C^+ Using Slater Orbitals. Int. J. Quantum Chem. 2011, 111, 1921.
24. Sims, J. S.; Hagstrom, S. A. Hylleraas-Configuration-Interaction Study of the ^1S Ground State of Neutral Beryllium. *Phys. Rev. A* **2011**, *83*, 032518.

25. Chakravorty, S. J.; Gwaltney, S. R.; Davidson, E. R.; Parpia, F. A.; Froese Fischer, C. Ground-State Correlation Energies for Atomic Ions With 3 to 18 Electrons. *Phys. Rev. A* **1993**, *47*, 3649.
26. Kramida, A.; Ralchenko, Y.; Reader, J.; NIST ASD Team. *NIST Atomic Spectra Database (ver. 5.1) [Online]*; National Institute of Standards and Technology: Gaithersburg, MD, 2013. Available http://physics.nist.gov/asd (30 June 2014).

FURTHER READING

27. Stevenson, R. *Multiplet Structure of Atoms and Molecules*; W. B. Saunders Company: Philadelphia and London, 1965.

SECTION III

Strong Correlation in Electronic Systems

CHAPTER THIRTEEN

Extension of the Configuration Interaction Monte Carlo Method to Atoms and Molecules

Alessandro Roggero*, Francesco Pederiva†,‡,1
*Institute for Nuclear Theory, University of Washington, Seattle, WA, United States
†University of Trento, Trento, Italy
‡INFN-TIFPA, Trento Institute for Fundamental Physics and Applications, Trento, Italy
[1]Corresponding author: e-mail address: pederiva@science.unitn.it

Contents

1. Introduction 242
2. Method 242
   2.1 CIMC Method 242
   2.2 Importance Functions and Recursion Relations 245
   2.3 Construction of the Single-Particle States 246
3. Results 247
   3.1 Atoms 247
   3.2 Molecules 250
4. Conclusions 252
Acknowledgment 252
References 252

Abstract

We present an extension of the configuration interaction Monte Carlo (CIMC) method to the computation of the ground-state properties of atoms and molecules. In particular we make use of orthonormalized Gaussian basis sets and compute the coupled clusters at the singles-doubles level (CCSD) wave function to be used as importance function in the imaginary-time propagation. We present a few results for first-row atoms and some simple molecules. In particular we will show a substantial independence of results when the CCSD wave function is truncated at second-order perturbation theory level, thereby confirming the possible use of CIMC as a viable accelerator of CC calculations given the more favorable scaling with the electron number.

Advances in Quantum Chemistry, Volume 76
ISSN 0065-3276
https://doi.org/10.1016/bs.aiq.2017.06.004

1. INTRODUCTION

Quantum Monte Carlo (QMC) methods in Fock space have become very popular in the last decade. The possibility of directly working within a space of antisymmetric functions suggested that this as a possible alternate way of better coping with the Fermion sign problem.[1–3]

A few years ago we proposed to improve the efficiency of the method by introducing importance sampling from a variational wave function expanded in the occupation number basis set. In particular we made use of the wave function provided by a coupled clusters (CC) calculation in order to compute the correlation energy in a uniform electron gas using a plane waves basis,[4] thereby effectively realizing a QMC propagation in momentum space. The technicalities were inspired by the shell model Monte Carlo methods used in nuclear physics and extended to condensed matter systems under the name of configuration interaction Monte Carlo (CIMC).[5]

CIMC calculations in bulk systems are particularly useful when in need of using nonlocal interactions, such as those naturally appearing in nuclear physics (e.g., Ref. 6), or those raising from an effective treatment of core electrons in large atoms, like in the standard pseudo-potentials for plane wave calculations widely used in solid state physics.[7]

In a previous edition of these proceedings[8] we showed how the method could be in principle extended to finite systems, namely a few first-row atoms, by making use of some Gaussian basis sets. Our results, however, were still affected by some technical problems, evidenced by the fact that there was no consistency (at least in some cases) between results projected from CC wave functions and the corresponding truncation at second-order perturbation theory.

In this chapter we will present novel results obtained from a refinement of the method, extended to the case of a few simple molecules. Although these calculations are still mostly at a level of a proof of concept, it is now clear that the method is perfectly functional also for standard quantum chemistry application.

2. METHOD

2.1 CIMC Method

In this section we will give a brief review of the CIMC method. To set the stage let us take a finite set $\mathcal{S}$ of single-particle (sp) states of size $\mathcal{N}_s$ and consider a general second-quantized fermionic Hamiltonian including two and possibly many-body interactions

$$H = \sum_{\alpha\beta\in\mathcal{S}} \epsilon_{\alpha\beta} a_\alpha^\dagger a_\beta + \sum_{\alpha\beta\gamma\delta\in\mathcal{S}} V_{\alpha\beta\gamma\delta} a_\alpha^\dagger a_\beta^\dagger a_\delta a_\gamma + \cdots . \tag{1}$$

In this expression, Greek letter indices indicate sp states (i.e., α is a collective label for all sp quantum numbers), the operator $a_\alpha^\dagger$ (a_α) creates (destroys) a particle in the sp state α and the $V_{\alpha\beta\gamma\delta}$ are general (antisymmetrized) two-body interaction matrix elements:

$$V_{\alpha\beta\gamma\delta} = \langle\alpha\beta|\hat{V}|\gamma\delta\rangle - \langle\alpha\beta|\hat{V}|\delta\gamma\rangle. \tag{2}$$

For an N-fermion system the resulting Fock space would be spanned by the full set of N-particle Slater determinants that can be generated using the sp orbitals $\alpha \in \mathcal{S}$. We will denote these Slater determinants in the occupation number basis by $|\mathbf{n}\rangle$, where $\mathbf{n} \equiv \{n_\alpha\}$ and $n_\alpha = 0, 1$ are occupation number of the single-particle orbital α satisfying $\sum_\alpha n_\alpha = N$. Notice that the formulation so far does not rely on any special choice of either the interaction or basis set. We can now use these states as a complete basis in our many-body Hilbert space and express a generic state in it as:

$$|\Psi\rangle = \sum_{\mathbf{n}} \langle\mathbf{n}|\Psi\rangle|\mathbf{n}\rangle \equiv \sum_{\mathbf{n}} \Psi(\mathbf{n})|\mathbf{n}\rangle \tag{3}$$

where the sum is carried out over all the possible basis vectors obtained using a finite set including $\mathcal{N}_S$ single-particle orbitals.

The core idea of diffusion Monte Carlo is to extract information on the ground state of the Hamiltonian H by evolving in imaginary-time a state $|\Psi\rangle$ nonorthogonal to the ground state itself:

$$\Psi_{\tau+\Delta\tau}(\mathbf{m}) = \sum_{\mathbf{n}} \langle\mathbf{m}|P|\mathbf{n}\rangle\Psi_\tau(\mathbf{n}), \tag{4}$$

with a suitable projection operator P as for instance:

$$\langle\mathbf{m}|P|\mathbf{n}\rangle = \langle\mathbf{m}|e^{-\Delta\tau(H-E_T)}|\mathbf{n}\rangle. \tag{5}$$

In order to interpret Eq. (4) as a stochastic process we, however, need the matrix elements of the projection operator to be positive definite and, apart from very special Hamiltonians, this cannot be achieved in practice. This fact gives rise to a sign problem. In our case we deal with this sign problem by using a trial state $|\Phi\rangle$ to define a one-parameter family of Hamiltonians $\widetilde{H}_\gamma$

such that the corresponding projector is positive definite $\forall\gamma > 0$. For this Hamiltonian the off diagonal matrix elements ($\mathbf{n}\neq\mathbf{m}$) are defined as:

$$\langle\mathbf{m}|\mathcal{H}_\gamma|\mathbf{n}\rangle = \begin{cases} -\gamma\langle\mathbf{m}|H|\mathbf{n}\rangle & \mathfrak{s}(\mathbf{m},\mathbf{n}) > 0 \\ \langle\mathbf{m}|H|\mathbf{n}\rangle & \text{otherwise} \end{cases}, \tag{6}$$

while the diagonal elements are given by:

$$\langle\mathbf{n}|\mathcal{H}_\gamma|\mathbf{n}\rangle = \langle\mathbf{n}|H|\mathbf{n}\rangle + (1+\gamma)\sum_{\substack{\mathbf{n}\neq\mathbf{m}\\ \mathfrak{s}(\mathbf{m},\mathbf{n})>0}} \langle\mathbf{m}|H|\mathbf{n}\rangle \tag{7}$$

For any $\gamma \geq 0$ the propagator

$$\langle\mathbf{n}|\mathcal{P}_\gamma|\mathbf{n}\rangle = 1 - \Delta\tau\frac{\langle\Phi|\mathbf{m}\rangle}{\langle\mathbf{n}|\Phi\rangle}\langle\mathbf{m}|\mathcal{H}_\gamma - E_T|\mathbf{n}\rangle \tag{8}$$

is free from the sign problem. This procedure is essentially equivalent to the fixed-node approximation in standard diffusion Monte Carlo. In fact, the energies obtained with this transformed Hamiltonians (which can be estimated with arbitrary accuracy due to the lack of the sign problem) satisfy the variational property:

$$E_0 \leq \langle\Psi_0^\gamma|\widetilde{H}_\gamma|\Psi_0^\gamma\rangle \leq \langle\Phi|H|\Phi\rangle \tag{9}$$

or, in words, the ground-state energy of the transformed Hamiltonians is bounded from below to the true ground-state energy of the original problem and from above by the expectation value of the original Hamiltonian with the trial state used to define the transformation. The energy will eventually be dependent only on the overall nodal (or phase, if complex) structure of the trial state $|\Phi\rangle$ used to construct the parametrized Hamiltonians. The enforcement of the upper bound-property for our energy estimates is important in order to guarantee that by improving the $|\Phi\rangle$ the energy will result systematically lower, without ever overshooting the true ground-state value.

Another interesting property of this construction is that not only the upper-bound property Eq. (9) is valid for positive values of γ, but also for any linear extrapolation to the negative γ region (in particular is valid for extrapolations to $\gamma = -1$). All the results shown below are obtained by performing calculations at two different values $\gamma = 0$, 1, performing the extrapolation toward $\gamma = -1$ and considering as best estimate the lowest energy among them. We note in passing that most of the time this corresponds to the extrapolated value.

2.2 Importance Functions and Recursion Relations

It becomes then clear that, in order for this to make any practical sense, we need a trial state which is both accurate and whose overlaps $\boldsymbol{\Phi}(\mathbf{n})$ can be computed quickly for any Slater determinant. In this work we use a trial state modeled using the CC ansatz:

$$|\Psi_{\rm CC}\rangle = e^{\hat{T}}|\Phi_{\rm HF}\rangle, \tag{10}$$

where $|\Phi_{\rm HF}\rangle$ is the Hartree–Fock ground state. In the above equation, correlations are introduced trough the excitation operator $\hat{T}$, which in CC theory is hierarchically divided as

$$\hat{T} = \hat{T}_1 + \hat{T}_2 + \hat{T}_3 + \cdots \tag{11}$$

where the index $1, 2, \ldots$ refers to the number of creation/annihilation operators that compose them. The first two terms that we will be using in this calculations can be written explicitly as:

$$\hat{T}_1 = \sum_{\alpha,\beta\in\mathcal{S}} t_\alpha^\beta a_\beta^\dagger a_\alpha \quad \hat{T}_2 = \frac{1}{4}\sum_{\alpha,\beta,\gamma,\delta\in\mathcal{S}} t_{\alpha\beta}^{\gamma\delta} a_\gamma^\dagger a_\delta^\dagger a_\alpha a_\beta \cdots, \tag{12}$$

and correspond to the truncation dubbed coupled cluster singles and doubles (CCSD). The amplitudes t_α^β and $t_{\alpha\beta}^{\gamma\delta}$ are usually obtained by solving the CCSD equations using the basis set of choice for our calculation. Since this procedure is quite computationally expensive, we have also explored the possibility of using the amplitudes given by first-order perturbation in the wave function.

Once the amplitudes are known we use a recursive algorithm to evaluate the overlap of the CC wave function with a general Slater determinant. To set the notation we will express a generic Slater-determinant state describing a M-particle–M-hole state as:

$$|\mathbf{m}\rangle = a_{p_1}^\dagger \ldots, a_{p_M}^\dagger a_{h_1} \ldots, a_{h_M}|\Phi_{\rm HF}\rangle \equiv |\Phi_{h_1,\ldots,h_M}^{p_1,\ldots,p_M}\rangle. \tag{13}$$

The required amplitude can then be expressed as a superposition of $M-2$ particle/hole states that can be generated from $\mathbf{m}$. Eventually (the proof is tedious but straightforward) one obtains

$$\langle \mathbf{m}|\Psi_{\text{CC}}\rangle = \sum_{\gamma=2}^{M}\sum_{\mu<\nu}^{M}(-1)^{\gamma+\mu+\nu}t_{h_1 h_\gamma}^{p_\mu p_\nu}\Psi_{\text{CC}}^{M-2}\begin{pmatrix} p_1,p_2,\ldots,p_{\mu-1},p_{\mu+1},\ldots,p_{\nu-1},p_{\nu+1},\ldots,p_M \\ h_2\ldots,h_{\gamma-1},h_{\gamma+1},\ldots,h_M \end{pmatrix} + \sum_{\mu=1}^{M}(-1)^{\mu+1}t_{h_1}^{p_\mu}\Psi_{\text{CC}}^{M-1}\begin{pmatrix} p_1,p_2,\ldots,p_{\mu-1},p_{\mu+1},\ldots,p_M \\ h_2,\ldots,h_M \end{pmatrix} \tag{14}$$

assuming $p_1 < p_2 < \cdots < p_M$ and $h_1 < h_2 < \cdots < h_M$. The normalization is fixed in such a way that $\langle \Phi_{\text{HF}}|\Psi_{\text{CC}}\rangle = 1$. In the above expression we use the alternative notation

$$\Psi_{\text{CC}}^{M}\begin{pmatrix} p_1,p_2,\ldots,p_M \\ h_1,\ldots,h_M \end{pmatrix} \equiv |\Phi_{h_1,\ldots,h_M}^{p_1,\ldots,p_M}\rangle \tag{15}$$

to indicate a Slater determinant containing M-particle (corresponding to $p_1,p_2,\ldots,p_M$) and M-holes (corresponding to $h_1,h_2,\ldots,h_M$) as in Eq. (13).

We see then that the cost of a single evaluation of the wave function overlap is, strictly speaking, independent on the size of the system and scales only with the number of particle–hole excitations present in the given state. This may clearly provide a substantial scaling improvement compared to full CC in favorable situations, in particular this probably allows for computing systems up to a few tenths of electrons without the need of large computational resources.

2.3 Construction of the Single-Particle States

The computation of finite systems (in this case atoms and molecules) requires the choice of some set of basis functions from which to build the Slater determinants. We will later discuss in the results section the specific bases employed in our calculations. Here we want to remark the fact that we want our many-body Slater determinants to be orthogonal among each other, this can be achieved efficiently by a proper orthogonalization of the single-particle states. In general Gaussian orbitals do not make an orthogonal basis and we have then to find useful linear combinations of the orbitals to be used in building the Slater determinants. One way to achieve this is by simply orthonormalize the orbitals among themselves or, even better, by finding the linear combinations which diagonalize a one-body reduction of the full Hamiltonian such as the Fock matrix

$$F_{\alpha\beta} = \epsilon_{\alpha\beta} + \sum_{\gamma} V_{\alpha\gamma\beta\gamma}. \tag{16}$$

In the previous contribution we used the first procedure, which was leading to some inconsistent results for the first-row atoms, while in this work we will start from the orthonormal basis computed by diagonalizing the Fock matrix for the converged configuration solving the self-consistent Hartree–Fock equations. This approach is particularly beneficial when orbital hybridization is important as in the case of molecules.

3. RESULTS

3.1 Atoms

After a few improvements in the computational procedure, we retested the results on the singlet $S = 0$ state of some first-row atoms in order to better assess the reachable accuracy. Since imaginary-time projection cannot build components outside the employed model space, but would rather give the best representation of the wave function within that model space, results will depend on the choice of the basis set. As a test we used for all cases considered the STO-3G basis set and the cc-pVDZ basis set from Ref. 9. We then computed the correlation energy from MP2 perturbation theory, CC theory with singles and doubles (CCSD) and finally CCSD(T) with the inclusion of triplet at perturbative level. All the results in both Hartree–Fock, CC theory and full CI (FCI) are obtained using the publicly available implementation PySCF.[10] These energies are then compared to the diffusion Monte Carlo predictions obtained using the CCSD ansatz Eq. (14) with amplitudes obtained with either:

- solution of the self-consistent CCSD equations (CIMC results)
- second-order Moeller–Plesset perturbation theory (CIMC[pt2])

Notice that the choice of the importance function provides in principle different upper bounds to the energy due to the different nodal structures.

For all the results we have attempted the extrapolation to $\gamma = -1$ as explained in the section above. The correlation energy is estimated taking as reference the converged HF result obtained with PySCF.

The comparison between CCSD/CCSD(T) and CIMC results is quite satisfactory for most of the atoms considered. For Be CIMC slightly improves on the correlation energy with respect to CCSD as computed by us, but the difference is very small (about 0.2%) and might be attributed to numerical uncertainties. The main interesting point is the fact that while the MP2 calculation recovers only about 47% of the correlation energy, the CIMC calculation using the MP2 amplitudes CIMC[pt2] is able to recover more than 99% of the correlation energy, although with a standard deviation which is about an order of magnitude larger: order 10^{-2} mH to be compared to 10^{-3} mH of the full CIMC. This is consistent with previous calculations for the homogeneous electron gas.[4]

A rather different situation is observed for the O and C atoms where CIMC[pt2] seems to degrade its performance. This is somewhat expected as in these cases we consider the excited singlet states which will have a more complex nodal structure than the one possessed by the ground state, we then expect the simple PT2 ansatz to struggle to reproduce such complexity. It is then reasonable to expect that the extrapolation in γ within CIMC cannot fully recover from the degraded nodal structure of the wave function used for the fixed-node procedure.

This is especially clear in the case of singlet state of oxygen, in this case the amplitudes obtained by solving the CC equations seem extremely good: the statistical error of the calculation is extremely small. As expected the behavior of the perturbed solution is much worse and the MP2 energy recovers only 31% of the total correlation energy. Projection with CIMC unfortunately manages to bring this value only at 59%. This suggests that the fixed-node bias in excited states is greatly reduced by employing more accurate amplitudes to describe the nodes, especially in critical situations like here when the state is open shell and becomes essentially multireference (Table 1).

The case of C is somewhat troublesome in that the CIMC correlation energy results 9% larger than the CC result. This behavior may be connected

Table 1 Correlation Energy (in Hartree) for a Few First-Row Atoms for the STO-3G Basis

Atom	**Be**	**C ($S = 0$)**	**O ($S = 0$)**
FCI	−0.05177463	−0.05660322	−0.047444445
MP2	−0.02435837	−0.02267991	−0.01459329
CCSD	−0.05177027	−0.05163551	−0.047444445
CCSD(T)	−0.05177027	−0.05163551	−0.047444445
CIMC	−0.051861(3)	−0.05606(9)	−0.047444444(1)
CIMC[pt2]	−0.05164(4)	−0.0491(3)	−0.0282(2)

Shown are results computed within Moeller–Plesset perturbation theory at second order (MP2), coupled cluster without (CCSD) and with triplets (CCSD(T)) by perturbation, CIMC projected from the CCSD wave function, and CIMC projected from the MP2 Wave function. Also shown are results obtained with full CI.

to the open shell structure of this state that will call for a multireference description. The wave function of CC theory indeed contains a large number of states but only very few of them will contribute to the estimate of the energy, which is given by the matrix element

$$E^{CC} = \langle \mathrm{HF}|\hat{H}|\Psi_{\mathrm{CC}}\rangle. \tag{17}$$

In fact only configurations with at most 2particle–2hole excitations will be connected to the Hartree–Fock state by a two-body Hamiltonian. In situations where $|\mathrm{HF}\rangle$ is indeed the dominant configuration this estimate will be rather similar to the actual expectation value, whose upper-bound we are estimating with CIMC. In more multireference situations, when this dominance of the Hartree–Fock state is not realized, CIMC should then be able to achieve a greater accuracy. For the case of Carbon in fact we see that the dominant effect producing the difference between CIMC and CIMC[pt2] is a substantially larger contribution of quadrupole excitations, most of which are neglected in the estimate at Eq. (17) (Table 2).

We can use the result of FCI to see where the issue may be present and we see indeed that the multireference estimate for the singlet C state is substantially lower than the converged CC result and is now very similar to the full CIMC estimate. The case of O is rather peculiar and even if it is still an open shell system now FCI matches exactly the CCSD result: this is probably an artifact coming from the smallness of the Hilbert space in the STO-3G basis.

All calculations have been repeated using a larger model space generated by the cc-pVDZ Gaussian basis set. In this case the results need to be analyzed with further care. In the case of the Be atom CIMC gives very consistent results, substantially independent on the starting point (CCSD or MP2).

Table 2 Same as Table 1, But Computed With the cc-pVDZ Basis Set

Atom	Be	C ($S = 0$)	O ($S = 0$)	Ne
FCI	−0.04507188	−0.11131084	−0.1677582	−0.19210558
MP2	−0.02633594	−0.06705581	−0.1208955	−0.18756719
CCSD	−0.04503138	−0.10339119	−0.15546308	−0.1908614
CCSD(T)	−0.0450694622	−0.1067876	−0.16299929	−0.1919173
CIMC	−0.04015(5)	−0.0819(1)	−0.1427(3)	−0.1641(6)
CIMC[pt2]	−0.04027(7)	−0.0881(1)	−0.1474(2)	−0.1626(4)

However, the correlation energy is about 9% smaller than the corresponding CC and FCI ones. The same kind of bias is found also for the open shell oxygen case while the CIMC estimate is considerably worse for the more difficult case of carbon and we are able to retrieve only about 77% of the correlation energy using CC amplitudes. This is consistent with the results obtained with the smaller basis. The quality of the Ne results is in between with some 86% of correlation retrieved with CIMC.

In almost all cases the CIMC[pt2] estimates are lower than the corresponding CIMC, especially for oxygen. Even if not conclusive, this suggests that the nodal structure becomes actually worse by performing the CC self-consistent iterations. This may particularly be true in systems with multireference character like C.

Overall the results seem to indicate that the CC wave function is struggling to reproduce the correct nodal structure of the ground state, it is interesting to note, however, that the CC energies seem very well converged (especially for the closed-shell systems as we expect). This means that the biased estimator Eq. (17) manages to artificially increase the accuracy of the calculation despite the fact that the wave function is not so good.

3.2 Molecules

In this section we present some preliminary results on CIMC calculations for a few small molecules. In particular we focused on H_2O (covalent bonds), LiH (ionic molecule), and Be_2 (a particularly difficult case due to the nearly degeneracy of *s* and *p* orbitals of Be atom) in the singlet state. We did not attempt to optimize the geometry and used a bond length of 1.78 Å for Be_2, 1.6365 Å for LiH while for water we place the oxygen at the origin and the two hydrogen atoms at (0, −0.757, 0.587) and (0, 0.757, 0.587), respectively.

These results should be intended as a further check for CIMC to produce correct projections of the ground state also in cases where the symmetry might in principle make the task much harder. In these calculations we used the same basis sets as in atoms (STO-3G and cc-pVDZ) on each single atom in order to build up the HF ground state and the corresponding orbitals.

Let us begin with the STO-3G case. In Table 3 on can see how water and LiH do not present particular difficulties. In both cases the difference between the CCSD and CCSD(T) correlation energies is less than 0.1%. CIMC and CIMC[pt2] once again are substantially close to each other indicating that the quality of the nodal surface provided by the CC calculations

Table 3 Correlation Energies (in Hartree) for a Few Simple Molecules Computed Using the STO-3G Basis Set

Atom	**H_2O**	**Be_2 ($S = 0$)**	**LiH**
FCI	−0.04958399	−0.1233246	−0.02108321
MP2	−0.03556682	−0.06220229	−0.01320542
CCSD	−0.04946747	−0.1201179	−0.02107219
CCSD(T)	−0.04953481	−0.12154144	−0.02108098
CIMC	−0.04934(8)	−0.1129(1)	−0.01989(1)
CIMC[pt2]	−0.0494(1)	−0.1096(2)	−0.02020(6)

Labels are the same as in Table 1.

Table 4 Same as Table 3, But Computed With the cc-pVDZ Basis Set

Atom	**H_2O**	**Be_2 ($S = 0$)**	**LiH**
MP2	−0.20401998	−0.074517227	−0.02274641
CCSD	−0.21334326	−0.1139897	−0.0310663
CCSD(T)	−0.2164033	−0.11958044	−0.03107542
CIMC	−0.1748(3)	−0.1106(2)	−0.02936(1)
CIMC[pt2]	−0.1764(5)	−0.1067(1)	−0.02953(4)

and by MP2 is comparable. The estimated upper-bounds are rather well behaved in all cases and the worst situation is realized in Be_2 where were able to recover only 93% of the correlation energy.

This is expected as both H_2O and LiH do not present particular features like unsaturated bonds which might make the determination of the correct nodal structure more difficult. A comparison with FCI also shows that Be_2 is the system with the larger multireference character further corroborating the expectation that the CC wave function will provide somewhat degraded nodes (Table 4).

When using the expanded model space using cc-pVDZ, the H_2O and LiH molecules show a somewhat improved behavior as compared to the simpler case and we are now able to recover more than 92% of the correlation energy for the singlet state of the beryllium dimer. This confirms the trend observed for the problematic case of the C atom: enlarging the basis size helps CIMC in retrieving more of the total correlation energy.

The cause of the worsening for water needs further investigation.

4. CONCLUSIONS

In this paper we review the CIMC method and discuss the extension of the algorithm to the case of atoms and, for the first time, some small molecules. The atoms considered both open shell (C, O) and closed shell (Be, Ne) and here we studied only the $S = 0$ singlet state. In general we obtain worse results, with errors of even 40% in extreme cases, for open shell atoms and more generally in states with a strong multireference character. This is at least in part to be expected and strongly suggests to devise better ways to obtain an approximate but accurate trial state than the CC ansatz.

In more favorable situations, like closed-shell atoms and the molecules with the exception of water, the results consistently retrieve around 90% or more of the total correlation energy. In these cases CIMC and CIMC [pt2] results are usually quite consistent indicating that the approximate amplitudes are more than reasonable. Due to the more slow scaling of CIMC this means that similar systems of much larger size, which are prohibitive for the expensive CC, could be accessed while maintaining a decent precision.

A higher level of accuracy is probably achievable only by efficiently reducing the extent of the sign problem using cancellation techniques (as in Refs. 1 and 3). We finally note that the advances reported in this manuscript open now the door for the exploration of finite nuclear systems which will provide a great test ground for the growing number of potential parametrizations.

ACKNOWLEDGMENT

The work of A.R. was supported by NSF Grant No. AST-1333607.

REFERENCES

1. Booth, G. H.; Thom, A. J. W.; Alavi, A. Fermion Monte Carlo Without Fixed Nodes: A Game of Life, Death, and Annihilation in Slater Determinant Space. *J. Chem. Phys.* **2009**, *131*, 054106.
2. Booth, G. H.; Grüneis, A.; Kresse, G.; Alavi, A. Towards an Exact Description of Electronic Wavefunctions in Real Solids. *Nature* **2013**, *493*, 365.
3. Petruzielo, F. R.; Holmes, A. A.; Changlani, H. J.; Nightingale, M. P.; Umrigar, C. J. Semistochastic Projector Monte Carlo Method. *Phys. Rev. Lett.* **2012**, *109*, 230201.
4. Roggero, A.; Mukherjee, A.; Pederiva, F. Quantum Monte Carlo With Coupled-Cluster Wave Functions. *Phys. Rev. B* **2013**, *88*, 115138.
5. Mukherjee, A.; Alhassid, Y. Configuration-Interaction Monte Carlo Method and Its Application to the Trapped Unitary Fermi Gas. *Phys. Rev. A* **2013**, *88*, 053622.

6. Roggero, A.; Mukherjee, A.; Pederiva, F. Quantum Monte Carlo Calculations of Neutron Matter With Nonlocal Chiral Interactions. *Phys. Rev. Lett.* **2014**, *112*, 221103.
7. Troullier, N.; Martins, J. L. Efficient Pseudopotentials for Plane-Wave Calculations. *Phys. Rev. B* **1991**, *43*, 1993.
8. Roggero, A.; Mori, P.; Mukherjee, A.; Pederiva, F. Configuration Interaction Monte Carlo With Coupled Clusters Wave Functions. *Adv. Quantum Chem.* **2016**, *73*, 315.
9. Schuchardt, K. L.; Didier, B. T.; Elsethagen, T.; Sun, L.; Gurumoorthi, V.; Chase, J.; Li, J.; Windus, T. L. Basis Set Exchange: A Community Database for Computational Sciences. *J. Chem. Inf. Model.* **2007**, *47*, 1045–1052. https://bse.pnl.gov.
10. Sun, Q.; Berkelbach, T. C.; Blunt, N. S.; Booth, G. H.; Guo, S.; Li, Z.; Liu, J.; McClain, J.; Sayfutyarova, E. R.; Sharma, S.; Wouters, S.; Chan, G. K. L. The Python-Based Simulations of Chemistry Framework (PySCF). ArXiv:1701.08223.

CHAPTER FOURTEEN

Time-Dependent Linear-Response Variational Monte Carlo

Bastien Mussard[*,†,1], Emanuele Coccia[*,‡,1], Roland Assaraf[*], Matthew Otten[§], Cyrus J. Umrigar[§,1], Julien Toulouse[*,1]

[*]Laboratoire de Chimie Théorique, Université Pierre et Marie Curie, Sorbonne Universités, CNRS, Paris, France
[†]Institut des Sciences du Calcul et des Données, Université Pierre et Marie Curie, Sorbonne Universités, Paris, France
[‡]Dipartimento di Scienze Fisiche e Chimiche, Universitá degli Studi dell'Aquila, L'Aquila, Italy
[§]Laboratory of Atomic and Solid-State Physics, Cornell University, Ithaca, NY, United States
[1]Corresponding authors: e-mail address: bastien.mussard@upmc.fr; emanuele.coccia@nano.cnr.it; cyrusumrigar@cornell.edu; julien.toulouse@upmc.fr

Contents

1. Introduction 256
2. Theory 257
   2.1 Wave Function Parametrization 257
   2.2 Linear Optimization Method 258
   2.3 Linear-Response Theory 259
   2.4 Realization in VMC 261
   2.5 Computational Details 264
3. Results 264
4. Conclusions and Perspectives 268
Acknowledgments 268
References 269

Abstract

We present the extension of variational Monte Carlo (VMC) to the calculation of electronic excitation energies and oscillator strengths using time-dependent linear-response theory. By exploiting the analogy existing between the linear method for wave function optimization and the generalized eigenvalue equation of linear-response theory, we formulate the equations of linear-response VMC (LR-VMC). This LR-VMC approach involves the first- and second-order derivatives of the wave function with respect to the parameters. We perform first tests of the LR-VMC method within the Tamm–Dancoff approximation using single-determinant Jastrow–Slater wave functions with different Slater basis sets on some singlet and triplet excitations of the beryllium atom. Comparison with reference experimental data and with configuration-interaction-singles (CIS) results shows that LR-VMC generally outperforms CIS for excitation energies and is thus a promising approach for calculating electronic excited-state properties of atoms and molecules.

Advances in Quantum Chemistry, Volume 76
ISSN 0065-3276
https://doi.org/10.1016/bs.aiq.2017.05.005

1. INTRODUCTION

Quantum Monte Carlo (QMC) methods[1–3] are a powerful and reliable alternative to wave function methods and density functional theory (DFT) for quantum chemistry calculations, thanks to their favorable scaling with system size and to their suitability for high-performance computing infrastructures, such as petascale architectures. Variational Monte Carlo (VMC)[4] combines Monte Carlo integration for computing the expectation value of the electronic Hamiltonian $\hat{H}$ and the variational principle for the ground state. VMC scales as N^{3-4} (where N is the number of electrons), similar to DFT scaling. The main drawback of any QMC approach is the very large prefactor in the scaling, preventing the systematic use of QMC in quantum chemistry calculations of medium- and large-size systems. This drawback is alleviated by performing massive parallel calculations on supercomputers.[5,6]

A fundamental role is played by the trial wave function, often written as a product of a determinantal part and a bosonic Jastrow factor[7] which depends on interparticle distances (with electron–nucleus, electron–electron, higher many-body terms, etc.). For example, one can use for the determinantal part a linear combination of configuration state functions (CSF, i.e., spatial- and spin-symmetry adapted linear combinations of Slater determinants of one-electron molecular orbitals),[8] or the antisymmetrized geminal power (AGP) ansatz (a single determinant of geminal pairing functions[9–11]). Furthermore, the optimization of the wave function is crucial for an accurate description of both static and dynamic electron correlation. The linear method[12–14] allows one to efficiently perform such an optimization for all the parameters of the wave function, using only the first-order derivatives of the wave function with respect to the parameters.

The calculation of excited-state properties of molecules (from prototypical models to complex organic dyes and biochromophores) still represents an open challenge for theoreticians. The two commonly used approaches are time-dependent DFT, which is not computationally demanding but sometimes lacks accuracy, and wave function methods, which are more accurate but very computationally demanding. QMC methods were originally formulated for ground-state problems and their extension to excited states is not straightforward. Relatively few applications of QMC for electronic excitations are present in literature, see, e.g., the singlet and triplet energies for the benchmark CH_2 diradical,[15] the low-lying singlet excited states of biochromophores,[16–19] the $n \rightarrow \pi^*$ transition in acrolein,[20,21] and the recent extension of the AGP ansatz for calculating excited-state energies.[22–26]

The basic idea of the present work stems from the formal analogy existing between the linear method for wave function optimization and time-dependent linear-response theory.[27] Indeed, the generalized eigenvalue equations of linear-response theory in the Tamm–Dancoff approximation (TDA) and those of the linear method at the ground-state minimum exactly coincide. Starting from this observation, we derive and implement the linear-response equations in VMC (LR-VMC). This represents an extension of the well-established time-dependent linear-response Hartree–Fock or multiconfiguration self-consistent field methods, taking into account both static and dynamic electron correlations.

The chapter is organized as follows. In Section 2, VMC and linear-response theory are briefly reviewed, and the LR-VMC method is presented and discussed in detail. Results of LR-VMC calculations in the TDA of some singlet and triplet excitations of the beryllium atom are reported and discussed in Section 3. Conclusions and perspectives for future work are given in Section 4.

2. THEORY

We first briefly review the form of the wave function that we use and the linear optimization method. We then derive the time-dependent linear-response equations and show how to implement them in VMC.

2.1 Wave Function Parametrization

We consider Jastrow–Slater-type wave functions parametrized as[12,14]

$$|\Psi(\mathbf{p})\rangle = \hat{J}(\boldsymbol{\alpha}) e^{\hat{\kappa}(\boldsymbol{\kappa})} \sum_{I=1}^{N_{\text{CSF}}} c_I |C_I\rangle, \tag{1}$$

where $\hat{J}(\boldsymbol{\alpha})$ is a Jastrow factor operator depending on a set of parameters $\boldsymbol{\alpha}$, $e^{\hat{\kappa}(\boldsymbol{\kappa})}$ is the orbital rotation operator depending on a set of orbital rotation parameters $\boldsymbol{\kappa}$, and $|C_I\rangle$ are CSFs with associated coefficients $\mathbf{c} = \{c_I\}$. The CSFs are linear combinations of Slater determinants of orbitals $|\phi_i\rangle$, which are expanded in a basis of Slater functions $\{|\chi_\mu\rangle\}$

$$|\phi_i\rangle = \sum_{\mu=1}^{N_{\text{basis}}} \lambda_{i\mu} |\chi_\mu\rangle. \tag{2}$$

The Slater functions are centered on the nuclei and their spatial representation is

$$\langle \mathbf{r}|\chi_\mu\rangle = N_n(\zeta)\ r^{n-1} \mathrm{e}^{-\zeta r}\ Y_{\ell,m}(\theta,\phi), \tag{3}$$

each function being characterized by a set of quantum numbers n, ℓ, m and an exponent ζ, $Y_{\ell,m}(\theta, \phi)$ are real spherical harmonics, and $N(\zeta)$ a normalization factor. The full set of parameters to consider is thus $\mathbf{p} = \{\boldsymbol{\alpha}, \mathbf{c}, \boldsymbol{\kappa}, \boldsymbol{\zeta}\}$ where $\boldsymbol{\zeta}$ stands for the set of exponents.

2.2 Linear Optimization Method

The linear optimization method[12–14] allows one to find the optimal parameters $\mathbf{p}$ using an iterative procedure. At each iteration, we consider the intermediate-normalized wave function

$$|\overline{\Psi}(\mathbf{p})\rangle = \frac{|\Psi(\mathbf{p})\rangle}{\langle \Psi_0|\Psi(\mathbf{p})\rangle} \tag{4}$$

where $|\Psi_0\rangle = |\Psi(\mathbf{p}^0)\rangle$ is the wave function for the parameters $\mathbf{p}^0$ at the current iteration (taken as normalized to unity, i.e., $\langle \Psi_0|\Psi_0\rangle = 1$), and we expand it to linear order in the parameter variations $\boldsymbol{\Delta}\mathbf{p} = \mathbf{p} - \mathbf{p}^0$,

$$|\overline{\Psi}_{\text{lin}}(\mathbf{p})\rangle = |\Psi_0\rangle + \sum_i \Delta p_i |\overline{\Psi}_i\rangle, \tag{5}$$

where $|\overline{\Psi}_i\rangle$ are the first-order derivatives of the wave function $|\overline{\Psi}(\mathbf{p})\rangle$

$$|\overline{\Psi}_i\rangle = \left(\frac{\partial|\overline{\Psi}(\mathbf{p})\rangle}{\partial p_i}\right)_{\mathbf{p}=\mathbf{p}^0} = |\Psi_i\rangle - \langle \Psi_0|\Psi_i\rangle|\Psi_0\rangle, \tag{6}$$

where $|\Psi_i\rangle = (\partial|\Psi(\mathbf{p})\rangle/\partial p_i)_{\mathbf{p}=\mathbf{p}^0}$ are the first-order derivatives of the original wave function $|\Psi(\mathbf{p})\rangle$. Using the intermediate-normalized wave function has the advantage that the derivatives in Eq. (6) are orthogonal to $|\Psi_0\rangle$, i.e., $\langle \Psi_0|\overline{\Psi}_i\rangle = 0$. We then determine the parameter variations $\boldsymbol{\Delta}\mathbf{p}$ by minimizing the corresponding energy

$$E_{\text{lin}} = \min_{\mathbf{p}} \frac{\langle \overline{\Psi}_{\text{lin}}(\mathbf{p})|\hat{H}|\overline{\Psi}_{\text{lin}}(\mathbf{p})\rangle}{\langle \overline{\Psi}_{\text{lin}}(\mathbf{p})|\overline{\Psi}_{\text{lin}}(\mathbf{p})\rangle}, \tag{7}$$

we update the parameters as $\mathbf{p}^0 \rightarrow \mathbf{p}^0 + \boldsymbol{\Delta}\mathbf{p}$, and iterate until convergence.

The minimization in Eq. (7) leads to the following generalized eigenvalue equation to be solved at each iteration

$$\begin{pmatrix} E_0 & \mathbf{g}_R^T/2 \\ \mathbf{g}_L/2 & \mathbf{H} \end{pmatrix} \begin{pmatrix} 1 \\ \Delta\mathbf{p} \end{pmatrix} = E_{\text{lin}} \begin{pmatrix} 1 & \mathbf{0}^T \\ \mathbf{0} & \mathbf{S} \end{pmatrix} \begin{pmatrix} 1 \\ \Delta\mathbf{p} \end{pmatrix}, \tag{8}$$

where $E_0 = \langle\Psi_0|\hat{H}|\Psi_0\rangle$ is the current energy, $g_{L,i} = 2\langle\overline{\Psi}_i|\hat{H}|\Psi_0\rangle$ and $g_{R,j} = 2\langle\Psi_0|\hat{H}|\overline{\Psi}_j\rangle$ are the left and right energy gradients (identical except on a finite Monte Carlo sample), and $H_{ij} = \langle\overline{\Psi}_i|\hat{H}|\overline{\Psi}_j\rangle$ is the Hamiltonian matrix in the basis of the first-order wave function derivatives, and $S_{ij} = \langle\overline{\Psi}_i|\overline{\Psi}_j\rangle$ is the overlap matrix in this basis.

Note that in Eq. (8), $\mathbf{0}$ and $\mathbf{0}^T$ stand for the zero column vector and the zero row vector, respectively.

2.3 Linear-Response Theory

Starting from the previously optimized wave function, we introduce now a time-dependent perturbation (e.g., interaction with an electric field) in the Hamiltonian

$$\hat{H}(t) = \hat{H} + \gamma\hat{V}(t), \tag{9}$$

where γ is a coupling constant. The approximate ground-state wave function $|\Psi(\mathbf{p}(t))\rangle$ evolves in time through its parameters $\mathbf{p}(t)$, which become now generally complex. As before, it is convenient to introduce the intermediate-normalized wave function

$$|\overline{\Psi}(\mathbf{p}(t))\rangle = \frac{|\Psi(\mathbf{p}(t))\rangle}{\langle\Psi_0|\Psi(\mathbf{p}(t))\rangle}, \tag{10}$$

where $|\Psi_0\rangle = |\Psi(\mathbf{p}^0)\rangle$ is the wave function for the initial parameters $\mathbf{p}^0$, again taken as normalized to unity (i.e., $\langle\Psi_0|\Psi_0\rangle = 1$). At each time, the time-dependent parameters $\mathbf{p}(t)$ can be found from the Dirac–Frenkel variational principle (see, e.g., Ref. 27)

$$\frac{\partial}{\partial p_i^*} \frac{\langle\overline{\Psi}(\mathbf{p}(t))|\hat{H}(t) - i\frac{\partial}{\partial t}|\overline{\Psi}(\mathbf{p}(t))\rangle}{\langle\overline{\Psi}(\mathbf{p}(t))|\overline{\Psi}(\mathbf{p}(t))\rangle} = 0. \tag{11}$$

To apply Eq. (11) in linear order in γ, we start by expanding the wave function $|\overline{\Psi}(\mathbf{p}(t))\rangle$ around $\mathbf{p}^0$ to second order in the parameter variations $\Delta\mathbf{p}(t) = \mathbf{p}(t) - \mathbf{p}^0$

$$|\overline{\Psi}(\mathbf{p}(t))\rangle = |\Psi_0\rangle + \sum_i \Delta p_i(t)|\overline{\Psi}_i\rangle + \frac{1}{2}\sum_{i,j} \Delta p_i(t)\Delta p_j(t)|\overline{\Psi}_{ij}\rangle + \cdots, \tag{12}$$

where $|\overline{\Psi}_i\rangle$ are the first-order derivatives of $|\overline{\Psi}(\mathbf{p})\rangle$ already introduced in Eq. (6), and $|\overline{\Psi}_{ij}\rangle$ are the second-order derivatives of the wave function $|\overline{\Psi}(\mathbf{p})\rangle$

$$\begin{aligned} |\overline{\Psi}_{ij}\rangle &= \left(\frac{\partial^2|\overline{\Psi}(\mathbf{p})\rangle}{\partial p_i \partial p_j}\right)_{\mathbf{p}=\mathbf{p}^0} \\ &= |\Psi_{ij}\rangle - \langle\Psi_0|\Psi_j\rangle|\Psi_i\rangle - \langle\Psi_0|\Psi_i\rangle|\Psi_j\rangle \\ &\quad + \left(2\langle\Psi_0|\Psi_i\rangle\langle\Psi_0|\Psi_j\rangle - \langle\Psi_0|\Psi_{ij}\rangle\right)|\Psi_0\rangle, \end{aligned} \tag{13}$$

where $|\Psi_{ij}\rangle = \left(\partial^2|\Psi(\mathbf{p})\rangle/\partial p_i \partial p_j\right)_{\mathbf{p}=\mathbf{p}^0}$ are the second-order derivatives of the original wave function $|\Psi(\mathbf{p})\rangle$. Again, the advantage of using the intermediate-normalized wave function is that the second-order derivatives are orthogonal to $|\Psi_0\rangle$, i.e., $\langle\Psi_0|\overline{\Psi}_{ij}\rangle = 0$. Plugging Eq. (12) into Eq. (11) and keeping only first-order terms in $\boldsymbol{\Delta}\mathbf{p}(t)$, in the limit of a vanishing perturbation ($\gamma \to 0$), we find

$$\mathbf{A}\,\boldsymbol{\Delta}\mathbf{p}(t) + \mathbf{B}\,\boldsymbol{\Delta}\mathbf{p}(t)^* = i\mathbf{S}\,\frac{\partial\boldsymbol{\Delta}\mathbf{p}(t)}{\partial t}, \tag{14}$$

with the matrices $A_{ij} = \langle\overline{\Psi}_i|\hat{H} - E_0|\overline{\Psi}_j\rangle = H_{ij} - E_0 S_{ij}$ where E_0 is the ground-state energy, $B_{ij} = \langle\overline{\Psi}_{ij}|\hat{H}|\Psi_0\rangle$, and $S_{ij} = \langle\overline{\Psi}_i|\overline{\Psi}_j\rangle$. If we look for free-oscillation solutions of the form

$$\boldsymbol{\Delta}\mathbf{p}(t) = \mathbf{X}e^{-i\omega_n t} + \mathbf{Y}^* e^{i\omega_n t}, \tag{15}$$

where ω_n corresponds to an excitation (or de-excitation) energy, we arrive at the linear-response equation in the form of a non-Hermitian generalized eigenvalue equation[27]

$$\begin{pmatrix} \mathbf{A} & \mathbf{B} \\ \mathbf{B}^* & \mathbf{A}^* \end{pmatrix}\begin{pmatrix} \mathbf{X}_n \\ \mathbf{Y}_n \end{pmatrix} = \omega_n \begin{pmatrix} \mathbf{S} & \mathbf{0} \\ \mathbf{0} & -\mathbf{S}^* \end{pmatrix}\begin{pmatrix} \mathbf{X}_n \\ \mathbf{Y}_n \end{pmatrix}. \tag{16}$$

The Tamm–Dancoff approximation (TDA) corresponds to neglecting the contributions from $\mathbf{B}$, leading to

$$\mathbf{A}\mathbf{X}_n = \omega_n \mathbf{S}\mathbf{X}_n. \tag{17}$$

At the ground-state minimum, i.e., when the energy gradient is zero, the generalized eigenvalue equation of the linear method in Eq. (8) is equivalent to the TDA equation (17) which directly gives excitation energies $\omega_n = E_{\text{lin}} - E_0$.

Finally, the oscillator strength f_n for the transition from the ground state to the excited state n (with excitation energy ω_n) can be easily extracted from the response vector $(\mathbf{X}_n, \mathbf{Y}_n)$

$$f_n = \frac{2}{3}\omega_n \sum_{\alpha=x,y,z} \left[(\mathbf{X}_n + \mathbf{Y}_n)^{\mathrm{T}} \boldsymbol{\mu}^\alpha \right]^2, \tag{18}$$

where $\boldsymbol{\mu}^\alpha$ is the vector containing the transition dipole moments for the component α (x, y, or z) between the ground-state wave function $|\Psi_0\rangle$ and the wave function derivative $|\overline{\Psi}_i\rangle$

$$\mu_i^\alpha = \langle \overline{\Psi}_i | \hat{\mu}^\alpha | \Psi_0 \rangle, \tag{19}$$

and $\hat{\mu}^\alpha$ is the electronic dipole operator.

2.4 Realization in VMC

We now give the expressions for performing linear-response calculations in VMC, referred to as LR-VMC, i.e., for calculating the expressions in Section 2.3 in a VMC run. For convenience, we also recall the expressions necessary for the linear optimization method.

The current ground-state energy is calculated as

$$E_0 = \langle E_{\mathrm{L}}(\mathbf{R}) \rangle, \tag{20}$$

where $E_{\mathrm{L}}(\mathbf{R}) = [H\Psi_0(\mathbf{R})]/\Psi_0(\mathbf{R})$ is the local energy and $\langle \ldots \rangle$ stands for an average on a finite Monte Carlo sample of points $\mathbf{R}_k$ distributed according to $\Psi_0(\mathbf{R})^2$, with $\mathbf{R} = (\mathbf{r}_1, \mathbf{r}_2, \ldots, \mathbf{r}_N)$ designating the electron coordinates. The left and right energy gradients are evaluated as

$$\begin{aligned} g_{\mathrm{L},i} &= 2\left\langle \frac{\overline{\Psi}_i(\mathbf{R})}{\Psi_0(\mathbf{R})} \frac{H\Psi_0(\mathbf{R})}{\Psi_0(\mathbf{R})} \right\rangle \\ &= 2\left[\left\langle \frac{\Psi_i(\mathbf{R})}{\Psi_0(\mathbf{R})} E_{\mathrm{L}}(\mathbf{R}) \right\rangle - \left\langle \frac{\Psi_i(\mathbf{R})}{\Psi_0(\mathbf{R})} \right\rangle \langle E_{\mathrm{L}}(\mathbf{R}) \rangle \right], \end{aligned} \tag{21}$$

and

$$g_{\mathrm{R},j} = 2\left\langle \frac{\Psi_0(\mathbf{R})}{\Psi_0(\mathbf{R})} \frac{H\overline{\Psi}_j(\mathbf{R})}{\Psi_0(\mathbf{R})} \right\rangle = 2\left[\left\langle \frac{\Psi_j(\mathbf{R})}{\Psi_0(\mathbf{R})} E_{\mathrm{L}}(\mathbf{R}) \right\rangle - \left\langle \frac{\Psi_j(\mathbf{R})}{\Psi_0(\mathbf{R})} \right\rangle \langle E_{\mathrm{L}}(\mathbf{R}) \rangle + \langle E_{\mathrm{L},j}(\mathbf{R}) \rangle \right], \tag{22}$$

where $E_{\mathrm{L},\,j}(\mathbf{R})$ is the first-order derivative of the local energy

$$E_{\mathrm{L},j}(\mathbf{R}) = \frac{H\Psi_j(\mathbf{R})}{\Psi_0(\mathbf{R})} - \frac{\Psi_j(\mathbf{R})}{\Psi_0(\mathbf{R})} E_{\mathrm{L}}(\mathbf{R}). \tag{23}$$

Note that, in the limit of an infinite sample, $\langle E_{\mathrm{L},j}(\mathbf{R}) \rangle = 0$ due to the hermiticity of the Hamiltonian, and therefore the left and right gradients become identical. The elements of the overlap matrix $\mathbf{S}$ are calculated as

$$S_{ij} = \left\langle \frac{\overline{\Psi}_i(\mathbf{R})}{\Psi_0(\mathbf{R})} \frac{\overline{\Psi}_j(\mathbf{R})}{\Psi_0(\mathbf{R})} \right\rangle = \left\langle \frac{\Psi_i(\mathbf{R})}{\Psi_0(\mathbf{R})} \frac{\Psi_j(\mathbf{R})}{\Psi_0(\mathbf{R})} \right\rangle - \left\langle \frac{\Psi_i(\mathbf{R})}{\Psi_0(\mathbf{R})} \right\rangle \left\langle \frac{\Psi_j(\mathbf{R})}{\Psi_0(\mathbf{R})} \right\rangle, \tag{24}$$

and the elements of the matrix $\mathbf{H}$ are evaluated as

$$\begin{aligned} H_{ij} &= \left\langle \frac{\overline{\Psi}_i(\mathbf{R})}{\Psi_0(\mathbf{R})} \frac{H\overline{\Psi}_j(\mathbf{R})}{\Psi_0(\mathbf{R})} \right\rangle \\ &= \left\langle \frac{\Psi_i(\mathbf{R})}{\Psi_0(\mathbf{R})} \frac{\Psi_j(\mathbf{R})}{\Psi_0(\mathbf{R})} E_{\mathrm{L}}(\mathbf{R}) \right\rangle - \left\langle \frac{\Psi_i(\mathbf{R})}{\Psi_0(\mathbf{R})} \right\rangle \left\langle \frac{\Psi_j(\mathbf{R})}{\Psi_0(\mathbf{R})} E_{\mathrm{L}}(\mathbf{R}) \right\rangle \\ &\quad - \left\langle \frac{\Psi_j(\mathbf{R})}{\Psi_0(\mathbf{R})} \right\rangle \left\langle \frac{\Psi_i(\mathbf{R})}{\Psi_0(\mathbf{R})} E_{\mathrm{L}}(\mathbf{R}) \right\rangle + \left\langle \frac{\Psi_i(\mathbf{R})}{\Psi_0(\mathbf{R})} E_{\mathrm{L},j}(\mathbf{R}) \right\rangle \\ &\quad - \left\langle \frac{\Psi_i(\mathbf{R})}{\Psi_0(\mathbf{R})} \right\rangle \langle E_{\mathrm{L},j}(\mathbf{R}) \rangle + \left\langle \frac{\Psi_i(\mathbf{R})}{\Psi_0(\mathbf{R})} \right\rangle \left\langle \frac{\Psi_j(\mathbf{R})}{\Psi_0(\mathbf{R})} \right\rangle \langle E_{\mathrm{L}}(\mathbf{R}) \rangle. \end{aligned} \tag{25}$$

The elements of the matrix $\mathbf{A}$ are then given by

$$\begin{aligned} A_{ij} = H_{ij} - E_0 S_{ij} &= \left\langle \frac{\Psi_i(\mathbf{R})}{\Psi_0(\mathbf{R})} \frac{\Psi_j(\mathbf{R})}{\Psi_0(\mathbf{R})} E_{\mathrm{L}}(\mathbf{R}) \right\rangle - \left\langle \frac{\Psi_i(\mathbf{R})}{\Psi_0(\mathbf{R})} \right\rangle \left\langle \frac{\Psi_j(\mathbf{R})}{\Psi_0(\mathbf{R})} E_{\mathrm{L}}(\mathbf{R}) \right\rangle \\ &\quad - \left\langle \frac{\Psi_j(\mathbf{R})}{\Psi_0(\mathbf{R})} \right\rangle \left\langle \frac{\Psi_i(\mathbf{R})}{\Psi_0(\mathbf{R})} E_{\mathrm{L}}(\mathbf{R}) \right\rangle + \left\langle \frac{\Psi_i(\mathbf{R})}{\Psi_0(\mathbf{R})} E_{\mathrm{L},j}(\mathbf{R}) \right\rangle \\ &\quad - \left\langle \frac{\Psi_i(\mathbf{R})}{\Psi_0(\mathbf{R})} \right\rangle \langle E_{\mathrm{L},j}(\mathbf{R}) \rangle - \left\langle \frac{\Psi_i(\mathbf{R})}{\Psi_0(\mathbf{R})} \frac{\Psi_j(\mathbf{R})}{\Psi_0(\mathbf{R})} \right\rangle \langle E_{\mathrm{L}}(\mathbf{R}) \rangle \\ &\quad + 2\left\langle \frac{\Psi_i(\mathbf{R})}{\Psi_0(\mathbf{R})} \right\rangle \left\langle \frac{\Psi_j(\mathbf{R})}{\Psi_0(\mathbf{R})} \right\rangle \langle E_{\mathrm{L}}(\mathbf{R}) \rangle, \end{aligned} \tag{26}$$

and the elements of the matrix **B** are

$$
\begin{aligned}
B_{ij} &= \left\langle \frac{\overline{\Psi}_{ij}(\mathbf{R})}{\Psi_0(\mathbf{R})} \frac{H\Psi_0(\mathbf{R})}{\Psi_0(\mathbf{R})} \right\rangle \\
&= \left\langle \frac{\Psi_{ij}(\mathbf{R})}{\Psi_0(\mathbf{R})} E_{\mathrm{L}}(\mathbf{R}) \right\rangle - \left\langle \frac{\Psi_{ij}(\mathbf{R})}{\Psi_0(\mathbf{R})} \right\rangle \langle E_{\mathrm{L}}(\mathbf{R}) \rangle \\
&\quad - \left\langle \frac{\Psi_i(\mathbf{R})}{\Psi_0(\mathbf{R})} \right\rangle \left\langle \frac{\Psi_j(\mathbf{R})}{\Psi_0(\mathbf{R})} E_{\mathrm{L}}(\mathbf{R}) \right\rangle - \left\langle \frac{\Psi_j(\mathbf{R})}{\Psi_0(\mathbf{R})} \right\rangle \left\langle \frac{\Psi_i(\mathbf{R})}{\Psi_0(\mathbf{R})} E_{\mathrm{L}}(\mathbf{R}) \right\rangle \\
&\quad + 2 \left\langle \frac{\Psi_i(\mathbf{R})}{\Psi_0(\mathbf{R})} \right\rangle \left\langle \frac{\Psi_j(\mathbf{R})}{\Psi_0(\mathbf{R})} \right\rangle \langle E_{\mathrm{L}}(\mathbf{R}) \rangle .
\end{aligned}
\tag{27}
$$

Finally, the expression of the transition dipole moment needed for calculating oscillator strengths is

$$
\begin{aligned}
\mu_i^{\alpha} &= \left\langle \frac{\overline{\Psi}_i(\mathbf{R})}{\Psi_0(\mathbf{R})} \mu^{\alpha}(\mathbf{R}) \right\rangle \\
&= \left\langle \frac{\Psi_i(\mathbf{R})}{\Psi_0(\mathbf{R})} \mu^{\alpha}(\mathbf{R}) \right\rangle - \left\langle \frac{\Psi_i(\mathbf{R})}{\Psi_0(\mathbf{R})} \right\rangle \langle \mu^{\alpha}(\mathbf{R}) \rangle ,
\end{aligned}
\tag{28}
$$

where $\mu^{\alpha}(\mathbf{R}) = -\sum_{k=1}^{N} r_{k,\alpha}$ is the α-component of electronic dipole moment.

In the linear optimization method, using the nonsymmetric estimator of the matrix **H** in Eq. (25) instead of a symmetrized one has the advantage of leading to the strong zero-variance principle of Nightingale and Melik-Alaverdian[28]: in the limit where the current wave function $|\Psi_0\rangle$ and its first-order derivatives $|\overline{\Psi}_i\rangle$ form a complete basis for expanding an exact eigenstate of the Hamiltonian, the parameter variations $\mathbf{\Delta p}$ and the associated energy E_{lin} are found from Eq. (8) with zero variance provided that the Monte Carlo sample size is larger than the number of parameters (see discussion in Ref. 14). Unfortunately, this strong zero-variance principle does not apply when solving the linear-response equation (16). However, in the limit where $|\Psi_0\rangle$ is an exact eigenstate of the Hamiltonian, the left energy gradient $g_{\mathrm{L},\,i}$ in Eq. (21) vanishes with zero variance, and thus the TDA linear equation (17) becomes equivalent to Eq. (8) for calculating excited-state energies even on a finite Monte Carlo sample. Therefore, in this case, the strong zero-variance principle applies to the calculation of the response vectors $\mathbf{X}_n$ and excitation energies ω_n.

2.5 Computational Details

The calculations shown here were performed using the QMC program CHAMP,[29] starting from Hartree–Fock calculations done with GAMESS.[30] Two Slater basis sets of different sizes were used, namely the VB1 and VB2 basis set from Ref. 31. The VB1 basis set has five s and one p Slater functions ([5s, 1p]), whereas the VB2 basis set has six s, two p, and one d Slater functions ([6s, 2p, 1d]). We use a flexible Jastrow factor consisting of the exponential of the sum of electron–nucleus, electron–electron and electron–electron–nucleus terms, written as systematic polynomial and Padé expansions,[32–34] with 4 electron-nucleus parameters, 5 electron-electron parameters, and 15 electron-electron-nucleus parameters. For each VMC calculation, 10^4 blocks were employed with 10^4 steps each. One block was used for equilibration of the VMC distribution.

3. RESULTS

The beryllium atom was used as a first test of the LR-VMC approach, since accurate experimental reference values for the excitation energies are available from Ref. 35. An accurate description of the Be ground state requires a multiconfigurational wave function for accounting for the near-degeneracy between the 2s and 2p orbitals. However, for these preliminary tests, we present only results of calculations using a Jastrow–Slater single-determinant wave function for the ground state using TDA linear-response theory. This choice is motivated by the fact that a direct comparison between the LR-VMC/TDA method and configuration-interaction-singles (CIS) calculations represents a simple but essential first step for validating our approach. We expect that LR-VMC/TDA to outperform CIS because the Jastrow factor in LR-VMC should account for a substantial part of electronic correlation, and we find this to be the case for most of the excitations studied. The results are presented both as errors with respect to the experimental values in Fig. 1 and as detailed excitation energies in Tables 1–4.

In Table 1 results for the singlet 2s3s (^1S) state are reported. The effect of the Slater basis set adopted is dramatic at the CIS level, as a reasonable agreement with the reference experimental value of 0.249 Hartree is found only when the VB2 basis set is used. LR-VMC/TDA values are labeled as follow: (j) designates the response of the Jastrow parameters only, while (j + o) is the response of both the Jastrow and orbital parameters. The response of the

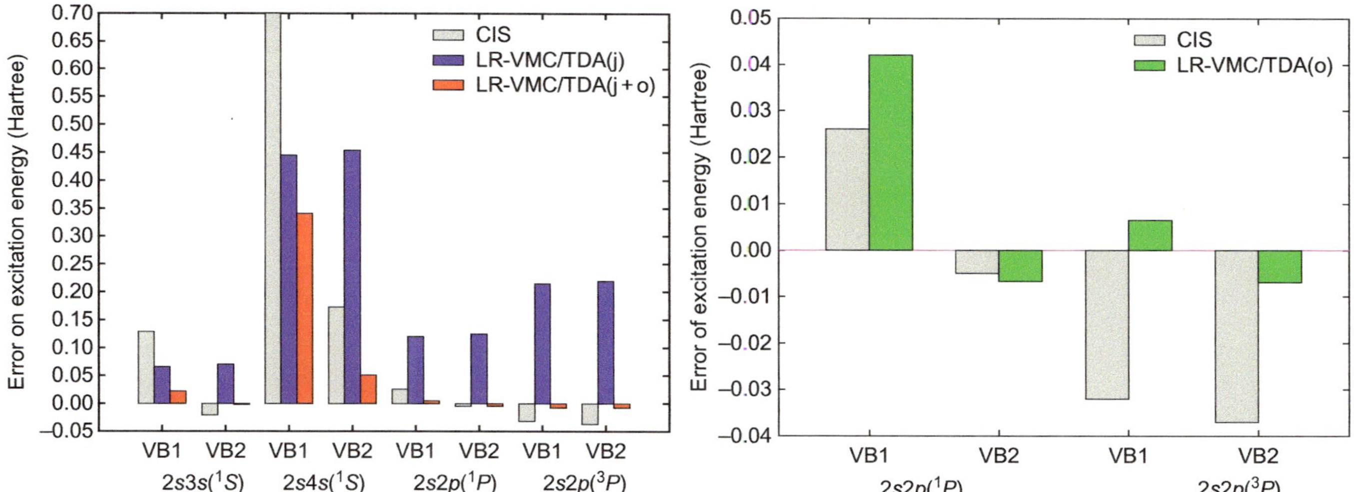

Fig. 1 Errors with respect to experimental values[35] of CIS (*grey*), and LR-VMC/TDA excitation energies including the response of the Jastrow parameters (*j*) (*blue*) and the response of the Jastrow and orbitals parameters (*j* + *o*) (*red*) for *S* excited states (*left*), and the response of the orbital parameters (*o*) (*green*) for *P* excited states (*right*). These results are also detailed in Tables 1–4.

Table 1 $2s3s$ (^1S) Excitation Energies (in Hartree) for the Beryllium Atom Calculated Using CIS and LR-VMC/TDA Including the Response of the Jastrow Parameters (j) and of the Jastrow and Orbital Parameters ($j + o$)

	CIS	LR-VMC/TDA(j)	LR-VMC/TDA($j + o$)	Exp.
VB1	0.378	0.2888(1)	0.2672(1)	0.249
VB2	0.228	0.2880(1)	0.2378(2)	0.249

The experimental value is taken from Ref. 35.

Table 2 $2s4s$ (^1S) Excitation Energies (in Hartree) for the Beryllium Atom Calculated Using CIS and LR-VMC/TDA Including the Response of the Jastrow Parameters (j) and of the Jastrow and Orbital Parameters ($j + o$)

	CIS	LR-VMC/TDA(j)	LR-VMC/TDA($j + o$)	Exp.
VB1	2.639	0.6106(1)	0.5578(2)	0.297
VB2	0.470	0.6104(1)	0.321(3)	0.297

The experimental value is taken from Ref. 35.

Jastrow factor substantially improves upon the CIS VB1 estimate, going from 0.378 to 0.2888(1) Hartree. The excitation energy improves further when the response of the orbital parameters is included in the LR-VMC/TDA calculation, yielding an error of around 0.02 Hartree with respect to the experimental value. Increasing the size of the Slater basis set, i.e., moving from VB1 to VB2, we obtain a fair agreement with the experimental data when both the Jastrow and the orbital parameters are included in the response (0.2378(2) Hartree).

The singlet $2s4s$ (^1S) excitation is higher in energy, and CIS fails to recover the experimental result of 0.297 Hartree, for both basis sets, as shown in Table 2. As already mentioned for the $2s3s$ excitation, the response of the Jastrow factor plays an important role for the VB1 basis set, reducing the error in the excitation energy by around 2 Hartree. Including the orbital parameters in the response lowers the excitation energy further to 0.5578(2), but this is still a large overestimate of the experimental value. With the VB2 basis, the LR-VMC/TDA($j + o$) calculation outperforms CIS, but a substantial error (>0.02 Hartree) still remains for this high-lying excitation. The failure of VB1 and, to a lesser extent, of VB2 is likely due to the poor description of the $4s$ orbital.

The extension of our proposed approach to P excitations is straightforward, with a relaxation of the spatial symmetry constraints in the orbital rotation parameters. Note that the Jastrow factor employed in this work only depends on interparticle distances, i.e., it has spherical symmetry, and therefore excited states with P symmetry cannot be represented with the wave function derivatives with respect to the Jastrow parameters. For this reason, only results concerning the response of the orbitals (o) are reported for the P excitations. In Table 3, results for the singlet $2s2p$ (^1P) state are given, which is the lowest energy excitation in the beryllium atom. The CIS calculations with the VB1 and VB2 basis sets show a fair agreement with the reference value of 0.194 Hartree, the CIS calculation using the VB2 basis set being only 5 mHartree below it. The LR-VMC/TDA(o) estimate is also close to the experimental reference when the VB2 basis set is employed (0.1873(2) Hartree), while for the VB1 basis set LR-VMC/TDA(o) greatly overestimates the excitation energy.

Similarly, our implementation of linear response allows us to easily compute triplet excitations by considering triplet orbital rotation parameters. The CIS calculation underestimates the correct excitation energy by more than 30 mHartree, while the LR-VMC/TDA(o) excitation energies are very close to the reference values of 0.100 Hartree (Table 4). The basis set effects are small is this case.

Finally, we computed the oscillator strength f (Table 5) corresponding to the singlet $2s2p$ (^1P) excitation, which is nonzero according to selection rules. The LR-VMC(o) oscillator strengths seem more sensitive to the basis set compared to the CIS oscillator strengths. Moreover, the inclusion of the Jastrow factor does not improve the oscillator strength.

Table 3 $2s2p$ (^1P) Excitation Energies (in Hartree) for the Beryllium Atom Calculated Using CIS and LR-VMC/TDA Including the Response of the Orbital Parameters (o)

	CIS	LR-VMC/TDA(o)	Exp.
VB1	0.220	0.2358(1)	0.194
VB2	0.189	0.1873(2)	0.194

The experimental value is taken from Ref. 35.

Table 4 $2s2p$ (^3P) Excitation Energies (in Hartree) for the Beryllium atom Calculated Using CIS and LR-VMC/TDA Including the Response of the Orbital Parameters (o)

	CIS	LR-VMC/TDA(o)	Exp.
VB1	0.068	0.1064(1)	0.100
VB2	0.063	0.0929(2)	0.100

The experimental value is taken from Ref. 35.

Table 5 Oscillator Strength *f* Corresponding to the $2s2p$ (^1P) Excitation of the Beryllium Atom Computed Using CIS and LR-VMC/TDA Including the Response of the Orbital Parameters (*o*)

	CIS	LR-VMC/TDA(*o*)	Exp.
VB1	0.648	0.435(1)	1.34(3)
VB2	0.669	0.57(2)	1.34(3)

The experimental value is taken from Ref. 36.

4. CONCLUSIONS AND PERSPECTIVES

In this work we have presented a formulation of time-dependent linear-response theory in the VMC framework using a Jastrow–Slater wave function. Compared to state-specific or state-average excited-state QMC methods, the advantage of this LR-VMC approach is that, after optimizing only one ground-state wave function, one can easily calculate several excitation energies of different spatial or spin symmetry. Compared to similar linear-response quantum chemistry methods, the presence of the Jastrow factor in LR-VMC allows one to explicitly treat part of the dynamical correlation. A disadvantage of the method is that the excitation energies are much more sensitive than the ground-state energy to the quality of the optimized ground-state wave function. This is true in other linear-response quantum chemistry methods as well, but is a bigger drawback in a method that employs stochastic optimization.

Using a Jastrow–Slater single-determinant wave function and the TDA, the LR-VMC method was shown to be more accurate that CIS for most of the excitation energies of the beryllium atom that were studied. The LR-VMC approach thus seems a promising method for calculating electronic excitation energies. In the near future, a systematic study on a set of molecules will be an essential step to further validate the proposed methodology, together with calculations using the full response equation beyond the TDA. Also, we will explore using multideterminant wave functions, larger basis sets, and including the wave function derivatives with respect to the exponents of the Slater functions.

ACKNOWLEDGMENTS

E.C. thanks University of L'Aquila for financial support and the Laboratoire de Chimie Théorique for computational resources. M.O. and C.J.U. were supported in part by NSF grant ACI-1534965.

REFERENCES

1. Hammond, B. L.; Lester, W. A., Jr.. Reynolds, P. J. *Monte Carlo Methods in Ab-Initio Quantum Chemistry*. World Scientific: Singapore, 1994.
2. Foulkes, W. M. C.. Mitas, L.; Needs, R. J.; Rajagopal, G. Quantum Monte Carlo Simulations of Solids. *Rev. Mod. Phys.* **2001**, *73*, 33–83. https://doi.org/10.1103/RevModPhys.73.33.
3. Toulouse, J.; Assaraf, R.; Umrigar, C. J. Introduction to the Variational and Diffusion Monte Carlo Methods. *Adv. Quantum Chem.* **2016**, *73*, 285–314.
4. Bressanini, D.; Reynolds, P. J. Between Classical and Quantum Monte Carlo Methods: "Variational" QMC. In: *Monte Carlo Methods in Chemical Physics*; Prigogine, I., Rice, S. A., Eds.; Advances in Chemical Physics, Vol. 105, 1998; pp 37–64.
5. Coccia, E.; Guidoni, L. Quantum Monte Carlo Study of the Retinal Minimal Model $C_5H_6NH_2^+$. *J. Comput. Chem.* **2012**, *33*, 2332–2339.
6. Coccia, E.; Varsano, D.; Guidoni, L. Ab Initio Geometry and Bright Excitation of Carotenoids: Quantum Monte Carlo and Many Body Green's Function Theory Calculations on Peridinin. *J. Chem. Theory Comput.* **2014**, *10*(1), 501–506.
7. Drummond, N. D.; Towler, M. D.; Needs, R. J. Jastrow Correlation Factor for Atoms, Molecules, and Solids. *Phys. Rev. B* **2004**, *70*, 235119. https://doi.org/10.1103/PhysRevB.70.235119.
8. Petruzielo, F. R.; Toulouse, J.; Umrigar, C. J. Approaching Chemical Accuracy With Quantum Monte Carlo. *J. Chem. Phys.* **2012**, *136*, 124116.
9. Casula, M.; Sorella, S. Geminal Wave Function With Jastrow Correlation: A First Application to Atoms. *J. Chem. Phys.* **2003**, *119*, 6500. https://doi.org/10.1063/1.1604379.
10. Casula, M.; Attaccalite, C.; Sorella, S. Correlated Geminal Wave Function for Molecules: An Efficient Resonating Valence Bond Approach. *J. Chem. Phys.* **2004**, *121*, 7110. https://doi.org/10.1063/1.1794632.
11. Zen, A.; Coccia, E.; Luo, Y.; Sorella, S.; Guidoni, L. Static and Dynamical Correlation in Diradical Molecules by Quantum Monte Carlo Using the Jastrow Antisymmetrized Geminal Power Ansatz. *J. Chem. Theory Comput.* **2014**, *10*, 1048–1061.
12. Toulouse, J.; Umrigar, C. J. Optimization of Quantum Monte Carlo Wave Functions by Energy Minimization. *J. Chem. Phys.* **2007**, *126*, 084102.
13. Umrigar, C. J.; Toulouse, J.; Filippi, C.; Sorella, S.; Hennig, R. G. Alleviation of the Fermion-Sign Problem by Optimization of Many-Body Wave Functions. *Phys. Rev. Lett.* **2007**, *98*, 110201. https://doi.org/10.1103/PhysRevLett.98.110201.
14. Toulouse, J.; Umrigar, C. J. Full Optimization of Jastrow-Slater Wave Functions With Application to the First-Row Atoms and Homonuclear Diatomic Molecules. *J. Chem. Phys.* **2008**, *128*, 174101.
15. Zimmerman, P. M.; Toulouse, J.; Zhang, Z.; Musgrave, C. B.; Umrigar, C. J. Excited States of Methylene From Quantum Monte Carlo. *J. Chem. Phys.* **2009**, *131*, 124103.
16. Valsson, O.; Filippi, C. Photoisomerization of Model Retinal Chromophores: Insight From Quantum Monte Carlo and Multiconfigurational Perturbation Theory. *J. Chem. Theory Comput.* **2010**, *6*, 1275.
17. Filippi, C.; Buda, F.; Guidoni, L.; Sinicropi, A. Bathochromic Shift in Green Fluorescent Protein: A Puzzle for QM/MM Approaches. *J. Chem. Theory Comput.* **2011**, *8*(1), 112–124.
18. Valsson, O.; Angeli, C.; Filippi, C. Excitation Energies of Retinal Chromophores: Critical Role of the Structural Model. *Phys. Chem. Chem. Phys.* **2012**, *14*, 11015.
19. Valsson, O.; Campomanes, P.; Tavernelli, I.; Rothlisberger, U.; Filippi, C. Rhodopsin Absorption From First Principles: Bypassing Common Pitfalls. *J. Chem. Theory Comput.* **2013**, *9*, 2441.

20. Toulouse, J.; Caffarel, M.; Reinhardt, P.; Hoggan, P. E.; Umrigar, C. J. Quantum Monte Carlo Calculations of Electronic Excitation Energies: The Case of the Singlet $n \rightarrow \pi^*$ (CO) Transition in Acrolein. In: *Advances in the Theory of Quantum Systems in Chemistry and Physics*; Hoggan, P. E., Brändas, E. J., Maruani, J., Piecuch, P., Delgado-Barrio, G., Eds.; *Progress in Theoretical Chemistry and Physics*; Vol. 22, 2012; pp 345–353.
21. Floris, F. M.; Filippi, C.; Amovilli, C. Electronic Excitations in a Dielectric Continuum Solvent With Quantum Monte Carlo: Acrolein in Water. *J. Chem. Phys.* **2014**, *140*, 034109.
22. Zen, A.; Coccia, E.; Gozem, S.; Olivucci, M.; Guidoni, L. Quantum Monte Carlo Treatment of the Charge Transfer and Diradical Electronic Character in a Retinal Chromophore Minimal Model. *J. Chem. Theory Comput.* **2015**, *11*, 992–1005. https://doi.org/10.1021/ct501122z.
23. Dupuy, N.; Bouaouli, S.; Mauri, F.; Sorella, S.; Casula, M. Vertical and Adiabatic Excitations in Anthracene From Quantum Monte Carlo: Constrained Energy Minimisation for Structural and Electronic Excited-State Properties in the JAGP Ansatz. *J. Chem. Phys.* **2015**, *142*, 214109.
24. Zhao, L.; Neuscamman, E. Equation of Motion Theory for Excited States in Variational Monte Carlo and the Jastrow Antisymmetric Geminal Power in Hilbert Space. *J. Chem. Theory. Comput.* **2016**, *12*, 3719–3726.
25. Zhao, L.; Neuscamman, E. An Efficient Variational Principle for the Direct Optimisation of Excited States. *J. Chem. Theory. Comput.* **2016**, *12*, 3436–3440.
26. Neuscamman, E. Variation After Response in Quantum Monte Carlo. *J. Chem. Phys.* **2016**, *145*, 081103.
27. McWeeny, R. *Methods of Molecular Quantum Mechanics*. Academic Press: Cambridge, MA, 1992.
28. Nightingale, M. P.; Melik-Alaverdian, V. Optimisation of Ground- and Excited-State Wave Functions and van der Waals Clusters. *Phys. Rev. Lett.* **2001**, *87*, 043401.
29. Umrigar, C. J., Filippi, C., Toulouse, J., Eds. CHAMP, A Quantum Monte Carlo Program Written. See http://www.physics.cornell.edu/cyrus/champ.html.
30. Schmidt, M. W.; Baldridge, K. K.; Boatz, J. A.; Elbert, S. T.; Gordon, M. S.; Jensen, J. H.; Koseki, S.; Matsunaga, N.; Nguyen, K. A.; Su, S. J.; Windus, T. L.; Dupuis, M.; Montgomery, J. A. General Atomic and Molecular Electronic Structure System. *J. Comput. Chem.* **1993**, *14*, 1347.
31. Ema, I.; Garcia de la Vega, J. M.. Ramirez, G.; Lopez, R.; Fernandez Rico, J.; Meissner, H.; Paldus, J. Polarized Basis Sets of Slater-Type Orbitals: H to Ne Atoms. *J. Comput. Chem.* **2003**, *24*, 859–868.
32. Umrigar, C. J. Unpublished.
33. Filippi, C.; Umrigar, C. J. Multiconfiguration Wave Functions for Quantum Monte Carlo Calculations of First-Row Diatomic Molecules. *J. Chem. Phys.* **1996**, *105*, 213.
34. Güçlü, A. D.; Jeon, G. S.; Umrigar, C. J.; Jain, J. K. Quantum Monte Carlo Study of Composite Fermions in Quantum Dots: The Effect of Landau-Level Mixing. *Phys. Rev. B* **2005**, *72*, 205327.
35. Kramida, A.; Martin, W. C. A Compilation of Energy Levels and Wavelengths for the Spectrum of Neutral Beryllium (Be I). *J. Phys. Chem. Ref. Data* **1997**, *26*, 1185.
36. Schnabel, R.; Kock, M. *f*-Value Measurement of the Be I Resonance Line Using a Nonlinear Time-Resolved Laser-Induced-Fluorescence Technique. *Phys. Rev. A* **2000**, *61*, 062506.

CHAPTER FIFTEEN

Quantum Monte Carlo Calculations for Industrial Catalysts: Accurately Evaluating the H_2 Dissociation Reaction Barrier on Pt(111)

Philip E. Hoggan[1]
Institute Pascal, UMR 6602 CNRS, Aubière Cedex, France
[1]Corresponding author: e-mail address: pehoggan@yahoo.com

Contents

1. Introduction 272
2. Setting Up the Model Geometries 273
3. QMC Methods 274
   3.1 Complex Jastrow Factors: Testing Against Real Factors 274
   3.2 Time-Step Bias, Other DMC Error Sources, and Specifications 275
4. Perspectives and Conclusions 276
Acknowledgments 277
References 277

Abstract

The yardstick of new first-principle approaches to key points on reaction paths at metal surfaces is chemical accuracy compared to reliable experiment. By this we mean that such values as the activation barrier are required to within 1 kcal/mol. Quantum Monte Carlo (QMC) is a promising (albeit lengthy) first-principles method for this and we are now beyond the dawn of QMC benchmarks for these systems, since hydrogen dissociation on Cu(111) has been studied with quite adequate accuracy in two improving QMC studies (Hoggan, 2015; Doblhoff-Dier et al., 2017).

Pt and Cu require the use of pseudo-potentials in these calculations and we show that those of Pt are less problematic than those for Cu, particularly for QMC work.

This work determines the activation barrier to hydrogen dissociation on Pt(111) using the bridging geometry. The method used is state-of-the-art (ab initio) QMC. This barrier agrees to better than chemical accuracy with a recent specific reaction parameter (SRP-DFT) calculation, fitted to measurement (Ghassemi et al., 2017).

The results give the dissociation barrier for hydrogen on Pt(111) as 5.4 kcal/mol (QMC and 6.2 by SRP-DFT) with a QMC standard error below 0.99 kcal/mol.

This is encouraging for establishing less well-known benchmark values of industrial reaction barriers on Pt(111).

Advances in Quantum Chemistry, Volume 76
ISSN 0065-3276
https://doi.org/10.1016/bs.aiq.2017.07.002

1. INTRODUCTION

Among metal surfaces that catalyse hydrogen dissociation, Cu(111) has been the subject of very recent benchmark studies with specific reaction parameter density functional theory (SRP-DFT) and state-of-the-art Quantum Monte Carlo (QMC), because reliable molecular beam measurements of the late barrier for this reaction are available.[1,2]

Experiment and SRP-DFT agree with QMC. Standard error in the QMC is close to chemical accuracy (i.e., just within a kcal/mol), and indications have been obtained in our earlier work for reducing the error.[2] The Pt(111) case is more favorable than copper, since the element can be described by trial wave functions with variance an order of magnitude less than in the copper case, which is adversely affected by the presence of a full 3d shell.

Industrial catalysts often use Pt(111) for certain reactions triggered by bond-breaking. This surface also catalyses bond dissociation in hydrogen. Unfortunately, Pt(111) does not yet benefit from accurate molecular beam results for the reaction. A recent DFT study[3] shows that there are several coordinations of the hydrogen molecule, leading to two main categories of transitions state. The "on-top" geometry has the molecule perpendicular to the surface plane, above a Pt atom. The reaction then appears to be almost barrier free. Such a barrier energy is inaccessible by QMC to date, being little more than the standard error in height. This work therefore focuses on the "bridge" orientation. The agreement is very good between the DFT study and the present QMC work, with the former needing a fitted parameter to ensure that the excesses of two functionals compensate each other and the latter, described here, being completely ab initio but rather time consuming. The QMC methods used in this work were developed in Refs. 1, 2, 4, and 5, where a complete description can be found.

The diffusion Monte Carlo (DMC) approach is unrivalled in accuracyamong first-principle methods.

QMC benchmarks the barrier for hydrogen dissociation on Cu(111) as 14.8 $\pm$ 1.5 kcal/mol.[2] The barrier on Cu(111) is significantly higher than on Pt(111). Accurate molecular beam experiments give 0.63 eV or 14.5 kcal/mol. The H_2 molecule dissociation on Pt(111) has a low experimental barrier which depends strongly on the molecular orientation and active site and ranges from some 0.06 to 0.420 eV. The higher barriers are for trigonal sites on the surface with minimal rearrangements and no defects.[3]

For an H_2 molecule parallel to the surface, the lowest experimental barrier is 0.27 eV.

Evaluating the bridging hydrogen dissociation barrier on Pt(111) is therefore likely to pose a stiff challenge to QMC. The lowest QMC standard error is just over 0.04 eV (which is termed chemical accuracy, i.e., error bars within 1 kcal/mol). This work shows hydrogen dissociation QMC barriers are reliable on Pt(111) for "bridge" orientations (standard error under 0.042 eV or 0.99 kcal/mol). For this barrier, the DMC value is obtained here as 5.4 kcal/mol or 0.23 eV, within standard error of experiment (0.27 c.f. 0.23 ± 0.04 eV).

2. SETTING UP THE MODEL GEOMETRIES

The reaction barrier heights in this work are obtained in first-principle calculations by stringently accurate ground-state QMC calculations on two geometries comprising the same atoms. This reference to the molecule–solid asymptote previously showed high accuracy.[6,7] The high-energy limit is the so-called transition state (TS). The low-energy asymptote corresponds to a geometry describing weakly physisorped hydrogen (at equilibrium bond length) on the Pt(111) slab.

The wave function used to initialize the QMC calculation is a single Slater determinant of DFT (Perdew, Burke, Ernzerhof) Kohn–Sham spin-orbitals, in view of the system size. Nevertheless, methods used here embedded a molecular portion of the active site and their geometry was then optimized within QMC.[4] This would allow a selected CI. Wave functions optimized by QMC from the DFT output (a hydrogen molecule and a platinum slab of the face-centered cubic lattice exposing a close-packed (111) face) are expanded in plane-waves.

The parallel orientation of the H_2 molecule between two Pt atoms of a trigonal site is described following a recent DFT and dynamics study, compared to experiment.[3] The TS is optimized with a hydrogen molecule parallel to the surface, above and between two Pt atoms (spanning 1.2 Å of a Pt–Pt linkage in the (111) surface like a bridge).

A slab thickness study[2] uses experimental top atom-layer spacing converging to bulk values in four layers. A five-layer slab model is used here since the wave function is much improved over that of four-layer slabs (little change for 6–12-layer slabs). Four hydrogen molecules are adsorbed on it and surface area is varied. First, a 2 by 2 mesh surface followed by a 4 by 4 mesh surface. Comparing these areas permits extrapolation to the infinite surface extent.

Note that for the cited work on copper, a significant overall reduction in barrier height was observed, whereas for Pt(111) the overall effect is very small. This shows that finite size effects are lower for Pt(111) than Cu(111) as confirmed by the relative single-particle contributions that are almost an order of magnitude lower for Pt (see Section 3.2).

To obtain a manageable number of valence electrons for heavy elements like platinum (and copper) a suitable pseudo-potential must be used. For Pt, we use a $Z=60$ pseudo-potential, leaving 18 valence electrons/atom.[4] Both the limitation of the trial wave function to a ground-state single determinant and the use of pseudo-potentials introduce deficiencies in its nodal structure. A novel complex Jastrow factor is used to remedy this as much as possible (Section 3.1).[8] Then, the fixed phase approximation preserves fermionic antisymmetry in DMC, i.e., configurations are distributed according to a density formed as the product of the trial wave function and its update. The source of error due to poor initial nodes is projected out by using a complex Jastrow factor to optimize the trial wave function at the variational Monte Carlo (VMC) stage. The wave function is complex for these slabs without inversion symmetry.

3. QMC METHODS

Two QMC methods are used which are detailed elsewhere.[9] Briefly, the first so-called VMC minimizes an observable, typically wave function variance, or total energy. The accurate diffusion Monte Carlo approach then uses a propagator method to solve the time-dependent Schrödinger equation of the system, cast as a diffusion–drift equation in imaginary time, which is then increased in small steps like a time variable.

3.1 Complex Jastrow Factors: Testing Against Real Factors

This complex Jastrow factor is a set of polynomials in the interparticle variables. It contains two independent parameter sets for its real and imaginary parts. This makes the trial wave function very flexible.

A VMC optimization of free-parameter Jastrow coefficients is first carried out. These polynomials in the two-particle and also three-particle (electron pair and nucleus) separation variables are taken to order 6–9 and coefficients treated as variational parameters. Energy minimization is used for fine-tuning following an initial variance minimization. Cut-off (range) parameters can be fixed after the variance minimization.

Results at the VMC level for the TS and asymptote geometries are compared. Total energy and variance values for the whole systems in the 2 by 2 by 5 mesh slab are tabulated in Table 1. This slab comprises 80 Pt–atoms and

Table 1 Total Energy and Variance in Atomic Units

Structure	VMC E_{tot} (c)	Var (c)	VMC E_{tot} (r)	Var (r)
ASY	−409.668	25.49	−409.297	30.12
TS	−409.479	27.66	−409.092	33.60

4 hydrogen molecules (for computational reasons, see Ref. 2). It contains 1448 electrons.

Table 1 reports energy minimization for the hydrogen and Pt(111) system by VMC.

The (c) denotes optimization with a complex Jastrow factor of order 6 and free spin state, to be used as a factor of the complex wave function, i.e., complex arithmetic is required. By (r) we denote optimization of two real Jastrow factors (of order 6 and free spin state), one for the complex wave function real components and the other for its imaginary components.

Note that, at the VMC level, minimum energies and variance of the TS structure show a more substantial improvement from use of the complex Jastrow factor. This is ascribed to intrinsic multideterminant character of the wave function of the TS, which would obviously be better represented by multiple determinant input. The complex Jastrow permits a partial cure of the limitations related to using single determinant (ground state) input by improving the nodes. The poor wave function quality is not so pronounced for the asymptote and the influence of the complex Jastrow factor is rather more modest.

Our DMC reaction barrier was evaluated using real Jastrows for comparison. The barrier height is 0.254 eV with standard error 0.12 eV when our result is 0.23 eV with standard error 0.042 eV.

3.2 Time-Step Bias, Other DMC Error Sources, and Specifications

The step-size (τ) chosen is 0.005 a.u. and was checked for possible bias by comparing three equivalent runs for τ, $\tau/2$, and 2τ. The energies extrapolate linearly to zero time-step falling within less than 1 meV, therefore, the chosen step-size will not lead to time-step bias.

The DMC algorithm used since February 2016 is less sensitive to time step.[10] This algorithm also recovers well from mediocre trial wave functions. The resulting DMC energies show little fluctuation for our chosen time step and do not become subvariational for nonoptimal Jastrow factors (but see earlier work for such adverse impact of suboptimal Jastrow factors).[5]

The software used for all calculations is CASINO[9] (running on 1024 cores) and 60,000 data points were collected for each geometry (for which 12 twists were used).[9] The bridging TS geometry is directly optimized within VMC. It is compared to an asymptote with the hydrogen molecules 8 Å from the surface. The clean metal surface has been shown to be a poor reference, due to the surface states and the fact that physisorped atoms and molecules have a symmetry-breaking role on these states, as shown in our previous work.[7]

Solid state QMC is currently coming of age and a number of metal–surface systems with interacting molecules have been studied since the activation barrier published by Pozzo and Alfè in 2008, for the H_2 molecule dissociating on Mg(0001).[6] This system does not yet have accurate (molecular beam experiment) measurements.

Applications to catalysis and many electronic devices require surfaces of transition metals. For (almost) all systems of interest, all-electron calculations are prohibitive in terms of cpu time. Pseudo-potentials are used. With this in mind, ensuing (statistical and also systematic) error was meticulously evaluated.[5] Those due to wave function and pseudo-potential (in particular non-locality, often neglected but significant when 3d electrons are studied) were quantified for transition metal binary compounds, involving first transition series elements.[5] The insight gained points to strategy for the improvement of metal slab design.[2] Nonlocality error and errors compounding the fixed-phase error are the most difficult to circumvent. The work on transition metal binary compounds and the subsequent benchmark for hydrogen dissociation on Cu(111)[2,5] use Casula T-moves. This is a modified DMC algorithm that guarantees the variation principle. In this work, we found little advantage in using T-moves.

Error assessment of platinum surfaces for adsorped molecules shows low trial wave function variance data. Problems encountered with elements involving partly or totally full 3d shells in Ref. 2 are less acute for Pt. Single-particle finite-size effect corrections are accounted for by twist averaging with the strategy developed in Ref. 1. This procedure will not be detailed in the present letter and accounts for a minor effect of 1% on the barrier. Additional corrections by the model periodic Coulomb method (described in Ref. 2) amount to less than 1/4.

4. PERSPECTIVES AND CONCLUSIONS

This QMC evaluation of the hydrogen dissociation barrier on Pt(111) in a bridge geometry has standard error within chemical accuracy (under

0.99 kcal/mol) and agrees with the SRP-DFT work[3] (note the QMC, achieves chemical accuracy, i.e., is within 1 kcal/mol of experiment). The functional used contains a parameter fitted to experiment and gives the measured barrier of 6.2 kcal/mol. The barrier of 5.4 kcal/mol from QMC is within standard error of experiment. Note the QMC barrier for the hydrogen dissociation at Cu(111) in Ref. 2 also falls below accurate molecular beam measurements for it.

The quality of trial wave functions in this work shows the advantage of projecting toward exact nodes using the complex Jastrow factor developed in Ref. 8. Further work for CI will be used in our embedding approach to the active site for catalysis. The VMC study summarized in Table 1 confirms this. From variance values, which are lower for Pt(111) than for Cu(111) we conclude that much of the pseudo-potential locality issue is specific to 3d electrons.

Bond dissociation at metal catalyst surfaces is often involved in industrial synthesis of chemicals. This work describes a prototype for future QMC benchmarks on the Pt(111) industrial catalysts and synthesis reactions.

ACKNOWLEDGMENTS

The author would like to thank the Regional Computing Facility (CRRI) on campus at Clermont Auvergne University for generous resources in calculation time. Thanks also go to my numerical analysts Laura and Marie Hoggan.

REFERENCES

1. Hoggan, P. E. *Quantum Monte Carlo Activation Barrier for Hydrogen Dissociation on Copper to Unprecedented Accuracy*; 2015. (submitted to ArXiv Mat. Cond. on 24 November 2015).
2. Doblhoff-Dier, K.; Meyer, J.; Hoggan, P. E.; Kroes, G. J. Quantum Monte Carlo Calculations on a Benchmark Molecule–Metal Surface Reaction: H2 + Cu(111). *J. Chem. Theory Comput.* **2017**, *13*(7), 3208–3219. https://doi.org/10.1021/acs.jctc.7b00344.
3. Ghassemi, E. N.; Wijzenbroek, M.; Somers, M. F.; Kroes, G. J. Chemically Accurate Simulation of Dissociative Chemisorption of D2 on Pt(111). *Chem. Phys. Lett.* **2017**, *683*, 329–335.
4. Absi, N.; Hoggan, P. E. Quantum Monte Carlo Investigation of Two Catalytic Reaction Paths for Hydrogen Synthesis on Pt(111). In: *Recent Progress in Quantum Monte Carlo*; ACS Symposium Series, Vol. 1234, 2016; pp 77–88. Chapter 5, Publication date: 1 December 2016 (web).
5. Doblhoff-Dier, K.; Meyer, J.; Hoggan, P. E.; Kroes, G. J.; Wagner, L. K. Diffusion Monte Carlo for Accurate Dissociation Energies of 3d Transition Metal Containing Molecules. *J. Chem. Theory Comput.* **2016**, *12*, 2583–2597. https://doi.org/10.1021/acs.jctc.6b00160. Publication date: 13 May 2016 (web).
6. Pozzo, M.; Alfè, D. Hydrogen Dissociation on Mg(0001) Studied Via Quantum Monte Carlo Calculations. *Phys. Rev. B* **2008**, *78*, 245313.

7. Hoggan, P. E.; Bouferguene, A. Relative Advantages of Quantum Monte Carlo Simulation for Changing Electron Correlation: CO Reactions on Copper and Platinum Catalysts. *Adv. Quantum Chem.* **2014**, *68*, 89–103.
8. Changlani, H. J.; Kinder, J. M.; Umrigar, C. J.; Chan, G. K. L. Approximating Strongly Correlated Wave Functions With Correlator Product States. *Phys. Rev. B.* **2009**, *80*, 245116.
9. Needs, R. J.; Towler, M. D.; Drummond, N. D.; Lopez Rios, P. Continuum Variational and Diffusion Quantum Monte Carlo Calculations. *J. Phys. Condens. Matter* **2010**, *22*, 023201. Casino website: www.tcm.phy.cam.ac.uk/mdt26/casino2.html.
10. Zen, A.; Sorella, S.; Gillan, M. J.; Michaelides, A.; Alf, D. Boosting the Accuracy and Speed of Quantum Monte Carlo: Size Consistency and Time Step. *Phys. Rev. B* **2016**, *93*, 241118(R). (R = Rapid Communications).

CHAPTER SIXTEEN

Density-Based Analysis of Spin-Resolved MP2 Method

Mateusz Witkowski, Szymon Śmiga[1], Ireneusz Grabowski
Institute of Physics, Faculty of Physics, Astronomy and Informatics, Nicolaus Copernicus University, Torun, Poland
[1]Corresponding author: e-mail address: szsmiga@fizyka.umk.pl

Contents

1. Introduction 279
2. Computational Details 281
3. Methodology and Results 282
   3.1 Proportionality of the Spin-Resolved MP2 Densities 282
   3.2 Two-Dimensional Scan 284
   3.3 Dipole Moments 288
4. Final Remarks 288
Acknowledgments 290
References 290

Abstract

The extensive study of the spin-resolved second-order Møller–Plesset method in the context of the electron density is performed. It was found the well-defined proportionality of the same- and opposite-spin parts of the MP2 correlated electronic density. We have rationalized the value of the scaling parameter used in the foundation of the SOS-MP2 (Jung et al., 2004) method from the density point of view. Our analysis is complemented by the calculations of the dipole moments using differently parameterized spin-resolved MP2 methods.

1. INTRODUCTION

Since 2003, when Grimme proposed the spin-component-scaled MP2 method,[1] a rapid development of spin-resolved (SR) methods is observed in the field of quantum chemistry. Many useful variants have been proposed, analyzed, and implemented[2] in different levels of approximations and different application areas.[3–7] However, those based on the second-order Møller–Plesset

Advances in Quantum Chemistry, Volume 76
ISSN 0065-3276
https://doi.org/10.1016/bs.aiq.2017.05.004

perturbation theory (MP2)[8] energy expression are the most interesting ones from point of view of a simplicity and computational cost.

Rescaling of the same-spin (SS) and opposite-spin (OS) part of the MP2 correlation energy proposed in Ref. 1 leads to a spin-resolved MP2 method (SR-MP2)

$$E_c^{\text{SR-MP2}} = c_{\text{OS}} E^{\text{MP2(OS)}} + c_{\text{SS}} E^{\text{MP2(SS)}} \tag{1}$$

with

$$E_c^{\text{MP2(OS)}} = \frac{1}{2} \sum_{\sigma, \tau \neq \sigma} \sum_{ijab} \frac{|(i_\sigma a_\sigma | j_\tau b_\tau)|^2}{\epsilon_{i\sigma} + \epsilon_{j\tau} - \epsilon_{a\sigma} - \epsilon_{b\tau}} \tag{2}$$

$$\begin{aligned} E_c^{\text{MP2(SS)}} = & \frac{1}{2} \sum_{\sigma} \sum_{ijab} \frac{|(i_\sigma a_\sigma | j_\sigma b_\sigma)|^2}{\epsilon_{i\sigma} + \epsilon_{j\sigma} - \epsilon_{a\sigma} - \epsilon_{b\sigma}} \\ & - \frac{1}{2} \sum_{\sigma} \sum_{ijab} \frac{(i_\sigma a_\sigma | j_\sigma b_\sigma)(a_\sigma j_\sigma | b_\sigma i_\sigma)}{\epsilon_{i\sigma} + \epsilon_{j\sigma} - \epsilon_{a\sigma} - \epsilon_{b\sigma}}, \end{aligned} \tag{3}$$

where c_{OS} and c_{SS} are the OS and SS coefficients, respectively. In the above a, b, and i, j label the virtual and occupied indexes, respectively, while σ and τ label the spin indexes. Such simple modification of the MP2 energy expression results in significant improvement in the accuracy of calculated properties which, in some cases, can be compared to those obtained from a more advanced methods such as the coupled-cluster single double and perturbative triple [CCSD(T)][9] approach. At the same time the SR-MP2 method retains the computational simplicity of its original counterpart as well as other important features like size-consistency and orbital-rotational invariance.

In the initial work[1] the $c_{\text{OS}} = 6/5$ and $c_{\text{SS}} = 1/3$ scaling parameters have been optimized with respect to a training set of reaction energies. Defined in this manner the "spin-component scaled" MP2 method (SCS-MP2) gives rather good improvement in the quality of atomization energy, barrier heights,[1,10] heats of formation,[11] or molecular geometries.[1,10,12] Another interesting variant of SR-MP2 method is the one, proposed by Jung et al. dubbed "scaled-opposite spin" MP2 (SOS-MP2).[10] In this simplified method, the SS contribution to the correlation energy in Eq. (1) is modeled by proper selection of the $c_{\text{OS}} = 1.3$ parameter, while $c_{\text{SS}} = 0$. In general, SOS-MP2 leads to a similar quality of results as the SCS-MP2 method. However, due to neglect of the SS term in Eq. (1), computational efficiency increases significantly, leading to reduction of the numerical cost from $\mathcal{O}(N^5)$ to $\mathcal{O}(N^4)$ or even less.[13,14] Many other useful parameterizations of Eq. (1) have been proposed, based on semiempirical[7,15–18] as well as theoretical considerations.[19,20] Furthermore, the spin-resolved concept was also applied in the context of the density functional theory (DFT) and ab initio DFT,[21,22] where Eq. (1) was utilized as the correlation energy functional in the double-hybrid[23–27] and second-order optimized effective potential (OEP) method,[28–31] respectively. The latter studies[28,30] are especially interesting because they have revealed that their exists almost linear proportionality between the OS and SS part of the correlation energies, Kohn–Sham correlation potentials and correlated electronic densities (although, the proportionality factors are different for different quantities). This suggests that similar proportionality of densities may also exist in the SR-MP2 method. To date, however, such an analysis was not carried out. Thus to fill this gap, in this chapter we devote special attention to the density analysis of spin-resolved methods.

The work is organized as follows. In Section 3.1 we discuss the proportionality of the spin-resolved densities and its implications. Next, in Section 3.2 we extend this analysis by presenting two-dimensional scans of the SR-MP2 densities with respect to the CCSD(T). Finally, in Section 3.3 we investigate the impact of these results on the dipole moments which are especially sensitive to the quality of the electron density.

2. COMPUTATIONAL DETAILS

The main quantity used in our analysis is the electron density calculated obtained from the MP2 and CCSD(T) methods. The densities are calculated from the relaxed density matrices[32–34] constructed using the Lagrangian approach.[35–37] In the case of SR-MP2 method the SS and OS parts of correlated densities have been obtained by a scaling of adequate parts in the Lagrangian.

Calculations have been carried out in NWChem software package.[38] The basis sets employed are the spherical cc-pVXZ and aug-cc-pVXZ basis sets, where X=(D,T,Q).[39–42] All electrons were included in the calculation of correlated quantities. The calculations were performed for several atomic: Ne, Ar, Kr, Zn, and molecular systems, i.e., Cl_2, CS, F_2, He_2, N_2, OCS, P_2, and those listed in Table 1. For the latter ones, the experimental geometries were taken from Ref. 43. As in our previous studies[16,17,28,30] the CCSD(T)[9] results were used as the reference benchmark data.

3. METHODOLOGY AND RESULTS

In this section we discuss several aspects related to the SR-MP2 method in the context of electron density. In particular we analyze the proportionality of the spin-resolved densities by examining its relative parts and the possibility of reconstruction of the CCSD(T) density by proper scaling of the SR-MP2 density. Due to large qualitative and quantitative resemblance of densities obtained within all basis sets, for the sake of clarity of the discussion, we report only those obtained with aug-cc-pVQZ basis set.

3.1 Proportionality of the Spin-Resolved MP2 Densities

In order to demonstrate all features of second-order spin-resolved densities we utilize the concept of difference total densities (DTD) distribution, defined with respect to the total Hartree–Fock (HF) density [$\rho^{\mathrm{HF}}(\mathbf{r})$]

$$\rho_c^{\mathrm{method}}(\mathbf{r}) = \rho^{\mathrm{method}}(\mathbf{r}) - \rho^{\mathrm{HF}}(\mathbf{r}), \quad (4)$$

where $\rho^{\mathrm{method}}(\mathbf{r})$ is the total density gained from correlated (e.g., MP2 or CCSD(T)) calculations. The DTD was successfully used as a testing tool in the development of new methods, especially in the context of DFT.[28,44–48] We note that in case of atoms, the $\rho_c(\mathbf{r})$ is closely related to another quantity, namely the difference radial-density (DRD) distributions,[44] defined as

$$\mathrm{DRD}^{\mathrm{method}}(r) = 4\pi r^2 \rho_c^{\mathrm{method}}(r). \quad (5)$$

Thus, the latter one is used for atoms. For the clarity of discussion, henceforth we will refer to both quantities as correlated densities.

In Fig. 1 we present the SS and OS parts of MP2 correlated densities obtained for Ne and Ar atoms and for the CO molecule. Similar results have been observed in the case of other systems (not reported here). One can immediately observe the clear point-by-point proportionality of the spin-resolved parts of the MP2 correlated densities. All minima and maxima positions are almost ideally aligned with each other, even in the molecular case. This in turn suggests that SS correlation effects can be effectively modeled by properly rescaling the OS contributions only, which is more favorable from the computational point of view. We remark, however, that the relative density scaling factor γ which links SS and OS correlated densities

$$\rho_c^{\mathrm{OS}}(\mathbf{r}) = \gamma \rho_c^{\mathrm{SS}}(\mathbf{r}) \quad (6)$$

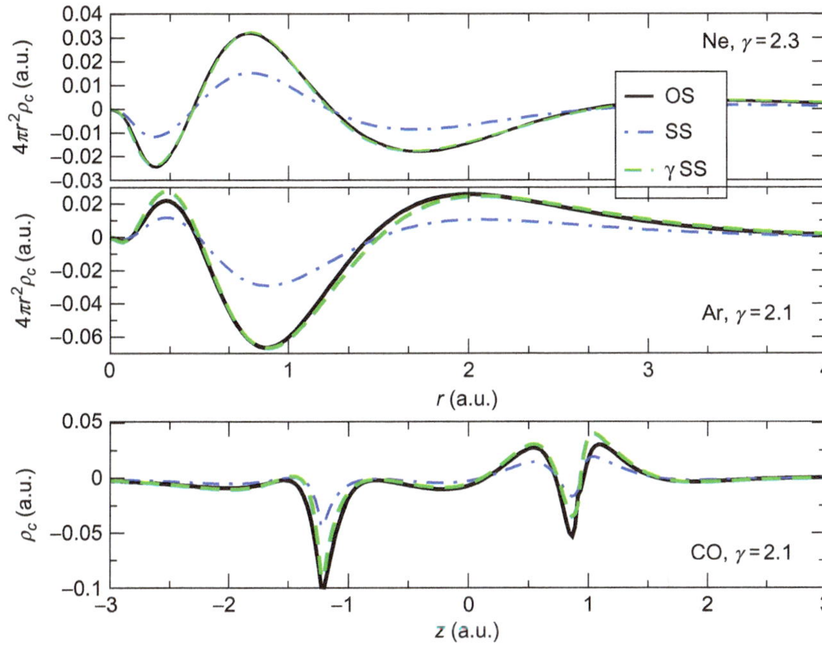

Fig. 1 Correlated density distributions for Ne (*top*), Ar (*middle*), and CO (*bottom*, plot along the molecular axis) calculated for SS and OS parts of MP2 density. On each plot, the value of γ factor rescaling SS correlated density has also been shown. See the text for details.

is not a general constant. It slightly depends on the choice of basis set as well as the molecular system itself. For example, in the case of the CO molecule for the aug-cc-pVDZ and aug-cc-pVQZ, γ changes from 2.3 to 2.1, respectively. Nevertheless, in general, for most systems and basis sets, the proportionality parameter ranges between the 1.7 and 2.5. For the systems presented in Fig. 1, the γ values are $\gamma=2.3$ for Ne and $\gamma=2.1$ for Ar and CO. We underline here that proportionality of resolved density is the general feature observed in all considered systems in our set, which is the first important finding of this study. Furthermore, above results confirm the previous findings,[2,16–18] based mostly on the analysis of SS and OS parts of the SR-MP2 correlation energy expression. The density-based analysis presented here is more demanding than those based on the correlation energy, because we scale not just one number (like in the correlation energy case) but the scaling holds for the whole space where the electron density is calculated. Hence, this kind of analysis can be crucial for understanding and rationalizing the behavior of spin-resolved-based methods.

3.2 Two-Dimensional Scan

In the previous section we have shown that SS and OS parts of the MP2 correlated densities display a very high degree of proportionality, and then that proper scaling of one term (e.g., the SS one) can reproduce general spatial behavior of another one (OS correlated density), with relatively good accuracy. In order to better understand and make possible use of this issue in its practical application, we perform more quantitative analysis. To this end, we conducted a full two-dimensional scan of the space spanned by the c_{OS} and c_{SS} parameters, analyzing, in each point of space the error indicator function (the integrated density differences—IDD[30]) defined as

$$\mathrm{IDD}(c_{\mathrm{OS}}, c_{\mathrm{SS}}) = \int d\mathbf{r} \left| \rho_c^{\mathrm{SR-MP2}}(\mathbf{r};\, c_{\mathrm{OS}}, c_{\mathrm{SS}}) - \rho_c^{\mathrm{CCSD(T)}}(\mathbf{r}) \right|. \qquad (7)$$

where $\rho_c^{\mathrm{CCSD(T)}}$ is the correlated CCSD(T) density. The first term in Eq. (7) is the correlated spin-resolved density calculated for the fixed pair of scaling coefficients as

$$\rho_c^{\mathrm{SR-MP2}}(\mathbf{r}; c_{\mathrm{OS}}, c_{\mathrm{SS}}) = c_{\mathrm{OS}}\ \rho_c^{\mathrm{OS-MP2}}(\mathbf{r}) + c_{\mathrm{SS}}\ \rho_c^{\mathrm{SS-MP2}}(\mathbf{r}). \qquad (8)$$

The c_{OS} and c_{SS} coefficients have been varied in the range between 0.0 and 2.0 with the 0.1 interval. The integration in Eq. (7) runs over whole three-dimensional space on a very accurate grid of spatial points.

One can note that the indicator defined by Eq. (7) tells explicitly how accurately the SR-MP2 method can reproduce the CCSD(T) reference density, i.e., we obtain ($\rho_c^{\mathrm{CCSD(T)}} \approx \rho_c^{\mathrm{SR-MP2}}$), for a fixed pair of coefficients. It also shows implicitly, the global qualitative relative proportionality of OS and SS parts of MP2 correlated densities. As before the scans have been done for all investigated systems. However, in Fig. 2, we report only the scan obtained for a CO molecular system which presents general trend observed in our results. Inspection of the figure confirms the observations made in Section 3.1, i.e., the proportionality of the OS and SS parts of the MP2 correlated densities. Furthermore we see that the SR-MP2 method can reproduce the reference CCSD(T) densities with quite good accuracy. We note that similar findings have been observe in Ref. 28, however, in the context of ab initio DFT. Nonetheless, several other features may be observed

(a) The IDD indicator shows the minima for a continuous set of parameter pairs, which can be obtained by a simple linear relation $c_{\mathrm{OS}} = \alpha c_{\mathrm{SS}} + \beta$, with α and β being system-dependent constants. This relation indicates the evident proportionality of the spin-resolved densities.

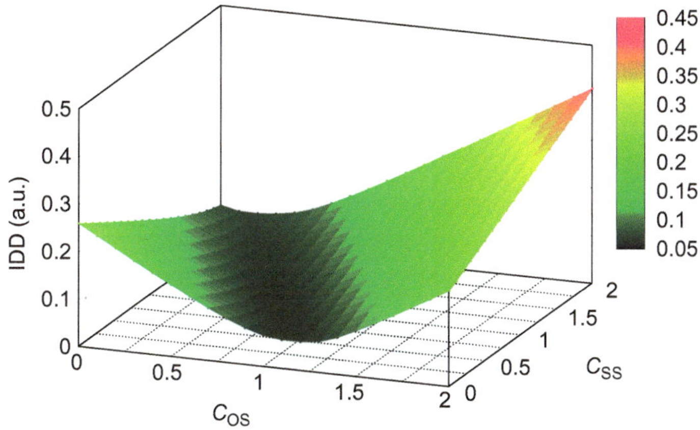

Fig. 2 The IDD scan in the function of c_{OS} and c_{SS} parameters for CO molecule.

(b) The performance of standard SCS-MP2 and SOS-MP2 methods, which yield almost identical IDD's, i.e., 0.07 and 0.065, respectively, is better than in the MP2 case where IDD = 0.09. The global minimum, with lower IDD = 0.06, is found for $c_{OS} = 1.1$ and $c_{SS} = 0.1$ pair of coefficients.

(c) The absolute minimum value of IDD is nonzero. This is due to the fact that the IDD is an integrated quantity which describes how correlation effects are reproduced in the density for all points in the space. This, in turn, means that the reference CCSD(T) density cannot be reproduced by a simple scaling of second-order spin-resolved densities. This may lead to the conclusion that wherever the MP2 method fails badly (e.g., in systems with multireference character) also the simple rescaled methods will lead to similar predictions.

(d) For the case where SS part of the density is neglected ($c_{SS} = 0$), it is always possible to find a value of the c_{OS} parameter such that the IDD error is minimized and its value is close to the global absolute minimum. This is a direct consequence of the proportionality found here and described in Section 3.1. In fact, the SS compensation term is proportional to the inverse of γ (present in Eq. (6)) and gives

$$\rho_c^{\mathrm{CCSD(T)}} \approx (c_{\mathrm{OS}} + \gamma^{-1} c_{\mathrm{SS}}) \rho_c^{\mathrm{OS-MP2}}(\mathbf{r}). \tag{9}$$

This, in turn, justifies assumed proportionality used in the foundation of the SOS-MP2 method. According to SOS-MP2 definition,[10] in order to compensate the lack of SS part of the correlation, one needs to increment the OS scaling factor of SCS-MP2[1] method by about 1/9. This means that the proportionality coefficient γ for SOS-MP2 method equals $10/3 \approx 3.3$ which is slightly larger than those obtained in the present work ($\gamma = 1.7 \div 2.5$).

Due to the latter feature we have also performed the one-dimensional scans along the c_{OS} axis with $c_{SS} = 0$. The plots obtained for Ne, Ar, and CO are presented in Fig. 3. Note that all plots have a parabolic profile with the clear minimum (indicated by blue-dotted vertical line) obtained at $c_{OS}^{min} = 1.22$ for Ne, $c_{OS}^{min} = 1.48$ for Ar and $c_{OS}^{min} = 1.14$ for CO. Similar results (not reported) have been obtained for the rest of the systems with the minima spreads in the range between $1.0 \div 1.5$. This feature shows evidently that the c_{OS} parameter for which the minimum IDD value is obtained (c_{OS}^{min}) is strictly system-dependent quantity. This, in turn, leads to the conclusion that the universal value of the parameter, which works with the same accuracy for all systems cannot be found. Thus, as in the SOS-MP2 method, it seems reasonable to consider an average optimal c_{OS}^{opt} parameter which in our case is

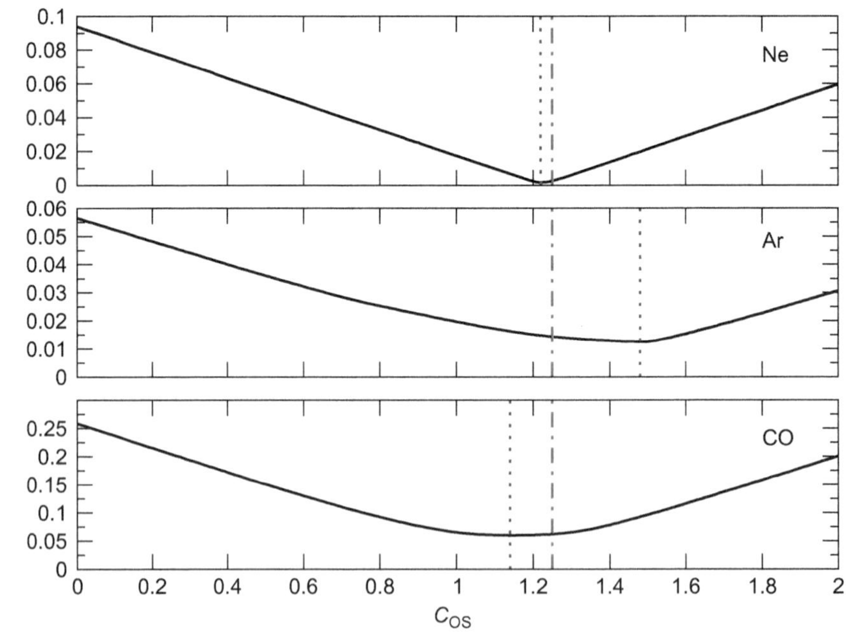

Fig. 3 The IDD scans along the c_{OS} axis ($c_{SS} = 0$) for Ne (*top*), Ar (*middle*), and CO (*bottom*). *Blue* (*dotted*) *line* indicates the minimum value of IDD error (Ne—$c_{OS}^{min} = 1.22$, Ar—$c_{OS}^{min} = 1.48$, CO—$c_{OS}^{min} = 1.14$) of the IDD error. *Red* (*dash-dotted*) *line* indicates the average optimal value of $c_{OS}^{opt} = 1.25$.

obtained as an overall average value of all $c_{\mathrm{OS}}^{\mathrm{min}}$ parameters minimizing IDD error, what gives $c_{\mathrm{OS}}^{\mathrm{opt}}=1.25$ (indicated in Fig. 3 by dash-dotted red line). Interestingly, this value is relatively close to the one proposed in SOS-MP2 method (i.e., $c_{\mathrm{OS}}=1.3$).[10] However, the correlation energies calculated with the $c_{\mathrm{OS}}^{\mathrm{opt}}=1.25$ coefficient (optimized for the correlated densities) are slightly worse than those obtained with the $c_{\mathrm{OS}}=1.3$ one, which was optimized to reproduce the CCSD(T) correlation energies. For the systems considered in this chapter, the MAE errors of correlation energies with respect to CCSD(T) data are 0.037 for SOS-MP2 and 0.054 for the SR-MP2 with $c_{\mathrm{OS}}^{\mathrm{opt}}=1.25$.

To conclude this section, in Fig. 4 we report the correlated densities calculated with optimal $c_{\mathrm{OS}}^{\mathrm{opt}}$ and minimizing $c_{\mathrm{OS}}^{\mathrm{min}}$ parameter compared to the reference CCSD(T) ones. Again, we see that for the $c_{\mathrm{OS}}^{\mathrm{min}}$ one can reproduce the reference CCSD(T) data with a very good accuracy. The correlated densities calculated using the $c_{\mathrm{OS}}^{\mathrm{opt}}$ parameter also perform very well, being slightly worse than in the case of $c_{\mathrm{OS}}^{\mathrm{min}}$ value. The SOS-MP2 correlated densities (not reported in Fig. 4) are in line with our optimal ones, which is not surprising due to relatively small difference in the c_{OS} parameter value.

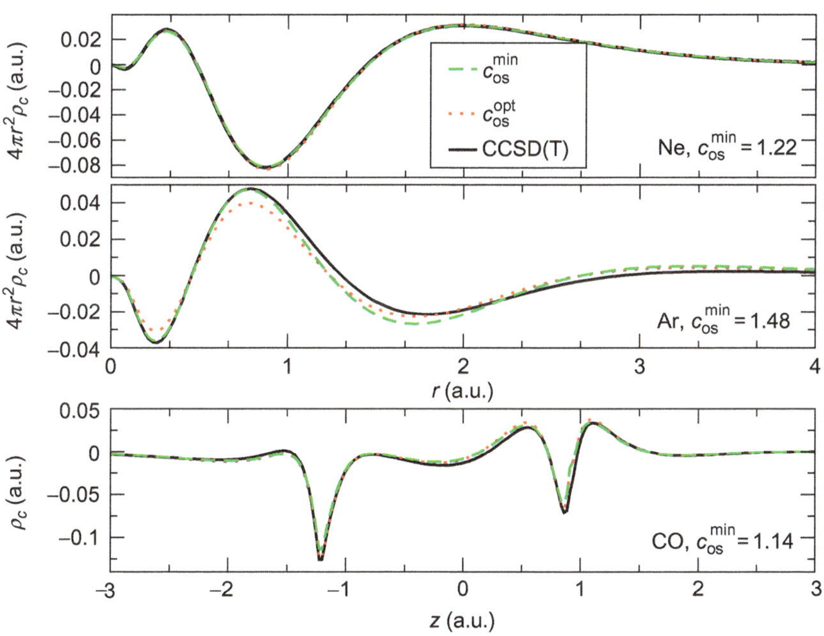

Fig. 4 The correlated densities calculated for OS-MP2 density with $c_{\mathrm{OS}}^{\mathrm{min}}$, $c_{\mathrm{OS}}^{\mathrm{opt}}=1.25$ scaling parameters for Ne (*top*), Ar (*middle*), and CO (*bottom*). For comparison the CCSD(T) correlated densities are also plotted.

3.3 Dipole Moments

To link the results of the density analysis conducted in the previous sections, with a more tangible quantity we present in Table 1 the assessment of the dipole moments calculated using different SR-MP2 methods (SCS-MP2, SOS-MP2, and with the c_{OS}^{opt} coefficient). The dipole moment is an important quantity characterizing the charge distribution of a molecular electronic system being directly related to the electronic density. Thus, it can be also used as a kind of measure of the quality of electron density.

For comparison, several other standard methods have been considered, i.e., HF and MP2. To assess the quality of the results, we report in Table 1 the root mean square errors (RMS), the mean absolute errors (MAE) and finally the mean absolute relative errors (MARE) calculated with respect to the CCSD(T) reference data. Note that only systems with nonzero dipoles are considered and the aug-cc-pVTZ basis set was used in the calculations.

At first glance one can note that all SR-MP2 methods show a reduction of all types of error compared to the standard MP2. The results obtained with optimized c_{OS} parameter are comparable with those from the SOS-MP2 method, yielding MAE and MARE of 0.26, 4.1% and 0.25, 3.6%, respectively. Slightly larger errors are observed for the SCS-MP2 method which yields MAE of 0.035 and MARE of 6.7%. Nevertheless, we note that the performance of all SR-MP2 methods is good, being roughly twice better than the standard MP2.

It must be noted, that in the case of NNO molecule the MP2 method predicts a qualitatively incorrect dipole moment −0.052 D compared with +0.115 D for CCSD(T) (the experimental value is +0.167 D[43]). The error is reduced in all SR-MP2 methods for which the correct sign and quantitatively accurate predictions of dipole moment are obtained.

The errors obtained for the different SR-MP2 methods are minor, which can be a result of the pretty flat minimum in the density scan (see Fig. 3). So, all the parameterizations used in SCS-MP2, SOS-MP2, and with our c_{OS}^{opt} parameter seem to work similarly and give the dipole moments of almost the same quality. Certainly because of the lower computational cost of the opposite part of the MP2 method, the one parameter variant (i.e., with $c_{SS} = 0$) should be used in practical calculations.

4. FINAL REMARKS

The density-based analysis presented in this work indicated several aspects related to SR-MP2 methods. First of all, we have shown the evident

Table 1 Dipole Moments (in Debye) Obtained for Several Spin-Resolved Methods

	CCSD(T)	HF	MP2	SR-MP2 SCS-MP2	SOS-MP2	This Work[a]
CH_2CF_2	1.416	1.559	1.352	1.398	1.420	1.426
ClF	0.893	1.114	0.893	0.909	0.917	0.925
CH_3CN	4.517	4.918	4.533	4.513	4.503	4.519
H_2S	0.973	1.091	0.999	0.988	0.982	0.986
HCOOH	1.379	1.641	1.351	1.358	1.363	1.373
HNO	1.615	1.945	1.617	1.618	1.621	1.634
HNO_3	2.194	2.558	2.122	2.138	2.148	2.164
CHF_3	1.466	1.635	1.473	1.480	1.483	1.489
SiO	3.048	3.713	2.955	3.014	3.042	3.068
SO	1.372	1.763	1.312	1.339	1.352	1.368
SO_2	1.653	2.026	1.571	1.614	1.635	1.650
N_2O_3	2.440	3.566	2.378	2.408	2.426	2.470
$SOCl_2$	1.493	1.943	1.411	1.457	1.492	1.509
FCN	2.191	2.333	2.226	2.195	2.180	2.186
H_2CO	2.360	2.841	2.355	2.372	2.380	2.398
H_2CS	1.674	2.165	1.632	1.642	1.646	1.666
H_2O	1.838	1.980	1.851	1.851	1.850	1.855
HCCF	0.713	0.895	0.683	0.727	0.748	0.754
HCl	1.089	1.209	1.120	1.107	1.101	1.105
HCN	3.007	3.293	3.030	3.029	3.028	3.038
HF	1.803	1.937	1.814	1.811	1.809	1.814
NH_3	1.507	1.611	1.521	1.524	1.525	1.528
CO	0.138	−0.254	0.287	0.247	0.228	0.209
NNO	0.115	0.618	−0.052	0.052	0.103	0.123
	Relative to CCSD(T)	RMS[b]	0.063	0.035	0.026	0.025
		MAE[c]	0.046	0.026	0.019	0.020
		MARE[d] [%]	12.6	6.7	4.1	3.6

[a]SOS-MP2 method with $c_{OS}^{opt} = 1.25$.
[b]Root mean square error.
[c]Mean absolute error.
[d]Mean absolute relative error.
The HF and CCSD(T) data are also included for comparison. The calculations have been performed in aug-cc-pVTZ basis set.

proportionality of the SS and OS parts of the MP2 correlated densities. This physical feature is important because it provides a rationale for the well-known SOS-MP2 method, which up to now have been based mainly on the empirical observations focusing only on the correlation energy. Moreover, we have shown that the c_{OS} is a system-dependent parameter which optimized for particular system leads to the significant improvement of the representation of correlation effects in density. This means that all optimal-parameter-based SR-MP2 methods (e.g., SCS-MP2, SOS-MP2) benefit rather from the statistical error cancelation than the actual better representation of correlation effects. Nonetheless, despite this small drawback, the spin-resolved methods can lead to relative good increase of accuracy of the results. Furthermore, due to relatively small numerical cost (especially the SOS variant), the method can be applied to relatively large molecular systems.

In conclusion, the analysis provides a valuable insight for practical and deeper understanding of the SR-MP2 methods what can be essential for the further development of SR-MP2 type approximation for different applications (e.g., noncovalent interaction as in Refs. 16 and 17). Additional work can be foreseen. For example, a similar type of analysis can be performed for other spin-resolved methods, e.g., SCS-CCSD.[3]

ACKNOWLEDGMENTS

This work was partially supported by the National Science Center under Grant No. DEC-2016/21/D/ST4/00903 and DEC-2013/11/B/ST4/00771.

Conflict of interest: The authors declare no competing financial interest.

REFERENCES

1. Grimme, S. Improved Second-Order Møller-Plesset Perturbation Theory by Separate Scaling of Parallel- and Antiparallel-Spin Pair Correlation Energies. *J. Chem. Phys.* **2003**, *118*(20), 9095–9102. https://doi.org/10.1063/1.1569242.
2. Grimme, S.; Goerigk, L.; Fink, R. F. Spin-Component-Scaled Electron Correlation Methods. *WIREs Comput. Mol. Sci.* **2012**, *2*(6), 886–906. https://doi.org/10.1002/wcms.1110.
3. Takatani, T.; Hohenstein, E. G.; Sherrill, C. D. Improvement of the Coupled-Cluster Singles and Doubles Method Via Scaling Same- and Opposite-Spin Components of the Double Excitation Correlation Energy. *J. Chem. Phys.* **2008**, *128*(12), 124111. https://doi.org/10.1063/1.2883974.
4. Hellweg, A.; Grun, S. A.; Hattig, C. Benchmarking the Performance of Spin-Component Scaled CC2 in Ground and Electronically Excited States. *Phys. Chem. Chem. Phys.* **2008**, *10*, 4119–4127. https://doi.org/10.1039/B803727B.
5. Pitonak, M.; Rezac, J.; Hobza, P. Spin-Component Scaled Coupled-Clusters Singles and Doubles Optimized Towards Calculation of Noncovalent Interactions. *Phys. Chem. Chem. Phys.* **2010**, *12*, 9611–9614. https://doi.org/10.1039/C0CP00158A.

6. Lochan, R. C.; Jung, Y.; Head-Gordon, M. Scaled Opposite Spin Second Order Møller-Plesset Theory With Improved Physical Description of Long-Range Dispersion Interactions. *J. Phys. Chem. A* **2005**, *109*(33), 7598–7605. https://doi.org/10.1021/jp0514426. PMID: 16834130.
7. Distasio, R. A., Jr.; Head-Gordon, M. Optimized Spin-Component Scaled Second-Order Møller-Plesset Perturbation Theory for Intermolecular Interaction Energies. *Mol. Phys.* **2007**, *105*(8), 1073–1083. https://doi.org/10.1080/00268970701283781.
8. Møller, C.; Plesset, M. S. Note on an Approximate Treatment for Many-Electron Systems. *Phys. Rev.* **1934**, *36*, 618–622.
9. Raghavachari, K.; Trucks, G. W.; Pople, J. A.; Head-Gordon, M. A Fifth-Order Perturbation Comparison of Electron Correlation Theories. *Chem. Phys. Lett.* **1989**, *157*(6), 479–483. https://doi.org/10.1016/S0009-2614(89)87395-6.
10. Jung, Y.; Lochan, R. C.; Dutoi, A. D.; Head-Gordon, M. Scaled Opposite-Spin Second Order Møller-Plesset Correlation Energy: An Economical Electronic Structure Method. *J. Chem. Phys.* **2004**, *121*(20), 9793–9802. https://doi.org/10.1063/1.1809602.
11. Grimme, S. Accurate Calculation of the Heats of Formation for Large Main Group Compounds With Spin-Component Scaled MP2 Methods. *J. Phys. Chem. A* **2005**, *109*(13), 3067–3077. https://doi.org/10.1021/jp050036j. PMID: 16833631.
12. Gerenkamp, M.; Grimme, S. *Spin-Component Scaled Second-Order Møller-Plesset Perturbation Theory for the Calculation of Molecular Geometries and Harmonic Vibrational Frequencies. Chem. Phys. Lett.* **2004**, *392*(1–3), 229–235. https://doi.org/10.1016/j.cplett.2004.05.063. http://www.sciencedirect.com/science/article/pii/S000926140400764X.
13. Maurer, S. A.; Kussmann, J.; Ochsenfeld, C. Communication: A Reduced Scaling J-Engine Based Reformulation of SOS-MP2 Using Graphics Processing Units. *J. Chem. Phys.* **2014**, *141*(5), 051106. https://doi.org/10.1063/1.4891797.
14. Song, C.; Martínez, T. J. M. Atomic Orbital-Based SOS-MP2 With Tensor Hypercontraction. II. Local Tensor Hypercontraction. *J. Chem. Phys.* **2017**, *146*(3), 034104. https://doi.org/10.1063/1.4973840.
15. Hill, J. G.; Platts, J. A. Spin-Component Scaling Methods for Weak and Stacking Interactions. *J. Chem. Theory Comput.* **2007**, *3*(1), 80–85. https://doi.org/10.1021/ct6002737.
16. Grabowski, I.; Fabiano, E.; Della Sala, F. A Simple Non-empirical Procedure for Spin-Component-Scaled MP2 Methods Applied to the Calculation of the Dissociation Energy Curve of Noncovalently-Interacting Systems. *Phys. Chem. Chem. Phys.* **2013**, *15*, 15485–15493. https://doi.org/10.1039/C3CP51431E.
17. Fabiano, E.; Della Sala, F.; Grabowski, I. *Accurate Non-Covalent Interaction Energies Via an Efficient MP2 Scaling Procedure. Chem. Phys. Lett.* **2015**, *635*, 262–267. https://doi.org/10.1016/j.cplett.2015.06.082. http://www.sciencedirect.com/science/article/pii/S0009261415005138.
18. Tan, S.; Barrera Acevedo, S.; Izgorodina, E. I. Generalized Spin-Ratio Scaled MP2 Method for Accurate Prediction of Intermolecular Interactions for Neutral and Ionic Species. *J. Chem. Phys.* **2017**, *146*(6), 064108. https://doi.org/10.1063/1.4975326.
19. Szabados, Á. Theoretical Interpretation of Grimme's Spin-Component-Scaled Second Order Møller-Plesset Theory. *J. Chem. Phys.* **2006**, *125*(21), 214105. https://doi.org/10.1063/1.2404660.
20. Fink, R. F. Spin-Component-Scaled Møller-Plesset (SCS-MP) Perturbation Theory: A Generalization of the MP Approach With Improved Properties. *J. Chem. Phys.* **2010**, *133*(17), 174113. https://doi.org/10.1063/1.3503041.
21. Grabowski, I.; Hirata, S.; Ivanov, S.; Bartlett, R. J. Ab Initio Density Functional Theory: OEP-MBPT(2). A New Orbital-Dependent Correlation Functional. *J. Chem. Phys.* **2002**, *116*(11), 4415–4425. https://doi.org/10.1063/1.1445117.

22. Bartlett, R. J.; Grabowski, I.; Hirata, S.; Ivanov, S. The Exchange-Correlation Potential in Ab Initio Density Functional Theory. *J. Chem. Phys.* **2005**, *122*(3), 034104. https://doi.org/10.1063/1.1809605.
23. Kozuch, S.; Martin, J. M. L. DSD-PBEP86: In Search of the Best Double-Hybrid DFT With Spin-Component Scaled MP2 and Dispersion Corrections. *Phys. Chem. Chem. Phys.* **2011**, *13*, 20104–20107. https://doi.org/10.1039/C1CP22592H.
24. Yu, F. Spin-Component-Scaled Double-Hybrid Density Functionals With Nonlocal van der Waals Correlations for Noncovalent Interactions. *J. Chem. Theory Comput.* **2014**, *10*(10), 4400–4407. https://doi.org/10.1021/ct500642x. PMID: 26588137.
25. Kozuch, S.; Martin, J. M. L. Spin-Component-Scaled Double Hybrids: An Extensive Search for the Best Fifth-Rung Functionals Blending DFT and Perturbation Theory. *J. Comput. Chem.* **2013**, *34*(27), 2327–2344. https://doi.org/10.1002/jcc.23391.
26. Bremond, E.; Savarese, M.; Sancho-Garcia, J. C.; Perez-Jimenez, Á. J.; Adamo, C. Quadratic Integrand Double-Hybrid Made Spin-Component-Scaled. *J. Chem. Phys.* **2016**, *144*(12), 124104. https://doi.org/10.1063/1.4944465.
27. Mezei, P. D.; Csonka, G. I.; Ruzsinszky, A.; Kállay, M. Construction of a Spin-Component Scaled Dual-Hybrid Random Phase Approximation. *J. Chem. Theory Comput.* **2017**, *13*(2), 796–803. https://doi.org/10.1021/acs.jctc.6b01140. PMID: 28052197.
28. Grabowski, I.; Fabiano, E.; Della Sala, F. Optimized Effective Potential Method Based on Spin-Resolved Components of the Second-Order Correlation Energy in Density Functional Theory. *Phys. Rev. B* **2013**, *87*, 075103. https://doi.org/10.1103/PhysRevB.87.075103.
29. Buksztel, A.; Śmiga, S.; Grabowski, I. The Correlation Effects in Density Functional Theory Along the Dissociation Path. In: *Electron Correlation in Molecules—Ab Initio Beyond Gaussian Quantum Chemistry*; Hoggan, P. E., Ozdogan, T., Eds.; *Advances in Quantum Chemistry*; Vol. 73; Academic Press: Oxford, 2016; pp. 263–283. https://doi.org/10.1016/bs.aiq.2015.07.002 (chapter 14).
30. Grabowski, I.; Fabiano, E.; Teale, A. M.; Śmiga, S.; Buksztel, A.; Della Sala, F. Orbital-Dependent Second-Order Scaled-Opposite-Spin Correlation Functionals in the Optimized Effective Potential Method. *J. Chem. Phys.* **2014**, *141*(2), 024113. https://doi.org/10.1063/1.4887097.
31. Śmiga, S.; Della Sala, F.; Buksztel, A.; Grabowski, I.; Fabiano, E. Accurate Kohn-Sham Ionization Potentials From Scaled-Opposite-Spin Second-Order Optimized Effective Potential Methods. *J. Comput. Chem.* **2016**, *37*(22), 2081–2090. https://doi.org/10.1002/jcc.24436.
32. Handy, N. C.; Schaefer, H. F., III. On the Evaluation of Analytic Energy Derivatives for Correlated Wave Functions. *J. Chem. Phys.* **1984**, *81*(11), 5031–5033.
33. Bartlett, R. J. Analytical Evaluation of Gradients in Coupled-Cluster and Many-Body Perturbation Theory. In: *Geometrical Derivatives of Energy Surfaces and Molecular Properties*, Jørgensen, P., Simons, J., Eds.; Reidel, Dordecht: The Netherlands, 1986; pp. 35–61.
34. Salter, E. A.; Trucks, G. W.; Bartlett, R. J. Analytic Energy Derivatives in Many-Body Methods. I. First Derivatives. *J. Chem. Phys.* **1989**, *90*(3), 1752–1766. https://doi.org/10.1063/1.456069.
35. Jørgensen, P.; Helgaker, T. Møller-Plesset Energy Derivatives. *J. Chem. Phys.* **1988**, *89*(3), 1560–1570.
36. Koch, H.; Jensen, H. J. A.; Jørgensen, P.; Helgaker, T.; Scuseria, G. E.; Schaefer, H. F., III Coupled Cluster Energy Derivatives. Analytic Hessian for the Closed-Shell Coupled Cluster Singles and Doubles Wave Function: Theory and Applications. *J. Chem. Phys.* **1990**, *92*(8), 4924–4940.
37. Hald, K.; Halkier, A.; Jørgensen, P.; Coriani, S.; Hättig, C.; Helgaker, T. A Lagrangian, Integral-Density Direct Formulation and Implementation of the Analytic CCSD and CCSD(T) Gradients. *J. Chem. Phys.* **2003**, *118*(7), 2985–2998.

38. Valiev, M.; Bylaska, E. J.; Govind, N.; Kowalski, K.; Straatsma, T. P.; Van Dam, H. J. J.; Wang, D.; Nieplocha, J.; Apra, E.; Windus, T. L.; de Jong, W. A. NWChem: A Comprehensive and Scalable Open-Source Solution for Large Scale Molecular Simulations. *Comput. Phys. Commun.* **2010**, *181*(9), 1477–1489. https://doi.org/10.1016/j.cpc. 2010.04.018. http://www.sciencedirect.com/science/article/pii/S0010465510001438.
39. Dunning, T. H., Jr. Gaussian Basis Sets for use in Correlated Molecular Calculations. I. The Atoms Boron Through Neon and Hydrogen. *J. Chem. Phys.* **1989**, *90*(2), 1007–1023. https://doi.org/10.1063/1.456153.
40. Woon, D. E. Benchmark Calculations With Correlated Molecular Wave Functions. V. The Determination of Accurate Ab Initio Intermolecular Potentials for He2, Ne2, and Ar2. *J. Chem. Phys.* **1994**, *100*(4), 2838–2850. https://doi.org/10.1063/1.466478.
41. Kendall, R. A.; Dunning, T. H., Jr.; Harrison, R. J. Electron Affinities of the First Row Atoms Revisited. Systematic Basis Sets and Wave Functions. *J. Chem. Phys.* **1992**, *96*(9), 6796–6806. https://doi.org/10.1063/1.462569.
42. Woon, D. E.; Dunning, T. H., Jr. Calculation of the Electron Affinities of the Second Row Atoms: Al-Cl. *J. Chem. Phys.* **1993**, *99*(5), 3730–3737. https://doi.org/10.1063/1.466148.
43. Johnson, R. D., III., Ed. *NIST Computational Chemistry Comparison and Benchmark Database*, NIST Standard Reference Database Number 101 Release 18, 2016, http://cccbdb.nist.gov/.
44. Jankowski, K.; Nowakowski, K.; Grabowski, I.; Wasilewski, J. Coverage of Dynamic Correlation Effects by Density Functional Theory Functionals: Density-Based Analysis for Neon. *J. Chem. Phys.* **2009**, *130*(16), 164102 (pages 9).
45. Jankowski, K.; Nowakowski, K.; Grabowski, I.; Wasilewski, J. Ab Initio Dynamic Correlation Effects in Density Functional Theories: A Density Based Study for Argon. *Theor. Chem. Acc.* **2010**, *125*(3–6), 433–444.
46. Grabowski, I.; Teale, A. M.; Śmiga, S.; Bartlett, R. J. Comparing Ab Initio Density-Functional and Wave Function Theories: The Impact of Correlation on the Electronic Density and the Role of the Correlation Potential. *J. Chem. Phys.* **2011**, *135*(11), 114111.
47. Śmiga, S.; Buksztel, A.; Grabowski, I. Density-Dependent Exchange-Correlation Potentials Derived From highly Accurate Ab Initio Calculations. In: *Proceedings of MEST 2012: Electronic Structure Methods With Applications to Experimental Chemistry*; Hoggan, P., Ed.; Vol. 68 Elsevier Academic Press Inc, 2014; pp. 125–151.
48. Grabowski, I.; Teale, A. M.; Fabiano, E.; Śmiga, S.; Buksztel, A.; Della Sala, F. A Density Difference Based Analysis of Orbital-Dependent Exchange-Correlation Functionals. *Mol. Phys.* **2014**, *112*(5–6), 700–710. https://doi.org/10.1080/00268976.2013.854424.

CHAPTER SEVENTEEN

Time-Dependent Configuration Interaction Using the Graphical Unitary Group Approach: Nonlinear Electric Properties

Patrick J. Lestrange*, Mark R. Hoffmann†, Xiaosong Li[1],*
*University of Washington, Seattle, WA, United States
†University of North Dakota, Grand Forks, ND, United States
[1]Corresponding author: e-mail address: xsli@uw.edu

Contents

1. Introduction 296
2. Theory 297
   2.1 Configuration Interaction 297
   2.2 Time-Dependent CI 299
   2.3 Dynamic (Hyper)Polarizabilities 300
3. Results 302
4. Conclusions 308
Acknowledgments 309
References 309

Abstract

Dynamic electric properties are most commonly determined by applying linear and nonlinear response theory. This is often a sequential process as each order of response depends on the solution for the previous lower order. Response theory is a perturbative approach and is not directly amenable to modeling time-resolved spectroscopies or experiments involving exotic pulse shapes. Nonperturbative interaction between a system and an electric field can be modeled explicitly in time. This makes it possible to more easily resolve higher-order properties and highly nonlinear processes. Time-dependent configuration interaction has asserted itself as a powerful tool for accurately modeling electronic dynamics. We have implemented time-dependent configuration interaction using the graphical unitary group approach in order to study the dynamics of open-shell systems while retaining spin as a good quantum number. This approach has been used to resolve linear and nonlinear electric properties of molecular systems. Important considerations when modeling dynamic electric properties in the time-domain are presented as well as comparisons to properties of broken symmetry solutions.

Advances in Quantum Chemistry, Volume 76
ISSN 0065-3276
https://doi.org/10.1016/bs.aiq.2017.06.003

1. INTRODUCTION

Intense atto- and femtosecond laser sources have opened up new ways to probe electronic and nuclear dynamics in real time.[1–5] Intricate experiments take advantage of these new laser sources and tease out previously inaccessible information about different molecular systems.[6–8] A more detailed understanding of linear and nonlinear responses to these strong electric fields will aid in the interpretation of these complex experiments.

Dynamical polarizabilities and hyperpolarizabilities are often calculated with time-dependent perturbation theory using sum-over-states expressions.[9–12] These techniques assume a much weaker field strength than those seen in typical attosecond experiments. Further, they are not able to account for the complicated pulse shapes that are often employed or the time-dependence between pump and probe pulses. An alternative approach would be to model the system explicitly in the time-domain under the influence of an external field and extract relevant properties from the simulation. This goes beyond the perturbative regime and captures all orders of response that could be described by perturbation theory all at once. It also allows for the use of pulse shapes and field strengths more representative of the experiment being modeled.

Real-time time-dependent density functional theory (RT-TDDFT),[13–18] due to its low computational cost and reasonable accuracy, has been used to model absorption spectra of large molecules,[19,20] linear and nonlinear electric properties,[21,22] and spin[16,23] and charge transport.[24,25] However, RT-TDDFT suffers from some well-documented problems associated with using adiabatic functionals of the density. This can lead to a time-dependence in the resonant frequencies as the density is propagated in time.[26–28] This is a particularly vexing problem that makes it difficult to model time-resolved spectroscopies or excited state properties although some groups have made recent progress in this area.[29,30]

Multiconfigurational wave functions propagated in time do not suffer from the same difficulties of RT-TDDFT. These techniques have been used to model time-resolved photoionization[31,32] and nonlinear responses[33–37] to intense electric fields. Time-dependent configuration interaction singles (TD-CIS) is the simplest and most commonly used multiconfigurational approach. Unfortunately, it tends to overestimate (hyper)polarizabilities due to a lack of electron correlation in the description of the ground state, although it is able to qualitatively describe high harmonic generation spectra.[33,36,37] Madsen and Miyagi recently gave a more detailed analysis of the role of multielectron excitations on the description of nonlinear properties with their series of time-dependent restricted-active-space self-consistent field (TD-RASSCF) approaches.[38] Their analysis highlighted the importance of including doubly excited configurations in the basis describing the time-evolving wave function. This work will focus on practical considerations of evaluating nonlinear electric properties in the time-domain for both closed and open-shell systems with the graphical unitary group approach (GUGA) to configuration interaction (GUGA-CI). We will also highlight the importance of including doubly excited configurations in the description of the time-evolving wave function.

2. THEORY

2.1 Configuration Interaction

Configuration interaction describes electronic wave functions as linear combinations of different electron configurations $\{\phi_i\}$

$$|\psi\rangle = \sum_i C_i|\phi_i\rangle. \tag{1}$$

CIS uses singly excited configurations from the reference state to describe excited electronic states, but does not improve the description of the reference ground state.[39] The expansion in Eq. (1) can be truncated at other levels of excitation leading to methods like CISD and CISDT where doubly and triply excited configurations are also included. These approaches can be used to describe multiconfigurational ground and excited electronic states. The bases used in the expansion are usually either excited Slater determinants[40–43] or spin-adapted configurations known as configuration state functions (CSFs).[44–49]

Excited Slater determinants can be used as the CI basis for any form of the mean-field reference including a restricted, unrestricted, or noncollinear wave function.[50–52] Any relaxation or breaking of a particular symmetry in the reference configuration will also be broken for the resulting CI wave functions. There are some advantages in using a broken symmetry wave function including most notably that breaking symmetry usually lowers the energy of the variational wave function. However, when a broken symmetry wave function is used as the reference configuration for response theory of CI, there can often be unphysical bright transitions to different spin states.[53] These unphysical transitions should usually be removed from calculated spectra and should influence calculated electric properties as well.

CSFs are spin-adapted combinations of determinants that can also be used as the basis in the CI expansion. They do require the use of a restricted reference configuration that is both an eigenfunction of $\hat{S}_z$ and $\hat{S}^2$. CSFs are an appealing alternative to Slater determinants since any CI space, including Full CI, will be smaller due to the lack of coupling between different spin states and the Hamiltonian being block-diagonalized into different spin states. For higher spin states, CIS on top of a restricted open-shell Hartree–Fock (ROHF) wave function will add a small amount of electron correlation to the description of the ground state.[54] This is of course different from what is seen when only singly excited Slater determinants are used or when the reference configuration is a closed-shell singlet. The combination of Slater determinants needed to create the excited low spin open-shell CSFs may include doubly excited determinants which would offer a small correction to the description of the ground state.

The time-independent nonrelativistic electronic Hamiltonian is defined as

$$\hat{H}_0 = \sum_{pq} \langle p|\hat{h}|q\rangle \hat{E}_{pq} + \frac{1}{2}\sum_{pqrs} (pq|rs)\left(\hat{E}_{pq}\hat{E}_{rs} - \delta_{qr}\hat{E}_{ps}\right) \tag{2}$$

where $\{p, q, r, s\}$ are orbital indices, $\langle p|\hat{h}|q\rangle$ and $(pq|rs)$ are one- and two-electron integrals, respectively, and $\hat{E}_{pq}$ is a spin-adapted excitation operator

$$\hat{E}_{pq} = \sum_{\tau} \hat{a}^{\dagger}_{p\tau}\hat{a}_{q\tau}. \tag{3}$$

The one- and two-body coupling coefficients between states must be calculated when constructing the CI matrix: $\langle \psi' |\hat{E}_{pq}|\psi\rangle$ and $\langle \psi' |\hat{E}_{pq}\hat{E}_{rs} - \delta_{qr}\hat{E}_{ps}|\psi\rangle$. These are simple to calculate when using orthogonal orbitals in determinants since they can only be 0 or $\pm$ 1. The couplings between CSFs are more complicated, however, since each one is constructed from a linear combination of determinants.

The GUGA[44–48,55] determines these coupling coefficients by finding patterns in a graph constructed from available spin couplings within different orbital occupations. The graph is a convenient representation of a distinct row table, which contains all unique rows of the Paldus arrays that define different spin-adapted configurations. Each path that traverses the graph is a unique CSF and the coupling coefficients are determined by the shape made by two different CSFs. This graph can be efficiently searched to build the Hamiltonian or, in the case of direct CI, the matrix–vector product

$$\sigma_i = \sum_j H_{ij} C_j \tag{4}$$

This technique can also be applied to multiconfigurational SCF (MCSCF) or any CI excitation level simply by changing the graph that is searched.

2.2 Time-Dependent CI

The dynamics of a time-evolving wave function are determined by solving the time-dependent Schrödinger equation

$$\mathrm{i}\frac{\partial \Psi(t)}{\partial t} = \hat{H}(t)\Psi(t). \tag{5}$$

The time-dependent Hamiltonian, in the length gauge,[56] is defined as

$$\hat{H}(t) = \hat{H}_0 - \hat{\mu} \cdot \mathbf{E}(t) \tag{6}$$

where the field free Hamiltonian is augmented by the dipole operator $\hat{\mu} = -\sum_i^N \mathbf{r}_i + \sum_A^{N_A} Z_A \mathbf{R}_A$ coupling to an external electric field $\mathbf{E}(t)$ in the dipole approximation.

The time-dependent wave function is described in the basis of CI states

$$\Psi(t) = \sum_i C_i(t)\psi_i. \tag{7}$$

The initial wave function at $t = 0$ can be any of the CI states or a superposition of several states. Without an external electric field, the propagation of the time-evolving coefficients is trivial

$$\mathbf{C}(t + \Delta t) = e^{-\mathrm{i}\hat{\mathbf{H}}_0 \Delta t}\mathbf{C}(t) \tag{8}$$

where $\hat{\mathbf{H}}_0$ is the diagonal field free Hamiltonian in the CI basis. This just causes the wave function to rotate through the complex plane, but does not change any observables. The split operator approach[57] can be used to separate the time-dependent and -independent parts of the Hamiltonian and describe the system under the influence of an oscillating electric field

$$\mathbf{C}(t+\Delta t) = \left[\prod_{q=x,y,z} \mathbf{U}_q^{\dagger} e^{\mathrm{i}E_q(t)\boldsymbol{\mu}_q\Delta t}\mathbf{U}_q\right] e^{-\mathrm{i}\hat{\mathbf{H}}_0\Delta t}\mathbf{C}(t) \tag{9}$$

where $\mathbf{U}_q$ is a unitary matrix that transforms the CI dipole matrix $\boldsymbol{\mu}_q$ so that it is diagonal. With this formalism, the dynamics of a time-dependent wave function can be determined in the basis of resolved CI states. It should be noted that the $\mathbf{U}_q$ matrices will be the dimension of the used CI basis and will constitute the memory bottleneck in these simulations.

2.3 Dynamic (Hyper)Polarizabilities

In these simulations, the field is turned on using a linear ramp envelope that grows to a maximum intensity over several optical cycles of the applied field $\left(k = \frac{2\pi}{\omega}\right)$. This type of time-dependent electric field can be written as

$$\mathbf{E}(t) = \begin{cases} \frac{t}{nk}\mathbf{A}\cos(\omega t) & \text{if } 0 \le t < no \\ \mathbf{A}\cos(\omega t) & \text{if } t \ge no \end{cases} \tag{10}$$

where $\mathbf{A}$ is the constant magnitude vector of the field and n is the number of optical cycles over which the field is turned on.

To compute the dynamic polarizabilities and hyperpolarizabilities from real-time simulations, a finite difference method can be applied to obtain the time-dependent dipole responses from the time-dependent electric dipole moment.[22] The TD-CI dipole moment is evaluated at every time step as

$$\mu_i(t) = \mathbf{C}^{\dagger}(t)\mathbf{D}_i\mathbf{C}(t) \tag{11}$$

where $i \in \{x, y, z\}$ and $\mathbf{D}_i$ is a matrix of dipoles between CI states. To review how this approach works, we must first notice that the time-evolving dipole moment can be expanded in a Taylor series to all orders of response.[58] Each order can be separated into its own expression that can be used to extract properties from the time-evolving dipole moment with a small error related to the field strengths used, $\mathcal{O}(A^4)$. The first-, second-, and third-order responses are determined as

$$\mu_{ij}^{(1)} = \frac{8[\mu_i(t, A_j) - \mu_i(t, -A_j)] - [\mu_i(t, 2A_j) - \mu_i(t, -2A_j)]}{12A_j} \tag{12}$$

$$\mu_{ijj}^{(2)} = \frac{16[\mu_i(t, A_j) + \mu_i(t, -A_j)] - [\mu_i(t, 2A_j) + \mu_i(t, -2A_j)] - 30\mu_i^0}{24A_j^2} \tag{13}$$

$$\mu_{ijjj}^{(3)} = \frac{-13[\mu_i(t, A_j) - \mu_i(t, -A_j)] + 8[\mu_i(t, 2A_j) - \mu_i(t, -2A_j)] - [\mu_i(t, 3A_j) - \mu_i(t, -3A_j)]}{48A_j^3} \tag{14}$$

where $\mu_i(t, \pm A_j)$ are the i-th components of the time-dependent dipole induced by a field with maximum intensity $\pm A_j$.

In an oscillatory electric field, such as Eq. (10), time-dependent dipole response functions have analytical expressions in terms of frequency-dependent (hyper)polarizabilities (see Ref. 22 for derivations),

$$\mu_{ij}^{(1)} = \alpha_{ij}(-\omega;\omega)\cos(\omega t) \tag{15}$$

$$\mu_{ijk}^{(2)} = \frac{1}{4}[\beta_{ijk}(-2\omega;\omega,\omega)\cos(2\omega t) + \beta_{ijk}(0;\omega,-\omega)] \tag{16}$$

$$\mu_{ijkl}^{(3)} = \frac{1}{24}[\gamma_{ijkl}(-3\omega;\omega,\omega,\omega)\cos(3\omega t) + 3\overline{\gamma}_{ijkl}(-\omega;\omega,\omega,-\omega)\cos(\omega t)] \tag{17}$$

The frequency-dependent polarizability $\alpha_{ij}(-\omega;\omega)$ is related to the linear refractive index. The first hyperpolarizabilities $\beta_{ijk}(-2\omega;\omega, \omega)$ and $\beta_{ijk}(0;\omega, -\omega)$ correspond to second-harmonic generation and optical rectification, respectively. The second hyperpolarizabilities $\gamma_{ijkl}(-3\omega;\omega,\omega,\omega)$ and $\overline{\gamma}_{ijkl}(-\omega;\omega,\omega,-\omega)$ related to third-harmonic generation and averaged degenerate four-wave mixing.

To compute the frequency-dependent (hyper)polarizabilities, we first carry out TD-CI simulations of electron dynamics in an oscillatory external electric field. Time-evolutions of the dipole moment are collected at different field strengths ($\pm A$, $\pm 2A$, $\pm 3A$). Time-dependent first-, second-, and third-order dipole responses are computed using Eqs. (12)–(14). At last, the frequency-dependent polarizability and hyperpolarizability values are then extracted by fitting different frequency-dependent response functions in Eqs. (15)–(17) to the simulation data of time-dependent dipole responses in Eqs. (12)–(14).

3. RESULTS

All simulations used a time step of 0.1 au (0.00242 fs). All CI calculations were performed in a development version of Gaussian[59] and all properties from response theory were obtained using the Dalton16 package.[60,61] Field strengths were kept sufficiently weak so that ionization would not need to be considered and the Keldysh parameters were close to one for all simulations.[62]

The dynamic polarizabilities and hyperpolarizabilities of H_2 at the experimental bond length of 0.7414 Å[63] were computed using TD-CIS and TD-CISD with the 6-31G basis set. This results in 4 configurations for CIS and 10 configurations for CISD. The molecule is oriented along the z-axis and an experimentally equivalent 632.8 nm light is propagated along the molecular axis. All even order properties are zero because molecular hydrogen has an inversion center, so only the odd order molecular properties are presented.

In these calculations, a monochromatic field is linearly ramped to a maximum intensity over several optical cycles. The field must be turned on slowly so that it remains an adiabatic perturbation to the system. If the field is turned on too quickly the system will be perturbed too far from the ground state, preparing a superposition of the ground and excited states. The time-evolving dipole would then report on how all states in the superposition respond to the field and not just the ground state. This can certainly be problematic when trying to determine the properties of a the ground state or any other single eigenstate.

We have compared the (hyper)polarizabilities extracted from simulations where the electric field is turned on over different periods of time. The field is ramped up to its maximum intensity over 1, 2, or 5 optical cycles and the electric properties are extracted from four additional optical cycles of propagation. The extracted polarizabilities α are largely unaffected by how quickly the field is turned on (Table 1). The quality of the fits to the 2nd hyperpolarizabilities γ signals are significantly improved by turning on the field more slowly (Fig. 1). This observation suggests that molecular hyperpolarizabilities are very sensitive to the adiabaticity of the simulation or the extent of excited state population in the wave function. This data indicates that turning on the field over 5 optical cycles is a sufficient length of time to achieve an adiabatic perturbation and will be used for all subsequent simulations.

Table 1 TD-CIS Polarizabilities and 2nd Hyperpolarizabilities of H_2 Under the Influence of 632.8 nm Light With the Field Turned on at Different Rates

	1 Optical Cycle	2 Optical Cycles	5 Optical Cycles
$\alpha_{zz}(-\omega;\omega)$	7.407 (0.999)	7.405 (0.999)	7.407 (0.999)
$\gamma_{zzzz}(-3\omega;\omega,\omega,\omega)$	−506.404 (0.783)	−515.418 (0.850)	−514.369 (0.992)
$\bar{\gamma}_{zzzz}(-\omega;\omega,\omega,-\omega)$	−454.117	−457.244	−457.202

R^2 values for fits are included in parentheses.

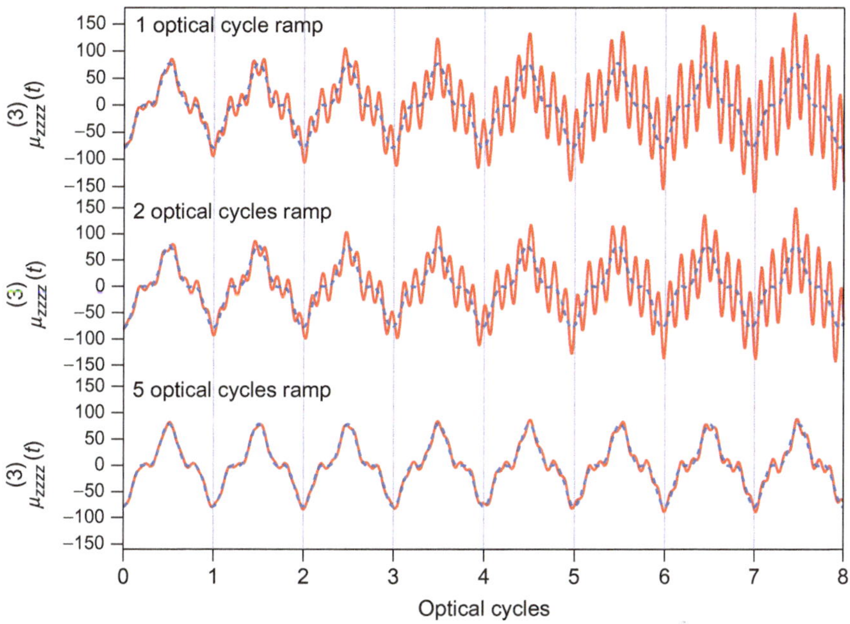

Fig. 1 The quality of the 2nd hyperpolarizability fits for H_2 if the field is turned on over different lengths of time.

Table 2 compares properties extracted from the real-time TD-CI simulations with those calculated from response theory. The fits to the simulation data are excellent with R^2 always above 0.99 (Fig. 2). The TD-CIS properties are consistently larger in magnitude than those calculated with response theory from a HF reference or with CISD (full CI for this system). A previous study by Saalfrank and coworkers also noted the tendency of TD-CIS to overestimate polarizabilities when compared with TD-CISD.[33] This behavior comes from two factors: the first excitation energy is lower for CIS than for CISD and the transition moments from

Table 2 Dynamic (Hyper)Polarizabilities of H_2 With $\omega = 632.8$ nm Modeled Using Various Methods and the 6-31G Basis Set

	Resp. (HF)	Resp. (CIS)	TD-CIS	Resp. (CISD)	TD-CISD	Exp.[64]
$\alpha_{zz}(-\omega;\omega)$	6.614	7.630	7.407	6.193	6.015	6.940
$\gamma_{zzzz}(-3\omega;\omega, \omega, \omega)$	−134.107		−514.369	−63.201	−63.206	
$\overline{\gamma}_{zzzz}(-\omega;\omega,\omega,-\omega)$	−118.138		−457.202	−69.062	−56.387	

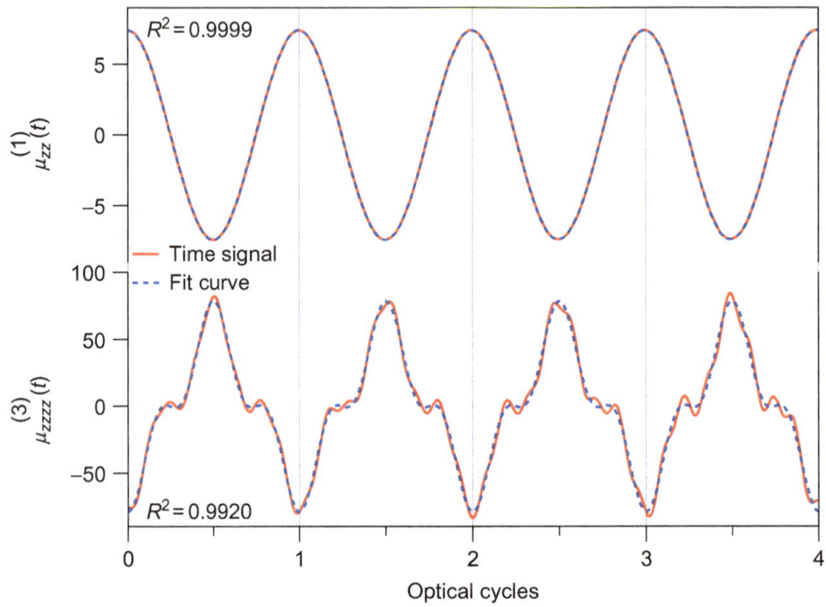

Fig. 2 Time evolution of the first- and third-order responses of H_2 modeled with TD-CIS using the 6-31G basis set. The fit curves and their R^2 values overlay the simulation data.

the ground state are larger for CIS. The same laser pulse thus perturbs the ground state further from equilibrium and induces a larger dipole moment. This behavior can also be rationalized by observing the relative changes in energy and populations of the ground and first excited state over the course of the simulations (Fig. 3). The change in energy and projection onto the first excited state is much greater for CIS than for CISD.

For most systems larger than H_2 it may not be computationally affordable to use the entire CI space as the basis describing the time-evolving wave function. We have investigated how different properties change as the number of CI states increases for a water molecule. The experimental geometry

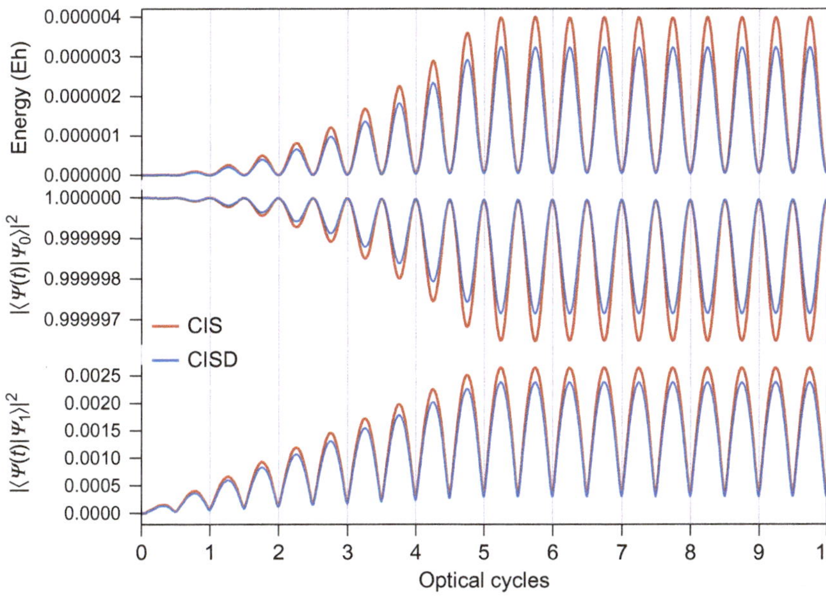

Fig. 3 Change in energy of the time-dependent H_2 wave function relative to the ground state energy and its projection onto the ground and first excited state modeled with CIS and CISD. The incident 631.8 nm pulse increases to a maximum intensity of 0.001 au over 5 optical cycles.

of gaseous water was used with $R_{O\text{-}H} = 0.957$ Å and $\alpha_{HOH} = 104.474$ degrees.[65] For these simulations, the system is oriented in the xy-plane with the molecular dipole along the z-axis.

Fig. 4 shows how electric properties change as the CI space is truncated and if only some of the lowest energy CIS states are used to describe the time-evolving wave function. The majority of the electric properties converge smoothly toward the result that would be obtained using the full CIS space. However, the third-order properties along the static dipole axis ($\gamma_{zzzz}(-3\omega;\omega, \omega, \omega)$ and $\overline{\gamma}_{zzzz}(-\omega;\omega, \omega, -\omega)$) do not converge as smoothly as the lower-order properties. It seems that as much as 2/3 of the total space may be required to converge the linear and nonlinear properties to within 5% of what would be calculated using the total CI space, consistent with what other studies have shown.[36] At that point all of the valence transitions have been included and only the oxygen K-edge transitions are remaining, which are energetically well-separated from other transitions and contribute very little in the determination of these properties. It may be advantageous for future high harmonic generation studies to also evaluate the convergence

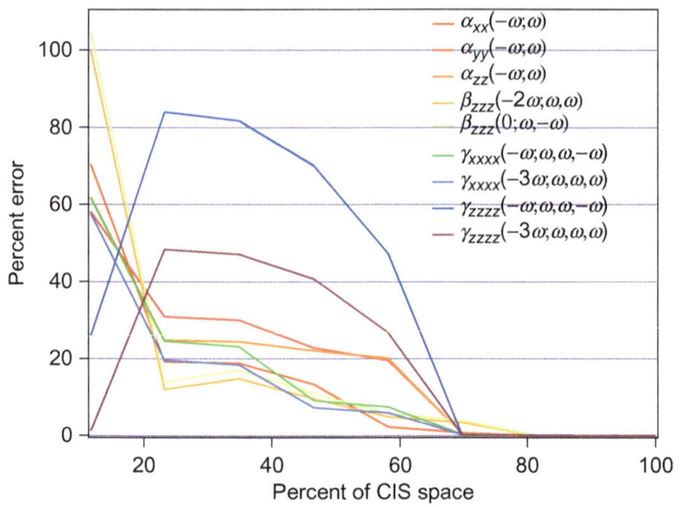

Fig. 4 Convergence of various properties of water as the number of CIS states is increased until it spans the full space.

of higher-order properties as well as that of the first-order polarizability when determining the basis for the time-evolving wave function. The vast majority of valence transitions may be required.

For the water molecule, the first-order properties calculated with TD-CIS are again consistently larger in magnitude than those calculated with TD-CISD (Table 3). Unfortunately, both methods and linear response HF underestimate the experimental polarizabilities although TD-CIS is closer. The deviation from experiment is due to several factors most notably the limited one-electron basis set used in these simulations. Certainly the larger basis sets would bring these computed properties closer to the experimental values. The CISD excitation energies for water are known to be too large compared to experimental values leading to the diminished response seen at all orders.[33]

An important benefit of the GUGA-CI approach is that it restricts the system to have a particular spin symmetry. This is in contrast to CI on top of an unrestricted Hartree–Fock (UHF) wave function in which all CI states will have broken $\hat{S}^2$ symmetry just like the reference configuration. We will illustrate that breaking spin symmetry and mixing with higher spin states influences the determination of dynamic electric properties. The TD-UCIS results used energies and transition dipoles from unrestricted CIS calculations. Table 4 presents the linear and nonlinear properties of

Table 3 Dynamic (Hyper)Polarizabilities of Water With $\omega = 514.5$ nm Modeled Using Various Methods and the 6-31+G* Basis Set

	Resp. (HF)	Resp. (CIS)	TD-CIS	Resp. (CISD)	TD-CISD	Exp.[66]
$\alpha_{xx}(-\omega;\omega)$	5.245	5.587	5.578	4.899	4.892	9.546
$\alpha_{yy}(-\omega;\omega)$	7.249	8.133	8.123	6.150	6.143	10.314
$\alpha_{zz}(-\omega;\omega)$	5.837	6.419	6.411	5.039	5.031	9.906
$\beta_{zzz}(-2\omega;\omega,\ \omega)$	−5.694	−12.461	−12.371	−2.563	−2.566	
$\beta_{zzz}(0;\omega,\ -\omega)$	−4.495	−10.434	−10.346	−2.077	−2.071	
$\gamma_{xxxx}(-3\omega;\omega,\ \omega,\ \omega)$	1901.647		1851.873		1078.802	
$\overline{\gamma}_{xxxx}(-\omega;\omega,\omega,-\omega)$	1032.180		1016.642		722.862	
$\gamma_{zzzz}(-3\omega;\omega,\ \omega,\ \omega)$	455.240		350.143		260.333	
$\overline{\gamma}_{zzzz}(-\omega;\omega,\omega,-\omega)$	278.554		159.873		189.269	

Table 4 Dynamic (Hyper)Polarizabilities of BeH With $\omega = 800$ nm Modeled Using Various Methods and the 6-31+G* Basis Set

	Resp. (ROHF)	TD-UCIS	TD-GUGA-CIS	TD-GUGA-CISD
$\alpha_{xx}(-\omega;\omega)$	36.359	37.834	38.720	28.633
$\alpha_{zz}(-\omega;\omega)$	20.566	20.758	20.732	20.307
$\beta_{zzz}(-2\omega;\omega,\ \omega)$	−221.381	−193.605	−211.705	−205.635
$\beta_{zzz}(0;\omega,\ -\omega)$	−151.097	−132.211	−143.883	−153.281
$\gamma_{zzzz}(-3\omega;\omega,\ \omega,\ \omega)$		41148.193	47154.336	38425.598
$\overline{\gamma}_{zzzz}(-\omega;\omega,\omega,-\omega)$		10735.933	13278.785	31112.808

the paramagnetic BeH molecule, a typically example of a spin-contaminated ground state wave function,[53] and Fig. 5 shows the time-evolutions of dipole responses. The molecule is fixed at its experimental geometry of 1.3426 Å.[67] The off-resonant field for BeH is representative of a Ti:Sapphire laser having a maximum intensity of 800 nm.

The first-order properties are virtually identical for TD-UCIS and TD-GUGA-CIS, but both methods significantly overestimate the polarizabilities compared to the TD-GUGA-CISD method. The calculated first hyperpolarizabilities are also similar between both single excitation methods and are comparable to the CISD results. The third-order properties do start

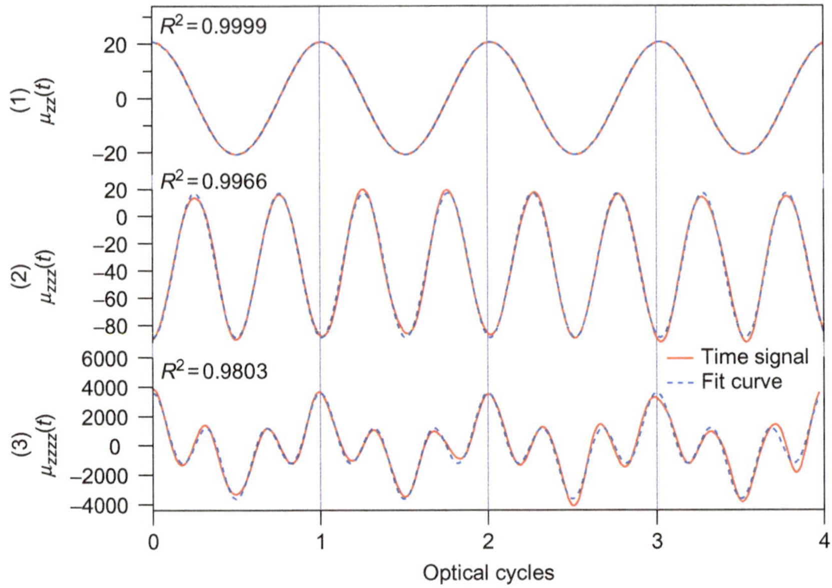

Fig. 5 Time evolution of the first-, second-, and third-order responses of BeH modeled with TD-GUGA-CIS using the 6-31+G* basis set. The fit curves and their R^2 values overlay the simulation data.

to deviate significantly between all time-dependent methods, particularly for $\overline{\gamma}_{zzzz}(-\omega;\omega,\omega,-\omega)$. The TD-GUGA-CISD results are about three times greater than the singly excited methods. Part of the difference between the UCIS and GUGA-CIS approaches is explained by the different reference states. The rest is due to the different eigenspectra for the approaches. Many of the excited states determined by the UCIS approach mix with neighboring quartet states and shift to different energies. Of course, this does not happen with the GUGA-CI approach and all states are true doublets.

4. CONCLUSIONS

We have developed a TD-GUGA-CI approach to simulate electron dynamics driven by external perturbations. Linear and nonlinear electric properties can be extracted from time-dependent configuration interaction calculations by fitting equations to the time signal. This approach is nonperturbative and has the potential to go beyond what is possible with traditional response theory since high-order properties can be determined all at once without needing to sequentially solve each response problem. The time over which an electric field is turned on was shown to have a significant impact on the quality of the fit. Although the extracted properties did vary much for H_2, this is not likely to be the case for more complicated systems. The quality of the CI space used to describe the time-evolving wave function should also be considered. The vast majority of valence excitations may be necessary to adequately converge linear and nonlinear properties. The TD-GUGA-CI approach forces open-shell systems to remain spin eigenfunctions. Results for the broken spin symmetry TD-UCIS method deviate from the GUGA-CI properties mostly due to changes in the eigenspectrum as states mix with other higher spin states. In short, TD-GUGA-CI is a powerful tool for modeling linear and nonlinear electric properties and has the potential to model more complex time-resolved spectra in the future.

ACKNOWLEDGMENTS

This work was supported by the National Science Foundation (CHE-1565520 to X.L.). Computations were facilitated through the use of advanced computational, storage, and networking infrastructure provided by the Hyak supercomputer system at the University of Washington, funded by the Student Technology Fee. M.R.H. wishes to acknowledge the ND EPSCoR for partial support.

REFERENCES

1. Krausz, F.; Ivanov, M. Attosecond Physics. *Rev. Mod. Phys.* **2009**, *81*, 163–234.
2. Spanner, M.; Mikosch, J.; Gijsbertsen, A.; Boguslavskiy, A. E.; Stolow, A. Multielectron Effects and Nonadiabatic Electronic Dynamics in Above Threshold Ionization and High-Harmonic Generation. *New J. Phys.* **2011**, *13*(9), 093010.
3. Boguslavskiy, A. E.; Mikosch, J.; Gijsbertsen, A.; Spanner, M.; Patchkovskii, S.; Gador, N.; Vrakking, M. J.; Stolow, A. The Multielectron Ionization Dynamics Underlying Attosecond Strong-Field Spectroscopies. *Science* **2012**, *335*(6074), 1336–1340.
4. Salières, P.; Maquet, A.; Haessler, S.; Caillat, J.; Taïeb, R. Imaging Orbitals With Attosecond and Ångström Resolutions: Toward Attochemistry?*Rep. Prog. Phys.* **2012**, *75*(6), 062401.
5. Zewail, A. H. Ultrafast Light and Electrons: Imaging the Invisible. In: *Optics in Our Time*, Al-Amri, M. D., El-Gomati, M., Zubairy, M. S., (Eds.). Springer: New York City, 2016; pp. 43–68.
6. Shafir, D.; Soifer, H.; Bruner, B. D.; Dagan, M.; Mairesse, Y.; Patchkovskii, S.; Ivanov, M. Y.; Smirnova, O.; Dudovich, N. Resolving the Time When an Electron Exits a Tunnelling Barrier. *Nature* **2012**, *485*(7398), 343–346.
7. Ott, C.; Kaldun, A.; Raith, P.; Meyer, K.; Laux, M.; Evers, J.; Keitel, C. H.; Greene, C. H.; Pfeifer, T. Lorentz Meets Fano in Spectral Line Shapes: A Universal Phase and Its Laser Control. *Science* **2013**, *340*, 716–720.
8. Shelby, M. L.; Lestrange, P. J.; Jackson, N. E.; Haldrup, K.; Mara, M. W.; Stickrath, A. B.; Zhu, D.; Lemke, H.; Chollet, M.; Hoffman, B. M.; Li, X.; Chen, L. X. Ultrafast Excited State Relaxation of a Metalloporphyrin Revealed by Femtosecond X-ray Absorption Spectroscopy. *J. Am. Chem. Soc.* **2016**, *138*, 8752–8764.
9. Orr, B. J.; Ward, J. F. Perturbation Theory of the Nonlinear Optical Polarization of an Isolated System. *Mol. Phys.* **1971**, *20*, 513.

10. Chen, G. H.; Mukamel, S.; Beljonne, D.; Bredas, J. L. The Coupled Electronic Oscillators vs the Sum-over-states Pictures for the Optical Response of Octatetraene. *J. Chem. Phys.* **1996**, *104*(14), 5406.
11. Bishop, D. M.; Kirtman, B.; Champagne, B. Differences Between the Exact Sum-over-states and the Canonical Approximation for the Calculation of Static and Dynamic Hyperpolarizabilities. *J. Chem. Phys.* **1997**, *107*(15), 5780–5787.
12. Tretiak, S.; Chernyak, V. Resonant Nonlinear Polarizabilities in the Time-Dependent Density Functional Theory. *J. Chem. Phys.* **2003**, *119*(17), 8809–8823.
13. Li, X.; Smith, S. M.; Markevitch, A. N.; Romanov, D. A.; Levis, R. J.; Schlegel, H. B. A Time-Dependent Hartree-Fock Approach for Studying the Electronic Optical Response of Molecules in Intense Fields. *Phys. Chem. Chem. Phys.* **2005**, 7, 233–239.
14. Lopata, K.; Govind, N. Modeling Fast Electron Dynamics With Real-Time Time-Dependent Density Functional Theory: Application to Small Molecules and Chromophores. *J. Chem. Theory Comput.* **2011**, 7(5), 1344–1355.
15. Lopata, K.; Govind, N. Near and Above Ionization Electronic Excitations With Non-Hermitian Real-Time Time-Dependent Density Functional Theory. *J. Chem. Theory Comput.* **2013**, *9*(11), 4939–4946.
16. Ding, F.; Goings, J. J.; Frisch, M. J.; Li, X. Ab Initio Non-relativistic Spin Dynamics. *J. Chem. Phys.* **2014**, *141*(21), 214111.
17. Goings, J. J.; Kasper, J. M.; Egidi, F.; Sun, S.; Li, X. Real Time Propagation of the Exact Two Component Time-Dependent Density Functional Theory. *J. Chem. Phys.* **2016**, *145*(10), 104107.
18. Williams-Young, D.; Goings, J. J.; Li, X. Accelerating Real-Time Time-Dependent Density Functional Theory With a Nonrecursive Chebyshev Expansion of the Quantum Propagator. *J. Chem. Theory Comput.* **2016**, *12*(11), 5333–5338.
19. Tussupbayev, S.; Govind, N.; Lopata, K.; Cramer, C. J. Comparison of Real-Time and Linear-Response Time-Dependent Density Functional Theories for Molecular Chromophores Ranging From Sparse to High Densities of States. *J. Chem. Theory Comput.* **2015**, *11*(3), 1102–1109.
20. Goings, J. J.; Li, X. An Atomic Orbital Based Real-Time Time-Dependent Density Functional Theory for Computing Electronic Circular Dichroism Band Spectra. *J. Chem. Phys.* **2016**, *144*(23), 234102.
21. Takimoto, Y.; Vila, F. D.; Rehr, J. J. Real-Time Time-Dependent Density Functional Theory Approach for Frequency-Dependent Nonlinear Optical Response in Photonic Molecules. *J. Chem. Phys.* **2007**, *127*(15), 154114.
22. Ding, F.; Kuiken, B. E. V.; Eichinger, B. E.; Li, X. An Efficient Method for Calculating Dynamical Hyperpolarizabilities Using Real-Time Time-Dependent Density Functional Theory. *J. Chem. Phys.* **2013**, *138*, 064104.
23. Peralta, J. E.; Hod, O.; Scuseria, G. E. Magnetization Dynamics From Time-Dependent Noncollinear Spin Density Functional Theory Calculations. *J. Chem. Theory Comput.* **2015**, *11*(8), 3661–3668.
24. Ding, F.; Chapman, C. T.; Liang, W.; Li, X. Mechanisms of Bridge-mediated Electron Transfer: A TDDFT Electronic Dynamics Study. *J. Chem. Phys.* **2012**, *137*, 22A512.
25. Petrone, A.; Lingerfelt, D. B.; Rega, N.; Li, X. From Charge-Transfer to a Charge-Separated State: A Perspective From the Real-Time TDDFT Excitonic Dynamics. *Phys. Chem. Chem. Phys.* **2014**, *16*(44), 24457–24465.
26. Thiele, M.; Gross , E. K. U.; Kümmel, S. Adiabatic Approximation in Nonperturbative Time-Dependent Density-Functional Theory. *Phys. Rev. Lett.* **2008**, *100*(15), 153004.
27. De Giovannini, U.; Brunetto, G.; Castro, A.; Walkenhorst, J.; Rubio, A. Simulating Pump-Probe Photoelectron and Absorption Spectroscopy on the Attosecond Timescale With Time-Dependent Density Functional Theory. *ChemPhysChem* **2013**, *14*(7), 1363–1376.

28. Provorse, M. R.; Habenicht, B. F.; Isborn, C. M. Peak-Shifting in Real-Time Time-Dependent Density Functional Theory. *J. Chem. Theory Comput.* **2015**, *11*(10), 4791–4802.
29. Fischer, S. A.; Cramer, C. J.; Govind, N. Excited State Absorption From Real-Time Time-Dependent Density Functional Theory. *J. Chem. Theory Comput.* **2015**, *11*(9), 4294–4303.
30. Fuks, J. I.; Luo, K.; Sandoval, E. D.; Maitra, N. T. Time-Resolved Spectroscopy in Time-Dependent Density Functional Theory: An Exact Condition. *Phys. Rev. Lett.* **2015**, *114*(18), 183002.
31. Greenman, L.; Ho, P. J.; Pabst, S.; Kamarchik, E.; Mazziotti, D. A.; Santra, R. Implementation of the Time-Dependent Configuration-Interaction Singles Method for Atomic Strong-Field Processes. *Phys. Rev. A* **2010**, *82*(2), 023406.
32. Krause, P.; Sonk, J. A.; Schlegel, H. B. Strong Field Ionization Rates Simulated With Time-Dependent Configuration Interaction and an Absorbing Potential. *J. Chem. Phys.* **2014**, *140*(17), 174113.
33. Krause, P.; Klamroth, T.; Saalfrank, P. Molecular Response Properties From Explicitly Time-Dependent Configuration Interaction Methods. *J. Chem. Phys.* **2007**, *127*(3), 034107.
34. Sonk, J. A.; Caricato, M.; Schlegel, H. B. TD-CI Simulation of the Electronic Optical Response of Molecules in Intense Fields: Comparison of RPA, CIS, CIS(D), and EOM-CCSD. *J. Phys. Chem. A* **2011**, *115*(18), 4678–4690.
35. Sonk, J. A.; Schlegel, H. B. TD-CI Simulation of the Electronic Optical Response of Molecules in Intense Fields II: Comparison of DFT Functionals and EOM-CCSD. *J. Phys. Chem. A* **2011**, *115*(42), 11832 11840.
36. Luppi, E.; Head-Gordon, M. Computation of High-Harmonic Generation Spectra of H_2 and N_2 in Intense Laser Pulses Using Quantum Chemistry Methods and Time-Dependent Density Functional Theory. *Mol. Phys.* **2012**, *110*(9–10), 909–923.
37. White, A. F.; Heide, C. J.; Saalfrank, P.; Head-Gordon, M.; Luppi, E. Computation of High-Harmonic Generation Spectra of the Hydrogen Molecule using Time-Dependent Configuration-Interaction. *Mol. Phys.* **2015**, *114*(7–8), 947–956.
38. Miyagi, H.; Madsen, L. B. Time-Dependent Restricted-Active-Space Self-Consistent-Field Theory for Laser-Driven Many-Electron Dynamics. II. Extended Formulation and Numerical Analysis. *Phys. Rev. A* **2014**, *89*, 063416.
39. Foresman, J. B.; Head-Gordon, M.; Pople, J. A. Toward a Systematic Molecular Orbital Theory for Excited States. *J. Phys. Chem.* **1992**, *96*, 135–149.
40. Handy, N. C. Multi-Root Configuration Interaction Calculations. *Chem. Phys. Lett.* **1980**, *74*(2), 280–283.
41. Knowles, P. J.; Handy, N. C. A New Determinant-Based Full Configuration Interaction Method. *Chem. Phys. Lett.* **1984**, *111*(4), 315–321.
42. Olsen, J.; Roos, B. O.; Jørgensen, P.; Jensen, H. J. A. Determinant Based Configuration Interaction Algorithms for Complete and Restricted Configuration Interaction Spaces. *J. Chem. Phys.* **1988**, *89*(4), 2185–2192.
43. Hohenstein, E. G.; Luehr, N.; Ufimtsev, I. S.; Martínez, T. J. An Atomic Orbital-Based Formulation of the Complete Active Space Self-Consistent Field Method on Graphical Processing Units. *J. Chem. Phys.* **2015**, *142*(22), 224103.
44. Paldus, J. Group Theoretical Approach to the Configuration Interaction and Perturbation Theory Calculations for Atomic and Molecular Systems. *J. Chem. Phys.* **1974**, *61*(12), 5321.
45. Shavitt, I. Graph Theoretical Concepts for the Unitary Group Approach to the Many-Electron Correlation Problem. *Int. J. Quantum Chem.* **1977**, *12*(S11), 131–148.
46. Shavitt, I. Matrix Element Evaluation in the Unitary Group Approach to the Electron Correlation Problem. *Int. J. Quantum Chem.* **1978**, *14*, 5–32.

47. Brooks, B. R.; Schaefer, H. F., III The Graphical Unitary Group Approach to the Electron Correlation Problem. Methods and Preliminary Applications. *J. Chem. Phys.* **1979**, *70*(11), 5092–5106.
48. Brooks, B. R.; Laidig, W. D.; Saxe, P.; Handy, N. C.; Schaefer, H. F., III The Loop-Driven Graphical Unitary Group Approach: A Powerful Method for the Variational Description of Electron Correlation. *Phys. Scr.* **1980**, *21*(3–4), 312–322.
49. Werner, H.-J.; Knowles, P. J. An Efficient Internally Contracted Multiconfiguration-Reference Configuration Interaction Method. *J. Chem. Phys.* **1988**, *89*(9), 5803.
50. Maurice, D.; Head-Gordon, M. Configuration Interaction With Single Substitutions for Excited States of Open-Shell Molecules. *Int. J. Quantum Chem.* **1995**, *29*, 361–370.
51. Yamaki, D.; Shigeta, Y.; Yamanaka, S.; Nagao, H.; Yamaguchi, K. MP2, Tamm-Dancoff, and RPA Methods Based on the Generalized HF Solution. *Int. J. Quantum Chem.* **2000**, *80*, 701–707.
52. Goings, J. J.; Ding, F.; Davidson, E. R.; Li, X. Approximate Singly Excited States From a Two-component Hartree-Fock Reference. *J. Chem. Phys.* **2015**, *143*(14), 144106.
53. Ipatov, A.; Cordova, F.; Doriol, L. J.; Casida, M. E. Excited-State Spin-Contamination in Time-Dependent Density-Functional Theory for Molecules With Open-shell Ground States. *J. Mol. Struct.* **2009**, *914*, 60–73.
54. Roemelt, M.; Neese, F. Excited States of Large Open-Shell Molecules: An Efficient, General, and Spin-Adapted Approach Based on a Restricted Open-Shell Ground State Wave function. *J. Phys. Chem. A* **2013**, *117*(14), 3069–3083.
55. Jiang, W.; Khait, Y. G.; Hoffmann, M. R. Configuration-Driven Unitary Group Approach for Generalized Van Vleck Variant Multireference Perturbation Theory. *J. Phys. Chem. A* **2009**, *113*(16), 4374–4380.
56. Ding, F.; Liang, W.; Chapman, C. T.; Isborn, C. M.; Li, X. On the Gauge Invariance of Nonperturbative Electronic Dynamics Using the Time-Dependent Hartree-Fock and Time-Dependent Kohn-Sham. *J. Chem. Phys.* **2011**, *135*(16), 164101.
57. Leforestier, C.; Bisseling, R. H.; Cerjan, C.; Feit, M. D.; Friesner, R.; Guldberg, A.; Hammerich, A.; Jolicard, G.; Karrlein, W.; Meyer, H.-D.; Lipkin, N.; Roncero, O.; Kosloff, R. A Comparison of Different Propagation Schemes for the Time Dependent Schrödinger Equation. *J. Chem. Phys.* **1991**, *94*(1), 59–80.
58. Butcher, P. N.; Cotter, D. The Constitutive Relation. In: *The Elements of Nonlinear Optics*. Cambridge University Press, Cambridge, 1990; pp. 12–36. https://doi.org/10.1017/CBO9781139167994.
59. Frisch, M. J.; Trucks, G. W.; Schlegel, H. B.; Scuseria, G. E.; Robb, M. A.; Cheeseman, J. R.; Scalmani, G.; Barone, V.; Mennucci, B.; Petersson, G. A.; Nakatsuji, H.; Caricato, M.; Li, X.; Hratchian, H. P.; Izmaylov, A. F.; Bloino, J.; Zheng, G.; Sonnenberg, J. L.; Liang, W.; Hada, M.; Ehara, M.; Toyota, K.; Fukuda, R.; Hasegawa, J.; Ishida, M.; Nakajima, T.; Honda, Y.; Kitao, O.; Nakai, H.; Vreven, T.; Montgomery, J. A., Jr.; Peralta, J. E.; Ogliaro, F.; Bearpark, M.; Heyd, J. J.; Brothers, E.; Kudin, K. N.; Staroverov, V. N.; Keith, T.; Kobayashi, R.; Normand, J.; Raghavachari, K.; Rendell, A.; Burant, J. C.; Iyengar, S. S.; Tomasi, J.; Cossi, M.; Rega, N.; Millam, J. M.; Klene, M.; Knox, J. E.; Cross, J. B.; Bakken, V.; Adamo, C.; Jaramillo, J.; Gomperts, R.; Stratmann, R. E.; Yazyev, O.; Austin, A. J.; Cammi, R.; Pomelli, C.; Ochterski, J. W.; Martin, R. L.; Morokuma, K.; Zakrzewski, V. G.; Voth, G. A.; Salvador, P.; Dannenberg, J. J.; Dapprich, S.; Parandekar, P. V.; Mayhall, N. J.; Daniels, A. D.; Farkas, O.; Foresman, J. B.; Ortiz, J. V.; Cioslowski, J.; Fox, D. J. *Gaussian Development Version Revision I.03+*. Gaussian Inc.: Wallingford, CT, 2014.
60. Aidas, K.; Angeli, C.; Bak, K. L.; Bakken, V.; Bast, R.; Boman, L.; Christiansen, O.; Cimiraglia, R.; Coriani, S.; Dahle, P.; Dalskov, E. K.; Ekström, U.; Enevoldsen, T.; Eriksen, J. J.; Ettenhuber, P.; Fernández, B.; Ferrighi, L.; Fliegl, H.; Frediani, L.;

Hald, K.; Halkier, A.; Hättig, C.; Heiberg, H.; Helgaker, T.; Hennum, A. C.; Hettema, H.; Hjertenaes, E.; Høst, S.; Høyvik, I.-M.; Iozzi, M. F.; Jansik, B.; Jensen, H. J. A.; Jonsson, D.; Jørgensen, P.; Kauczor, J.; Kirpekar, S.; Kjaergaard, T.; Klopper, W.; Knecht, S.; Kobayashi, R.; Koch, H.; Kongsted, J.; Krapp, A.; Kristensen, K.; Ligabue, A.; Lutnaes, O. B.; Melo, J. I.; Mikkelsen, K. V.; Myhre, R. H.; Neiss, C.; Nielsen, C. B.; Norman, P.; Olsen, J.; Olsen, J. M. H.; Osted, A.; Packer, M. J.; Pawlowski, F.; Pedersen, T. B.; Provasi, P. F.; Reine, S.; Rinkevicius, Z.; Ruden, T. A.; Ruud, K.; Rybkin, V. V.; Sałek, P.; Samson, C. C. M.; de Merás, A. S.; Saue, T.; Sauer, S. P. A.; Schimmelpfennig, B.; Sneskov, K.; Steindal, A. H.; Sylvester-Hvid, K. O.; Taylor, P. R.; Teale, A. M.; Tellgren, E. I.; Tew, D. P.; Thorvaldsen, A. J.; Thøgersen, L.; Vahtras, O.; Watson, M. A.; Wilson, D. J. D.; Ziolkowski, M.; Ågren, H. The Dalton Quantum Chemistry Program System. *WIREs Comput. Mol. Sci.* 2014, *4*(3), 269–284.

61. Dalton, A Molecular Electronic Structure Program, Release Dalton 2016.2, 2016. see http://daltonprogram.org.
62. Keldysh, L. V. Ionization in the Field of a Strong Electromagnetic Wave. *Sov. Phys. JETP* **1965**, *20*(5), 1307–1314.
63. Herzberg, G.; Howe, L. L. The Lyman Bands of Molecular Hydrogen. *Can. J. Phys.* **1959**, *37*, 636–659.
64. Bridge, N. J.; Buckingham, A. D. The Polarization of Laser Light Scattered by Gases. *Proc. R. Soc. A* **1966**, *295*, 334–349.
65. Hasted, J. B. Liquid water: Dielectric properties. In: *Water a Comprehensive Treatise*; Franks, F. (Ed.). Plenum, New York, 1972; pp. 255–309.
66. Murphy, W. F. The Rayleigh Depolarization Ratio and Rotational Raman Spectrum of Water Vapor and the Polarizability Components for the Water Molecule. *J. Chem. Phys.* **1977**, *67*(12), 5877–5882.
67. Huber, K. P.; Herzberg, G. *Constants of Diatomic Molecules*. Van Nostrand Reinhold: New York, 1979.

CHAPTER EIGHTEEN

Hybrid Treatments Based on Determinant Seniority Numbers and Spatial Excitation Levels in the Configuration Interaction Framework

Diego R. Alcoba[*,†,1], Alicia Torre[‡], Luis Lain[‡], Ofelia B. Oña[§], Gustavo E. Massaccesi[¶], Pablo Capuzzi[*,†]

[*]Facultad de Ciencias Exactas y Naturales, Universidad de Buenos Aires, Buenos Aires, Argentina
[†]Instituto de Física de Buenos Aires, Consejo Nacional de Investigaciones Científicas y Técnicas, Buenos Aires, Argentina
[‡]Facultad de Ciencia y Tecnología, Universidad del País Vasco, Bilbao, Spain
[§]Instituto de Investigaciones Fisicoquímicas Teóricas y Aplicadas, Universidad Nacional de La Plata, CCT La Plata, Consejo Nacional de Investigaciones Científicas y Técnicas, La Plata, Argentina
[¶]Ciclo Básico Común, Universidad de Buenos Aires, Buenos Aires, Argentina
[1]Corresponding author: e-mail address: qfxaldad@lg.ehu.es

Contents

1. Introduction 316
2. Theory 317
 2.1 Commutation Relations 317
 2.2 Spin-Adapted Hybrid Methods 319
3. Results and Discussion 321
4. Concluding Remarks 330
Acknowledgments 331
References 331

Abstract

In this work we project the Hamiltonian of an N-electron system onto a set of N-electron determinants cataloged by their seniority numbers and their excitation levels with respect to a reference determinant. We show that, in open-shell systems, the diagonalization of the N-electron Hamiltonian matrix leads to eigenstates of the operator $\hat{S}^2$ when the excitation levels are counted in terms of spatial orbitals instead of spin-orbitals. Our proposal is based on the commutation relations between the N-electron operators seniority number and spatial excitation level, as well as between these operators and the spin operators $\hat{S}^2$ and $\hat{S}_z$. Energy and $\langle\hat{S}^2\rangle$ expectation values of molecular

Advances in Quantum Chemistry, Volume 76
ISSN 0065-3276
https://doi.org/10.1016/bs.aiq.2017.05.003

systems obtained from our procedure are compared with those arising from the standard hybrid configuration interaction methods based on seniority numbers and spin-orbital-excitation levels. We analyze the behavior of these methods, evaluating their computational costs and establishing their usefulness.

1. INTRODUCTION

The full configuration interaction (FCI) method provides the exact determination of electronic energies in N-electron systems but its practical use is limited to systems with few electrons and small basis sets, due to its high computational cost. To reduce this cost, configuration interaction (CI) methods have been proposed, in which only a limited number of N-electron Slater determinants is used to project the Hamiltonian onto the N-electron space.[1–5] This selection has been made following mainly two criteria. In the excitation-based scheme, the Hamiltonian is projected onto an N-electron determinant set constituted by the reference determinant and those possessing up to a given integer of different occupation numbers (counted in terms of spin-orbitals or spatial orbitals) with respect to that reference. Another criterion is to select N-electron Slater determinants using their seniority number, i.e., the number of unpaired electrons involved in each determinant.[6–11] The excitation-based methods have been qualified as more efficient to describe systems presenting essentially dynamic correlation while the seniority-based procedures are preferred[7] for systems exhibiting high static correlation. More recently, hybrid methods, combining both criteria, have been proposed to select the determinants.[12–14] This type of method has proven to be useful to describe systems which present both dynamic and static correlation,[7, 13–17] as well as dissociation or bond-breaking processes, where the system can show mainly dynamic correlation near the equilibrium geometries and static correlation at stretched arrangements. However, to our knowledge, the reported results have been restricted to closed-shell systems and the spin-orbital excitation version.

The seniority number-based CI treatments provide spin-adapted wave functions for all type of system, with any number of electrons (even or odd), and for any spin symmetry.[8,18] However, in open-shell systems, the ordinary excitation-based CI procedures may lead to spin-contaminated wave functions when the excitations are described in terms of spin-orbitals. In Ref. 19 we have recently proven that the spin contamination in the excitation-based CI treatments can be avoided counting the excitation levels in terms of spatial orbitals instead of spin-orbitals. We have proposed an N-electron spin-free excitation operator which suitably evaluates the spatial excitation number. In this work we point out that spin contamination is also detected in open-shell system wave functions obtained from hybrid methods if the selection criterion is the spin-orbital one. Our aim is to extend the hybrid CI methodology to the study of open-shell systems, to ascertain the ability and applicability conditions of these procedures to describe a broader set of systems. Our proposal guarantees eigenstates without spin contamination for any N-electron Hamiltonian matrix generated by projection of the corresponding Hamiltonian operator onto a set of N-electron Slater determinants selected according to their seniority numbers and their spatial excitation numbers.

This work has been organized as follows. Section 2 describes the N-electron seniority number operator and the excitation number operators that evaluate the excitation levels of an N-electron Slater determinant with respect to a reference determinant. We analyze the commutation relation between these operators, predicting the features of the linear combinations of Slater determinants, which depend on the nature of the determinant set used to project the Hamiltonian. We also describe in this section the commutation relations of these operators with the operators $\hat{S}^2$ and $\hat{S}_z$. Section 3 reports the computational details and numerical results in selected open-shell systems, confirming the predictions provided by the commutation relations and showing spin-contaminated wave functions in the hybrid methods arising from the spin-orbital excitation version. We propose several procedures within the hybrid methodology, studying the quality of their results and their computational costs. Finally, in Section 4 we point out the conclusions that can be drawn from this work.

2. THEORY

2.1 Commutation Relations

Let $a^{\dagger}_{i^\sigma}$ and a_{i^σ} be the fermion creation and annihilation operators, respectively, corresponding to a spin-orbital i^σ; σ denotes the spin coordinate, α or β, and i is one of the K spatial orbitals of an orthonormal set $\{i, j, k, l, \ldots\}$. The excitation level, in terms of spin-orbitals (*so*), of a given N-electron determinant with respect to a reference determinant (*ref*) is evaluated by means of the N-electron operator

$$\hat{X}^{so} = \hat{N} - \sum_{i^{\alpha} \in ref} a^{\dagger}_{i^{\alpha}} a_{i^{\alpha}} - \sum_{i^{\beta} \in ref} a^{\dagger}_{i^{\beta}} a_{i^{\beta}} \tag{1}$$

where $\hat{N}$ is the particle number operator. The expectation value of this operator $\langle \hat{X}^{so} \rangle$ for a Slater determinant constitutes the excitation level of the analyzed determinant in terms of spin-orbitals. Alternatively, an N-electron spatial-orbital (*sp*) excitation level operator can be defined by means of the spin-free first- and second-order replacement operators $\hat{E}^{i}_{j} = \sum_{\sigma} a^{\dagger}_{i^{\sigma}} a_{j^{\sigma}}$ and $\hat{E}^{ik}_{jl} = \sum_{\sigma_1,\sigma_2} a^{\dagger}_{i^{\sigma_1}} a^{\dagger}_{k^{\sigma_2}} a_{l^{\sigma_2}} a_{j^{\sigma_1}}$, respectively[20,21]; this N-electron excitation operator has been expressed as Ref. 19

$$\hat{X}^{sp} = \hat{N} - \sum_{i \in ref} \hat{E}^{i}_{i} + \frac{1}{2} \sum_{i \in sref} \hat{E}^{ii}_{ii} \tag{2}$$

In this equation, the index i runs over all the reference singly or doubly occupied orbitals in the sum $\sum_{i \in ref}$, while that index only runs over the reference singly occupied (*sref*) spatial orbitals in the sum $\sum_{i \in sref}$. The expectation value $\langle \hat{X}^{sp} \rangle$ for a given determinant evaluates its excitation level in the spatial-orbital procedure with respect to a given reference. Both operators $\hat{X}^{so}$ and $\hat{X}^{sp}$ coincide when the reference determinant is closed-shell. On the other hand, the N-electron seniority number operator has been formulated by means of spin-free replacement operators as Refs. 8, 12, and 18

$$\hat{\Omega} = \sum_{i} (\hat{E}^{i}_{i} - \hat{E}^{ii}_{ii}) \tag{3}$$

Its expectation value $\langle \hat{\Omega} \rangle$ for an N-electron determinant evaluates the number of unpaired electrons involved.

The commutation relations between the spin operators $\hat{S}^2$ and $\hat{S}_z$ and the operators $\hat{\Omega}$ and $\hat{X}^{sp}$ have been previously reported,[18,19] concluding that $[\hat{S}^2, \hat{\Omega}] = 0$; $[\hat{S}^2, \hat{X}^{sp}] = 0$; $[\hat{S}_z, \hat{\Omega}] = 0$; and $[\hat{S}_z, \hat{X}^{sp}] = 0$. These relations have easily been shown by means of the spin-free replacement operator product rules,[22,23] expressing the spin-free $\hat{S}^2$ operator as Refs. 24 and 25

$$\hat{S}^2 = \frac{1}{2} \sum_{i,j,k,l} \left[\frac{4-N}{2(N-1)} \delta_{ij}\delta_{kl} - \delta_{il}\delta_{jk} \right] \hat{E}^{ik}_{jl} \tag{4}$$

in which the δ symbol means the Kronecker delta. Using those operator product rules or the well-known commutation rules of the fermion operators, one also shows

$$[\hat{\Omega}, \hat{X}^{sp}] = 0 \quad (5)$$

and

$$[\hat{\Omega}, \hat{X}^{so}] = 0 \quad (6)$$

which allow us to decompose a space of determinants of a given seniority number into different subspaces characterized by their excitation levels, or vice versa. However, unlike the $\hat{X}^{sp}$ operator, for an open-shell reference one finds

$$[\hat{S}^2, \hat{X}^{so}] \neq 0 \quad (7)$$

This relation and the $[\hat{S}^2, \hat{X}^{sp}] = 0$ one[19] justify that the spin-orbital-based and the spatial-orbital-based determinant selection schemes lead to different spin features in the resulting CI wave functions.

2.2 Spin-Adapted Hybrid Methods

Eqs. (2) and (3) mean that a given Slater determinant possesses integer eigenvalues $X^{sp} = \langle \hat{X}^{sp} \rangle$ and $\Omega = \langle \hat{\Omega} \rangle$ for the corresponding $\hat{X}^{sp}$ and $\hat{\Omega}$ operators, respectively. Moreover, in agreement with Eq. (5) linear combinations of determinants of identical eigenvalue X^{sp} and those with identical Ω eigenvalue can also be eigenfunctions of the $\hat{S}^2$ and $\hat{S}_z$ operators. This feature is not maintained in the case of the $\hat{X}^{so}$ operator which, according to Eq. (7), does not commute with the $\hat{S}^2$ operator. These properties ensure the construction of spin-adapted configurations of Slater determinants with quantum numbers S, S_z, X^{sp}, and Ω. Likewise, these linear combinations of Slater determinants with different parameters X^{sp} and Ω but with identical quantum numbers S and S_z (spin-degenerated functions) can be combined, yielding more general eigenfunctions of the $\hat{S}^2$ and $\hat{S}_z$ operators.

An N-electron nonrelativistic Hamiltonian, $\hat{H}$, of chemical interest possesses pairwise interactions satisfying the well-known commutation relations $[\hat{H}, \hat{S}^2] = 0$ and $[\hat{H}, \hat{S}_z] = 0$. Within the CI treatment methodology, this Hamiltonian is usually projected onto an N-electron Slater determinant set of $S_z \geq 0$ quantum number, leading to an N-electron Hamiltonian matrix. If the determinants of the set onto which the Hamiltonian is projected have been cataloged according to the parameters X^{sp} or Ω, diagonalizing this matrix one finds spin-adapted eigenstates corresponding to all possible spin quantum numbers $S = S_z, \ldots, S_{\max}$ ($S_{\max} = \min\left(\frac{N}{2}, K - \frac{N}{2}\right)$). Contrary to this situation, one can obtain spin-contaminated states, in systems described by open-shell reference determinants, if the excitation levels of the determinants arise from the expectation value of the operator in Eq. (1). The N-electron determinant set can be the union of sets corresponding to several values of X^{sp} and Ω, where the determinants common to both subsets are only taken into account once. A preliminary study of this procedure, limited to closed-shell systems, has been reported in Ref. 12 attempting to describe systems which present dynamic and static correlation. These union hybrid configuration interaction methods hereafter will be denoted by CI-$\bigcup_{X=0,\ldots,X_q^{sp}}^{\Omega=\Omega_1,\ldots,\Omega_p}$. In this notation, $\Omega_1, \ldots, \Omega_p$ is a sequence with extremes Ω_1 and Ω_p and some intermediate values meaning that all determinants of these seniority numbers have been selected, and similarly for the sequence of excitation levels $0, \ldots, X_q^{sp}$ which starts with $X^{sp} = 0$ and ends at a given X_q^{sp} value. Obviously, the case CI-$\bigcup_{X=0,\ldots,X_{\max}^{sp}}^{\Omega=2S,\ldots,\Omega_{\max}}$ corresponds to the FCI treatment, which requires the use of all Ω values, from the minimum $\Omega_{\min} = 2S$ to the maximum $\Omega_{\max} = \min(N, 2K - N)$;[12] it contains the total number of possible determinants and, consequently, it also includes all possible excitations, from the zeroth level to the highest one, $X_{\max}^{sp} = \min(N, K - 2S, 2K - N)$ for the substates $S_z = S$. Alternatively, the selection of the determinants can be limited to the intersection of sets defined by the series of parameters X^{sp} and Ω, what will be denoted by CI-$\bigcap_{X=0,\ldots,X_q^{sp}}^{\Omega=\Omega_1,\ldots,\Omega_p}$; this kind of methods has been proposed for closed-shell systems in Ref. 14, trying to reduce the computational cost of the seniority number CI methodology. Similarly to the CI-$\bigcup_{X=0,\ldots,X_{\max}^{sp}}^{\Omega=2S,\ldots,\Omega_{\max}}$ method, the CI-$\bigcap_{X=0,\ldots,X_{\max}^{sp}}^{\Omega=2S,\ldots,\Omega_{\max}}$ one coincides with the FCI treatment. These procedures offer a wide variety of possibilities whose behaviors and computational costs deserve to be analyzed in the case of open-shell systems. Moreover, the results for the energy and $\langle \hat{S}^2 \rangle$ quantity arising from the proposed spatial selection procedure must be compared with those obtained when the excitation levels are selected according to the expectation values of the operator in Eq. (1).

3. RESULTS AND DISCUSSION

We have selected several N-electron Slater determinant sets cataloged by the parameters $\Omega = \langle\hat{\Omega}\rangle$ and $X = \langle\hat{X}^{sp}\rangle$ (in the spatial-orbital treatment) as well as by $\Omega = \langle\hat{\Omega}\rangle$ and $X = \langle\hat{X}^{so}\rangle$ (in the spin-orbital one), to construct the projected N-electron Hamiltonian matrices for open-shell systems in ground states. These N-electron matrices have been expressed in the canonical Hartree–Fock molecular orbitals and the molecular orbitals minimizing the energy of all the CI-expansions (as is well known, the CI methods are not invariant to arbitrary orbital rotations[7] except in the FCI limit case). The excitation levels of the determinants have been calculated with respect to those restricted open-shell Hartree–Fock (ROHF) Slater ones when the canonical molecular orbitals have been used; otherwise, the references have been built from the optimized orbital basis sets. To obtain an affordable computational cost we have used the one-electron minimal STO-3G atomic basis sets. Calculations have been carried out with all electrons and all orbitals correlated. The lowest energies of each spin symmetry and the expectation values $\langle\hat{S}^2\rangle$ have been described at optimized FCI/STO-3G geometries and at stretched ones (in one of the systems). The standard one- and two-electron integrals required to formulate the N-electron Hamiltonians of these species have been obtained from a modified version of the PSI 3.3 package.[26] The determination of energy and $\langle\hat{S}^2\rangle$ values of all CI methods described comes from our own codes, based on Ref. 27.

In Tables 1 and 2 we report results corresponding to doublet (CO^+ molecule-ion) and triplet (CH_2 diradical) states, respectively; these results arise from optimized geometries and canonical Hartree–Fock molecular orbital basis sets. In these calculations we have assumed the largest Abelian subgroup of the point group describing the full symmetry of each compound, which is selected by the PSI 3.3 package ($C_{2v} \subset C_{\infty v}$ or $D_{2h} \subset D_{\infty h}$). In both tables we specify FCI results on the first line which, as mentioned in previous section, arise from the projection of the Hamiltonians onto sets of all possible determinants of $S_z = S$ value constructed with the used one-electron basis set; these sets can be described either in terms of the seniority number parameter $\Omega = \Omega_{min}, \ldots, \Omega_{max}$ or in terms of the excitation one $X = 0, \ldots, X_{max}$. Both sets are identical and consequently their intersection set is again the same. The results reported in lines 2, 3, and 4 of these tables correspond to the configuration interaction at single excitations

Table 1 Number of Determinants and Energies (in E_h) Arising From CI Methods for the Doublet System $CO^+(^2\Sigma^+)$ at FCI/STO-3G Optimized Geometry

	Spatial-Orbital Treatment		Spin-Orbital Treatment	
CI Method	**No. Determ.**	**Energy**	**No. Determ.**	**Energy**
$\bigcap_{X=0-X_{max}}^{\Omega=1-\Omega_{max}}$ (FCI)	25,200	−110.961561	25,200	−110.961561
$\bigcap_{X=0,1}^{\Omega=1-\Omega_{max}}$ (CIS)	64	−110.808622	46	−110.802231
$\bigcap_{X=0-2}^{\Omega=1-\Omega_{max}}$ (CISD)	946	−110.946612	703	−110.946384
$\bigcap_{X=0-3}^{\Omega=1-\Omega_{max}}$ (CISDT)	5,300	−110.952460	4,220	−110.951380
$\bigcap_{X=0-X_{max}}^{\Omega=1}$ ($\Omega = 1$)	840	−110.871825	840	−110.871825
$\bigcap_{X=0-X_{max}}^{\Omega=1,3}$ ($\Omega = 1, 3$)	8,400	−110.910133	8,400	−110.910133
$\bigcap_{X=0,1}^{\Omega=1}$	10	−110.797197	10	−110.797197
$\bigcap_{X=0-2}^{\Omega=1}$	64	−110.869018	64	−110.869018
$\bigcap_{X=0-3}^{\Omega=1}$	190	−110.869482	190	−110.869482
$\bigcap_{X=0,1}^{\Omega=1,3}$	64	−110.808622	46	−110.802231
$\bigcap_{X=0-2}^{\Omega=1,3}$	496	−110.901978	433	−110.901975
$\bigcap_{X=0-3}^{\Omega=1,3}$	1,900	−110.905897	1,790	−110.905009
$\bigcup_{X=0,1}^{\Omega=1}$	894	−110.885796	876	−110.878489
$\bigcup_{X=0-2}^{\Omega=1}$	1,722	−110.950882	1,479	−110.950662
$\bigcup_{X=0-3}^{\Omega=1}$	5,950	−110.955263	4,870	−110.954101
$\bigcup_{X=0,1}^{\Omega=1,3}$	8,400	−110.910133	8,400	−110.910133
$\bigcup_{X=0-2}^{\Omega=1,3}$	8,850	−110.956276	8,670	−110.956093
$\bigcup_{X=0-3}^{\Omega=1,3}$	11,800	−110.957628	10,900	−110.957203

Results correspond to Hartree–Fock molecular orbitals arising from standard STO-3G atomic basis sets. Molecular orbital occupation numbers in the reference-ROHF determinant: $\sigma^2\sigma^2\sigma^2\sigma^2\pi^2\pi^2\sigma^1\pi^0\pi^0\sigma^0$; $E_{ROHF} = -110.797197E_h$; $\Omega_{max} = 7$; $X_{max} = 7$.

(CIS), at single and double excitations (CISD), and at single, double, and triple ones (CISDT), respectively. The subsets of determinants described by these excitations, $X = 0, 1$, $X = 0–2$, and $X = 0–3$ are included in the whole set $\Omega = \Omega_{min}, \ldots, \Omega_{max}$; consequently, in our formalism, they

Table 2 Number of Determinants and Energies (in E_h) Arising From CI Methods for the Triplet System $CH_2(^3B_1)$ at FCI/STO-3G Optimized Geometry

CI Method	Spatial-Orbital Treatment			Spin-Orbital Treatment		
	No. Determ.	Energy (A)	Energy (B)	No. Determ.	Energy (A)	Energy (B)
$\bigcap_{X=0-X_{max}}^{\Omega=2-\Omega_{max}}$ (FCI)	735	−38.474746	−38.474746	735	−38.474746	−38.474746
$\bigcap_{X=0,1}^{\Omega=2-\Omega_{max}}$ (CIS)	35	−38.438238	−38.455016	23	−38.434237	−38.451219
$\bigcap_{X=0-2}^{\Omega=2-\Omega_{max}}$ (CISD)	234	−38.474020	−38.474281	171	−38.473982	−38.474005
$\bigcap_{X=0-3}^{\Omega=2-\Omega_{max}}$ (CISDT)	558	−38.474433	−38.474745	475	−38.474292	−38.474704
$\bigcap_{X=0-X_{max}}^{\Omega=2}$ $(\Omega = 2)$	210	−38.449096	−38.468084	210	−38.449096	−38.468084
$\bigcap_{X=0-X_{max}}^{\Omega=2,4}$ $(\Omega = 2, 4)$	630	−38.461601	−38.474726	630	−38.461601	−38.474726
$\bigcap_{X=0,1}^{\Omega=2}$	11	−38.430850	−38.434888	11	−38.430850	−38.434888
$\bigcap_{X=0-2}^{\Omega=2}$	57	−38.448995	−38.465874	57	−38.448995	−38.465874
$\bigcap_{X=0-3}^{\Omega=2}$	129	−38.449019	−38.467744	129	−38.449019	−38.467744
$\bigcap_{X=0,1}^{\Omega=2,4}$	35	−38.438238	−38.455016	23	−38.434237	−38.451219
$\bigcap_{X=0-2}^{\Omega=2,4}$	189	−38.461134	−38.473913	153	−38.461121	−38.473779
$\bigcap_{X=0-3}^{\Omega=2,4}$	453	−38.461486	−38.474668	401	−38.461381	−38.474644
$\bigcup_{X=0,1}^{\Omega=2}$	234	−38.457048	−38.470934	222	−38.452784	−38.469557

Continued

Table 2 Number of Determinants and Energies (in E_h) Arising From CI Methods for the Triplet System $CH_2(^3B_1)$ at FCI/STO-3G Optimized Geometry—cont'd

	Spatial-Orbital Treatment			Spin-Orbital Treatment		
CI Method	**No. Determ.**	**Energy (A)**	**Energy (B)**	**No. Determ.**	**Energy (A)**	**Energy (B)**
$\bigcup_{X=0-2}^{\Omega=2}$	387	−38.474382	−38.474532	324	−38.474344	−38.474412
$\bigcup_{X=0-3}^{\Omega=2}$	639	−38.474746	−38.474746	556	−38.474595	−38.474739
$\bigcup_{X=0,1}^{\Omega=2,4}$	630	−38.461601	−38.474726	630	−38.461601	−38.474726
$\bigcup_{X=0-2}^{\Omega=2,4}$	675	−38.474745	−38.474746	648	−38.474716	−38.474745
$\bigcup_{X=0-3}^{\Omega=2,4}$	735	−38.474746	−38.474746	704	−38.474737	−38.474746

Results correspond to (A) Hartree–Fock molecular orbitals arising from standard STO-3G atomic basis sets; (B) orbitals which minimize the energy of all CI-expansions by means of orbital rotations (using all orbitals of the basis set). Molecular orbital occupation numbers in the reference-ROHF determinant: $a_1^2 a_1^2 b_2^2 b_1^1 a_1^1 b_2^0 a_1^0$; $E_{ROHF} = -38.430850\ E_h$; $\Omega_{max} = 6$; $X_{max} = 5$.

can be represented by the corresponding intersection set. Similarly, the results in lines 5 and 6 arise from a pure seniority number CI approach and correspond to the lower values of the **Ω** parameter. All these values have been included in those tables to be compared with those obtained from hybrid CI methods of intersection and union types reported in the rest of the lines.

The results shown in Table 1 (a doublet system) present a spin contamination in most of the states described within the spin-orbital scheme reaching a value $\langle \hat{S}^2 \rangle = 0.78604$ in the CIS method while the spatial-orbital treatments always yield values $\langle \hat{S}^2 \rangle = 0.75000$. The energy values of the pure spin states are lower than the spin-contaminated counterparts, what shows the approximated methods yielding spin-adapted wave functions are more suitable, without a significant increase of computational cost. These features can also be observed in Table 2, in which we describe the methylene radical in its triplet ground state reaching up $\langle \hat{S}^2 \rangle = 2.00852$ values for spin contamination in the spin-orbitals treatments and with $\langle \hat{S}^2 \rangle =$ 2.00000 values for the spatial-orbital methods. The number of Slater determinants contained in the intersection sets defined by the parameters Ω and X is markedly lower than in their counterpart pure or union set CI methods. Consequently, the energies obtained from the intersection methods are higher, since they arise from shorter determinant expansions. The CI-$\bigcap_{X=0,1}^{\Omega=1}$, CI-$\bigcap_{X=0,1}^{\Omega=2}$, and CI-$\bigcap_{X=0,1}^{\Omega=3}$ methods for doublet and triplet, respectively, require the addition of very few Slater determinants to the reference one; they do not alter the restricted open-shell Hartree–Fock energies (E_{ROHF}) and do not yield spin-contaminated wave functions, meaning such configurations do not contribute to the approximate wave function expansion. All these results also indicate that the counterpart union methods CI-$\bigcup_{X=0,1}^{\Omega=1}$, CI-$\bigcup_{X=0,1}^{\Omega=2}$, and CI-$\bigcup_{X=0,1}^{\Omega=3}$ lower the energy with respect to the ROHF one. As shown in Table 1, the CI-$\bigcap_{X=0-3}^{\Omega=1,3}$ method yields energies close to those of the pure seniority number CI-$\bigcap_{X=0-X_{max}}^{\Omega=1,3}$ (Ω = 1, 3) method, at a drastically lower computational cost. In Table 2 we show the results for the triplet system obtained from the methods CI-$\bigcap_{X=0-3}^{\Omega=2,4}$. These results are also close to those obtained from the pure seniority number CI-$\bigcap_{X=0-X_{max}}^{\Omega=2,4}$ (Ω = 2, 4) one. All these results point out the practical usefulness of the intersection methods; they approach their pure seniority number CI counterparts (including static correlation) at a considerable reduced computational cost. We must highlight the results arising from the methods CI-$\bigcup_{X=0-2}^{\Omega=\Omega_{\min}}$ and CI-$\bigcup_{X=0-3}^{\Omega=\Omega_{\min}}$ in relation to the FCI ones. In the case of doublet system CO^+ (Table 1) the methods CI-$\bigcup_{X=0-2}^{\Omega=1}$ and CI-$\bigcup_{X=0-3}^{\Omega=1}$ yield energy values (spatial-orbital version) -110.950882 E_h and -110.955263 E_h while the FCI gives -110.961561 E_h, with computational costs 1722, 5950, and 25,200 determinants, respectively. The numerical determinations shown in Table 2 for triplet state yield similar behaviors if we compare the results arising from the CI-$\bigcup_{X=0-2}^{\Omega=2}$, CI-$\bigcup_{X=0-3}^{\Omega=2}$, and FCI methods. These results show the improvement of the numerical values obtained from the pure minimum seniority number CI methods when the set of determinants projecting the Hamiltonian is extended with the set of lower excitation levels.

To complement this type of study we have gathered, in Table 2, other results corresponding to the methylene radical at optimized geometry. These results have been obtained performing a minimization of the energy for each CI expansion by means of orbital rotations. A comparison of these results with their counterparts also reported in Table 2 shows that greatest energy differences are shown mainly by the pure and hybrid seniority-based methods. This behavior has also been observed for the other molecular systems studied. All these results confirm the strong dependence of these seniority-based methods on the used basis set, which has been pointed out in other works for closed-shell systems,[7,9,12,14] as well as the possibility of improving this type of results if optimized orbitals are used instead of those Hartree–Fock ones.

In closed-shell systems, the hybrid treatments have proven to be specially useful to describe dissociation or bond-breaking processes presenting both dynamic and static correlation, depending on their geometrical arrangements.[12] To explore the behavior of the proposed spin contamination-free methods along dissociation paths, in the open-shell case, we have chosen the linear hydrogen chain H_7 in its ground state, which is a benchmark system for testing new methods and represents a very challenging case of breaking many bonds. We report, in Table 3, results arising from optimized orbital basis sets for that system at several internuclear distances R_{H-H}. We have calculated the nonparallelity error (NPE) for each method (the difference between the maximum and minimum deviation from the FCI energy values in the interval of bond distances R_{H-H} studied). The H_7 chain presents predominantly static correlation in its symmetrically stretched geometries and, consequently, the seniority number CI methods provide a more suitable

Table 3 Number of Determinants, Energies (in E_h) and Nonparallelity Errors (NPE) Arising From CI Methods for the Symmetric Dissociation of the Ground State Linear Chain $H_7(^2\Sigma_u^+)$ in the Spatial-Orbital Treatment

CI Method	No. Determ.	Energy				NPE
		0.5 Å	1.0 Å	1.5 Å	2.0 Å	
$\bigcap_{X=0-X_{max}}^{\Omega=1-\Omega_{max}}$ (FCI)	1225	−2.377684	−3.735490	−3.484820	−3.319563	—
$\bigcap_{X=0,1}^{\Omega=1-\Omega_{max}}$ (CIS)	34	−2.359550	−3.673101	−3.311253	−2.982748	0.3187
$\bigcap_{X=0-2}^{\Omega=1-\Omega_{max}}$ (CISD)	259	−2.377345	−3.728787	−3.430805	−3.165486	0.1537
$\bigcap_{X=0-3}^{\Omega=1-\Omega_{max}}$ (CISDT)	744	−2.377660	−3.734540	−3.483909	−3.267026	0.0525
$\bigcap_{X=0-4}^{\Omega=1-\Omega_{max}}$ (CISDTQ)	1119	−2.377684	−3.735485	−3.484810	−3.319559	<0.0001
$\bigcap_{X=0-X_{max}}^{\Omega=1}$ ($\Omega = 1$)	140	−2.358448	−3.696174	−3.448042	−3.304486	0.0242
$\bigcap_{X=0-X_{max}}^{\Omega=1,3}$ ($\Omega = 1, 3$)	770	−2.377000	−3.734178	−3.483144	−3.318770	0.0010
$\bigcap_{X=0,1}^{\Omega=1}$	7	−2.342917	−3.619286	−3.237604	−2.840664	0.4441
$\bigcap_{X=0-2}^{\Omega=1}$	34	−2.357674	−3.684688	−3.394301	−3.156930	0.1426
$\bigcap_{X=0-3}^{\Omega=1}$	70	−2.358425	−3.693982	−3.417059	−3.167314	0.1330
$\bigcap_{X=0-4}^{\Omega=1}$	115	−2.358448	−3.695781	−3.440819	−3.276667	0.0248
$\bigcap_{X=0,1}^{\Omega=1,3}$	34	−2.359550	−3.673101	−3.311253	−2.982748	0.3187
$\bigcap_{X=0-2}^{\Omega=1,3}$	169	−2.375968	−3.724878	−3.425913	−3.161032	0.1568

Continued

Table 3 Number of Determinants, Energies (in E_h) and Nonparallelity Errors (NPE) Arising From CI Methods for the Symmetric Dissociation of the Ground State Linear Chain $H_7(^2\Sigma_u^+)$ in the Spatial-Orbital Treatment—cont'd

CI Method	No. Determ.	Energy				NPE
		0.5 Å	1.0 Å	1.5 Å	2.0 Å	
$\bigcap_{X=0-3}^{\Omega=1,3}$	439	−2.376920	−3.732269	−3.453935	−3.241648	0.0772
$\bigcap_{X=0-4}^{\Omega=1,3}$	664	−2.377002	−3.733937	−3.478972	−3.298642	0.0202
$\bigcup_{X=0,1}^{\Omega=1}$	167	−2.369921	−3.714751	−3.452424	−3.306457	0.0246
$\bigcup_{X=0-2}^{\Omega=1}$	365	−2.377412	−3.731666	−3.466279	−3.308474	0.0183
$\bigcup_{X=0-3}^{\Omega=1}$	814	−2.377675	−3.735126	−3.484022	−3.310441	0.0091
$\bigcup_{X=0-4}^{\Omega=1}$	1144	−2.377684	−3.735482	−3.484816	−3.316080	0.0035
$\bigcup_{X=0,1}^{\Omega=1,3}$	770	−2.377000	−3.734178	−3.483144	−3.318770	0.0010
$\bigcup_{X=0-2}^{\Omega=1,3}$	860	−2.377640	−3.735104	−3.484242	−3.318876	0.0006
$\bigcup_{X=0-3}^{\Omega=1,3}$	1075	−2.377683	−3.735472	−3.484660	−3.319092	0.0005
$\bigcup_{X=0-4}^{\Omega=1,3}$	1225	−2.377684	−3.735490	−3.484820	−3.319563	0.0000

Results correspond to molecular orbitals arising from standard STO-3G atomic basis sets. For all CI-expansions the energy is minimized by performing orbital rotations (using all orbitals of the basis set). $E_{ROHF} = -2.332387\ E_h$ ($R_{H-H} = 0.5$ Å), $E_{ROHF} = -3.619286\ E_h$ ($R_{H-H} = 1.0$ Å), $E_{ROHF} = -3.208888\ E_h$ ($R_{H-H} = 1.5$ Å), $E_{ROHF} = -2.790254\ E_h$ ($R_{H-H} = 2.0$ Å). $\Omega_{max} = 7$. $X_{max} = 6$.

description. This behavior is confirmed by the results shown in Table 3 in which the values of the NPE quantity arising from the pure seniority number CI methods ($\bigcap_{X=0-X_{\max}}^{\Omega=1}(\Omega = 1)$ and $\bigcap_{X=0-X_{\max}}^{\Omega=1,3}(\Omega = 1,\ 3)$) are much lower than those obtained from the pure excitation CI ones ($\bigcap_{X=0,1}^{\Omega=1-\Omega_{\max}}$(CIS), $\bigcap_{X=0-2}^{\Omega=1-\Omega_{\max}}$(CISD), and $\bigcap_{X=0-3}^{\Omega=1-\Omega_{\max}}$(CISDT)). In the intersection methods the low excitation levels restrict the number of determinants involved in a determined Ω value, yielding higher values of the NPE quantity than the corresponding pure seniority number CI methods although they require a considerable lower computational cost, due to their polynomial scaling with orbital basis set size. It must be noted that determinant subspaces corresponding to quadruple excitations must be included in these CI expansions to properly describe the many-bond-breaking process involved in the symmetric stretching of this molecular system. Fig. 1 shows that at near-equilibrium bond distances the intersection method proposed (CI-$\bigcap_{X=0-4}^{\Omega=1}$) yields results very close to those obtained from the seniority-based procedure CI-$\bigcap_{X=0-X_{\max}}^{\Omega=1}(\Omega = 1)$, at lower computational cost; however, at stretched configurations, the accuracy of the results begins

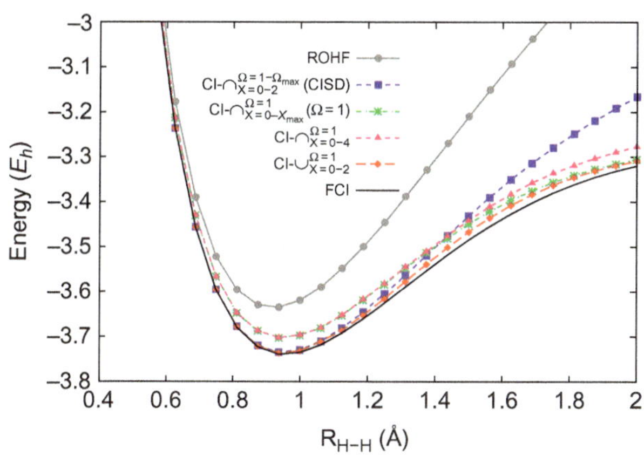

Fig. 1 Potential energy curve representing the symmetric dissociation of the H_7 linear chain for the ground state arising from the intersection hybrid method CI-$\bigcap_{X=0-4}^{\Omega=1}$ and union hybrid method CI-$\bigcup_{X=0-2}^{\Omega=1}$ and comparison with those obtained from the ROHF, CI-pure-excitation-based CISD, CI-pure-seniority-based $\Omega = 1$, and FCI methods. Results correspond to molecular orbitals arising from standard STO-3G atomic basis sets. For all CI-expansions the energy is minimized by performing orbital rotations (using all orbitals of the basis set).

to deteriorate due to the increasing static correlation. The union hybrid methods provide very low NPE values, e.g., the CI-$\bigcup_{X=0-2}^{\Omega=1}$ method gives NPE $= 0.0183$ with 365 determinants, which is a quarter of those required in the FCI method (1225). Fig. 1 also allows one to show that the union method (CI-$\bigcup_{X=0-2}^{\Omega=1}$) incorporates the dynamic correlation missing in the results of the pure seniority-based method (CI-$\bigcap_{X=0-X_{\max}}^{\Omega=1}(\Omega = 1)$) providing better results at near-equilibrium bond distances, while at the same time it leads to close results to that method at stretched distances, where the strong correlation dominates.

4. CONCLUDING REMARKS

We have proposed to project an N-electron Hamiltonian by means of sets of N-electron Slater determinants selected by hybrid procedures, based on seniority number and excitation number schemes, within the CI methodology framework. We show that the selection of determinants with excitation number defined in terms of spatial orbitals ensures spin contamination-free wave functions in open- and closed-shell systems, while the spin-orbital criterion can lead to spin contamination. Our proposal is based on the commutation relation between the N-electron seniority number operator and the recently reported spatial excitation one. We have tested this theoretical achievement performing numerical determinations in doublet and triplet systems, proving that the energies obtained from our determinant-union-set hybrid methods improve on those arising from the individual seniority number and spin-orbital-excitation-based treatments. An analysis of closeness to the exact energies (the FCI ones) shows that the pure seniority-level methods can be improved at a reasonable increase of computational cost, if the set of N-electron determinants which projects the Hamiltonian is extended to those possessing low excitation levels or vice versa, although in this latest case its cost is higher. Moreover, we have found energy results close to the pure seniority number CI treatments, with an important reduction of the computational cost, when one uses hybrid methods whose N-electron Slater determinants belong to the intersection sets between the excitation levels and the seniority number ones. Other possibilities of using union and intersection hybrid methods are currently being explored in our laboratories.

ACKNOWLEDGMENTS

D.R.A., G.E.M., and P.C. acknowledge support from the Universidad de Buenos Aires (Argentina) through research Grant No. 20020150100157BA. D.R.A. acknowledges the Consejo Nacional de Investigaciones Científicas y Técnicas (Argentina) for research Grant No. PIP 11220130100377CO. D.R.A., G.E.M., and O.B.O. thank the Consejo Nacional de Investigaciones Científicas y Técnicas (Argentina) for financial support of the Project No. 2013-1401PCB. A.T. and L.L. acknowledge the Universidad del Pais Vasco (Spain) for the Grant No. EHU16/10. O.B.O. acknowledges the Consejo Nacional de Investigaciones Científicas y Técnicas (Argentina) for research Grant No. PIP 11220130100311CO. P.C. thanks the Consejo Nacional de Investigaciones Científicas y Técnicas (Argentina) for research Grant No. PIP 11220150100442CO.

REFERENCES

1. Bytautas, L.; Ivanic, J.; Ruedenberg, K. Split-Localized Orbitals Can Yield stronger Configuration Interaction Convergence Than Natural Orbitals. *J. Chem. Phys* **2003**, *119*(16), 8217–8224.
2. Bunge, C. F. Selected Configuration Interaction With Truncation Energy Error and Application to the Ne Atom. *J. Chem. Phys.* **2006**, *125*(1), 014107.
3. Evangelista, F. A. Adaptive Multiconfigurational Wave Functions. *J. Chem. Phys* **2014**, *140*(12), 124114.
4. Alcoba, D. R.; Torre, A.; Lain, L.; Massaccesi, G. E.; Oña, O. B.; Ayers, P. W.; Van Raemdonck, M.; Bultinck, P.; Van Neck, D. Performance of Shannon-Entropy Compacted N-Electron Wave Functions for Configuration Interaction Methods. *Theor. Chem. Acc.* **2016**, *135*(6), 153.
5. Böhm, K.-H.; Auer, A. A.; Espig, M. Tensor Representation Techniques for Full Configuration Interaction: A Fock Space Approach Using the Canonical Product Format. *J. Chem. Phys.* **2016**, *144*(24), 244102.
6. Karafiloglou, P. An Efficient Generalized Polyelectron Population Analysis in Orbital Spaces: The Hole-Expansion Methodology. *J. Chem. Phys* **2009**, *130*(16), 164103.
7. Bytautas, L.; Henderson, T. M.; Jiménez-Hoyos, C. A.; Ellis, J. K.; Scuseria, G. E. Seniority and Orbital Symmetry as Tools for Establishing a Full Configuration Interaction Hierarchy. *J. Chem. Phys* **2011**, *135*(4), 044119.
8. Alcoba, D. R.; Torre, A.; Lain, L.; Massaccesi, G. E.; Oña, O. B. Seniority Number in Spin-Adapted Spaces and Compactness of Configuration Interaction Wave Functions. *J. Chem. Phys.* **2013**, *139*(8), 084103.
9. Limacher, P. A.; Kim, T. D.; Ayers, P. W.; Johnson, P. A.; De Baerdemacker, S.; Van Neck, D.; Bultinck, P. The Influence of Orbital Rotation on the Energy of Closed-Shell Wavefunctions. *Mol. Phys.* **2014**, *112*(5–6), 853–862.
10. Boguslawski, K.; Tecmer, P.; Limacher, P. A.; Johnson, P. A.; Ayers, P. W.; Bultinck, P.; De Baerdemacker, S.; Van Neck, D. Projected Seniority-Two Orbital Optimization of the Antisymmetric Product of One-reference Orbital Geminal. *J. Chem. Phys.* **2014**, *140*(21), 214114.
11. Lain, L.; Torre, A.; Alcoba, D. R.; Oña, O. B.; Massaccesi, G. E. A Study of the Compactness of Wave Functions Based on Shannon Entropy Indices: A Seniority Number Approach. *Theor. Chem. Acc.* **2015**, *134*(7), 85.
12. Alcoba, D. R.; Torre, A.; Lain, L.; Oña, O. B.; Capuzzi, P.; Van Raemdonck, M.; Bultinck, P.; Van Neck, D. A Hybrid Configuration Interaction Treatment Based on Seniority Number and Excitation Schemes. *J. Chem. Phys.* **2014**, *141*(24), 244118.

13. Bytautas, L.; Scuseria, G. E.; Ruedenberg, K. Seniority Number Description of Potential Energy Surfaces: Symmetric Dissociation of Water, N_2, C_2, and Be_2. *J. Chem. Phys.* **2015**, *143*(9), 094105.
14. Van Raemdonck, M.; Alcoba, D. R.; Poelmans, W.; De Baerdemaker, S.; Torre, A.; Lain, L.; Massaccesi, G. E.; Van Neck, D.; Bultinck, P. Polynomial Scaling Approximations and Dynamic Correlation Corrections to Doubly Occupied Configuration Interaction Wave Functions. *J. Chem. Phys* **2015**, *143*(10), 104106.
15. Mentel, Ł. M.; van Meer, R.; Gritsenko, O. V.; Baerends, E. J. The Density Matrix Functional Approach to Electron Correlation: Dynamic and Nondynamic Correlation Along the Full Dissociation Coordinate. *J. Chem. Phys.* **2014**, *140*(21), 214105.
16. Stein, T.; Henderson, T. M.; Scuseria, G. E. Seniority Zero Pair Coupled Cluster Doubles Theory. *J. Chem. Phys.* **2014**, *140*(21), 214113.
17. Poelmans, W.; Van Raemdonck, M.; Verstichel, B.; De Baerdemaker, S.; Torre, A.; Lain, L.; Massaccesi, G. E.; Alcoba, D. R.; Bultinck, P.; Van Neck, D. Variational Optimization of the Second-Order Density Matrix Corresponding to a Seniority-Zero Configuration Interaction Wave Function. *J. Chem. Theory Comput.* **2015**, *11*(9), 4064–4076.
18. Alcoba, D. R.; Torre, A.; Lain, L.; Massaccesi, G. E.; Oña, O. B. Configuration Interaction Wave Functions: A Seniority Number Approach. *J. Chem. Phys.* **2014**, *140*(23), 234103.
19. Alcoba, D. R.; Torre, A.; Lain, L.; Massaccesi, G. E.; Oña, O. B.; Capuzzi, P. Spin Contamination-Free N-Electron Wave Functions in the Excitation-Based Configuration Interaction Treatment. *J. Chem. Phys.* **2016**, *145*(1), 014109.
20. Torre, A.; Lain, L.; Millan, J. Calculation of Traces of p-Order Replacement Operators Over N-Electron Spin-Adapted Spaces. *Phys. Rev. A* **1993**, *47*(2), 923–928.
21. Lain, L.; Torre, A. Direct Computation of Traces of p-Order Replacement Operators Over N-Electron Spin-Adapted Spaces. *Phys. Rev. A* **1995**, *52*(3), 2446–2448.
22. Planelles, J.; Valdemoro, C.; Karwowski, J. Symmetric-Group Approach to the Study of the Traces of p-Order Reduced-Density Operators and of Products of These Operators. *Phys. Rev. A* **1990**, *41*(5), 2391–2397.
23. Valdemoro, C.; Torre, A.; Lain, L. Recent Theoretical Developments of the Spin Adapted Reduced Hamiltonian Theory and Its Application to the Study of Atoms. In: *Computational Chemistry: Structure, Interactions and Reactivity*; Fraga, S. Ed.; Elsevier: Amsterdam, 1992; pp. 90–113.
24. Torre, A.; Lain, L. Representation of the Spin Operator in the Spin-Free Second Quantized Approach. *J. Mol. Struct. (THEOCHEM)* **1998**, *426*(1–3), 25–28.
25. Alcoba, D. R.; Torre, A.; Lain, L.; Bochicchio, R. C. Determination of Local Spins by Means of a Spin-Free Treatment. *J. Chem. Theory Comput.* **2011**, 7(11), 3560–3566.
26. Crawford, T. D.; Sherrill, C. D.; Valeev, E. F.; Fermann, J. T.; King, R. A.; Leininger, M. L.; Brown, S. T.; Janssen, C. L.; Seidl, E. T.; Kenny, J. P.; Allen, W. D. PSI3: An Open-Source Ab Initio Electronic Structure Package. *J. Comput. Chem.* **2007**, *28*(9), 1610–1616.
27. Knizia, G.; Chan, G. K.-L. Density Matrix Embedding: A Simple Alternative to Dynamical Mean-Field Theory. *Phys. Rev. Lett.* **2012**, *109*(18), 186404.

CHAPTER NINETEEN

Potential Energy Curves for the LiK^+ and NaK^+ Molecular Ions With the Coupled Cluster Method

Monika Musiał[1], Anna Bewicz, Patrycja Skupin, Stanisław A. Kucharski
Institute of Chemistry, University of Silesia in Katowice, Katowice, Poland
[1]Corresponding author: e-mail address: musial@ich.us.edu.pl

Contents

1. Introduction 334
2. Synopsis of the Theory 335
3. Results and Discussion 337
   3.1 Atomic Calculations 338
   3.2 LiK^+ Molecular Ion 341
   3.3 NaK^+ Molecular Ion 344
4. Conclusions 346
Acknowledgment 347
References 347

Abstract

The EA-EOM (electron-attachment equation-of-motion) coupled cluster approach provides a description of the states obtained by attachment of a single electron to the reference system. If the reference is assumed to be a doubly ionized cation then the results relate to the cation. In the current work the above scheme is applied to the calculations of potential energy curves for the LiK^+ and NaK^+ molecular ions adopting as a reference system the doubly ionized structure, i.e., LiK^{+2} and NaK^{+2}. Such computational strategy benefits from the fact that the closed shell reference (LiK^{+2} or NaK^{+2}) dissociates into the closed shell fragments ($LiK^{+2} \Rightarrow Li^+ + K^+$, $NaK^{+2} \Rightarrow Na^+ + K^+$). This is advantageous since the RHF (restricted Hartree–Fock) function can be used as a reference in the whole range of interatomic distances. This scheme offers a *first principle* method without any model or effective potential parameters for the description of the bond-breaking processes. Moreover, the scalar relativistic effects are included by adding appropriate terms of the DK (Douglas–Kroll) Hamiltonian to the one-electron integrals.

Advances in Quantum Chemistry, Volume 76
ISSN 0065-3276
https://doi.org/10.1016/bs.aiq.2017.07.006

1. INTRODUCTION

Alkali metal diatomics are attractive objects both for the theoretical and experimental studies. The same applies to their cationic derivatives. Most of the theoretical works focused on the alkali metal diatomics replace the effect of the inner-shell electrons with the model or effective core potentials (ECP). Within this approach the neutral diatomics is treated as a two-electron system for which the full configuration interaction (FCI) treatment is easily accomplished (reduces to the CISD (CI singles and doubles)). The ionic system (AB^+) within the same approach reduces to the single electron moving within model or ECP potentials. In our approach realized in the recent work[1–5] we represent a different computational philosophy relying on a rigorous ab initio approach with all electrons included both at the Hartree–Fock (HF) and correlated level. In addition in our studies of the excited or ionized states based on the EOM (equation-of-motion) or Fock space (FS) multireference coupled cluster (MRCC) approach[6] we managed to use the RHF (restricted HF) reference which resulted in the significant simplification of the computational treatment.

In the present study we focus on the description of the electronic states of two molecular ions engaging potassium atom: LiK^+ and NaK^+. At first sight one may conclude that in order to use the RHF reference in the study of the cationic system one should employ the IP(ionization potential)-EOM scheme. This means that the reference RHF function determined for closed shell AB system would be used in the IP-EOM calculations for the AB^+ ion. However, the AB reference convenient in the calculations close to the equilibrium geometry turns out to be inadequate for the potential energy curve (PEC) evaluation since for the large internuclear distances the RHF function does not represent properly the reference. In such a situation in the calculations of PEC for AB^+ systems we have adopted a different strategy[7–9] aimed at such redefinition of the reference that the closed shell approach can be applied both to the dimer and also to its dissociation products.[1,4] In such case the RHF reference function can be used for the whole range of interatomic distances.

However, the situation mentioned above may occur in the cases when we dissociate a doubly positive ion (in our case LiK^{+2} and NaK^{+2}) instead of the neutral molecules LiK or NaK. Since in the former case we obtain closed shell products (Li^+ and K^+, Na^+, and K^+), isoelectronic with noble gas atoms, He, Ne, and Ar, respectively. This can be done if we are able to describe the molecular ion on the basis of the doubly ionized reference. Now we see that in order to do that we have to apply the EA (electron attachment) approach rather than IP since the former is capable to describe system after attachment of a single electron to the reference. Generally, the EA calculations can be illustratively written as: $AB \xrightarrow{EA} AB^-$ which in our case takes the form: $AB^{+2} \xrightarrow{EA} AB^+$. Thus, out of several possible routes to obtain the PECs for the AB^+-type molecular ions we choose the one which is based exclusively on the RHF approach. The second route includes a direct dissociation of the open shell AB^+ system using ROHF (restricted open shell HF) or UHF formalism. For both schemes we encounter some convergence problems when trying to reach the converged EOM solution for the desired electronic state. Finally, the third option relies on using IP approach for the neutral molecule AB based on the UHF reference for large interatomic distances with similar as in previous case limitations around critical geometries.

In the framework of the coupled cluster[10–13] method we can consider EA scheme formulated within the EOM theory.[14–19] The EA-EOM-CC technique seems particularly suitable for the bond-breaking problem owing to its size-extensivity which is crucial for the correct reproduction of the atomic excitation energies in the dissociation limit. In our previous work this method was applied successfully for the alkali metal molecular ions: Li_2^+,[7] Na_2^+,[8] and K_2^+.[9] Contrary to the previous studies of LiK^+ and NaK^+,[20–26] based on the effective potential methods, in this work we apply the size-extensive EA-EOM-CCSD method with all electrons correlated. This is a *first principle* computational scheme able to reproduce PECs for selected states of LiK^+ and NaK^+ without introduction of any auxiliary parameters.

In the current study we focus on calculations of the seven lowest lying electronic states (1–5$^2\Sigma^+$ and 1–2$^2\Pi$) of LiK^+ and NaK^+ ions using the EA-EOM-CCSD DK2 method. These are the first calculations of such type for these cations with the scalar relativistic effects included via DK2[27] (Douglas–Kroll second order). In the previous works[7–9] we focused on the homonuclear dimers, here we apply the same strategy to the heteronuclear diatomics.

2. SYNOPSIS OF THE THEORY

The electron-attached states, $|\Psi_k\rangle$, are parameterized via equation-of-motion coupled cluster theory[14–19] in a linear fashion as:

$$|\Psi_k\rangle = R(k)|\Psi_0\rangle \quad k = 1, 2, \ldots \tag{1}$$

with

$$R(k) = R_1(k) + R_2(k) + \cdots \tag{2}$$

$$= \sum_a r^a(k)a^\dagger + \frac{1}{2}\sum_{ab}\sum_i r_i^{ab}(k)a^\dagger b^\dagger i + \cdots \tag{3}$$

where the standard convention for indices is used, i.e., a, b, ... (i, j, ...) refer to virtual (occupied) one-particle levels. The ground state wave function is expressed as:

$$\Psi_0 = e^T|\Phi_0\rangle \tag{4}$$

and T is a cluster operator responsible for excitations; Φ_0 is a reference function.

After simple algebra we obtain eigenvalue problem:

$$\bar{H}R(k) = \omega_k R(k) \tag{5}$$

where ω_k is the energy change connected with the electron attachment process; $\bar{H}$ is the similarity transformed Hamiltonian defined, in terms of connected diagrams, as:

$$\bar{H} = e^{-T}He^T = \left(He^T\right)_c \tag{6}$$

We may summarize steps in the EA-EOM-CC calculations: (i) do CC calculations for the reference system, (ii) construct $\bar{H}$ elements using converged T amplitudes, and (iii) once amplitudes have been converged diagonalize the $\bar{H}$ matrix in the subspace of the determinants containing $N + 1$ electrons (N refers to the number of electrons in $|\Psi_0\rangle$) to obtain the sought eigenvalues.

Most of the results is obtained with the EA-EOM-CCSD (S = singles, D = doubles) model but exceptionally we also apply a more advanced model, i.e., full EA-EOM-CCSDT (T = triples). It means solving the CC equations for the reference state at the CCSDT level and next solving the EA-EOM-CC equations assuming that the $R(k)$ operator includes also the R_3 operator defined as:

$$R_3 = \frac{1}{12}\sum_{abc}\sum_{ij} r_{ij}^{abc}(k)a^\dagger b^\dagger c^\dagger ji \tag{7}$$

3. RESULTS AND DISCUSSION

All calculations are done using GAMESS[28] and ACES2[29] packages supplemented with the local version of the EA-EOM-CC module.[16] The results are collected in Figs. 1 and 2 and in Tables 1–4. The computed spectroscopic constants are obtained with help of the LEVEL-8 program of Le Roy.[30] In all electron attachment calculations the orbitals used are obtained by the RHF solution for the LiK^{+2} and NaK^{+2} systems. We used the POL1,[31] unANO-RCC+, i.e., uncontracted ANO-RCC[32] with additional diffuse functions,[3,8,9] Sapporo-QZP+diffuse[33] (acronym: SPK-AQZP) and relativistic Sapporo-DK-QZP+diffuse (acronym: SPKRAQZP)[33] basis functions. In all cases the spherical harmonic polarization functions were used with all electrons correlated. The size of the basis sets for LiK^+ is 64 (POL1), 158 (SPKRAQZP and SPK-AQZP) and 248 (unANO-RCC+); 72 (POL1), 207 (SPKRAQZP and SPK-AQZP), and 258 (unANO-RCC+) for NaK^+.

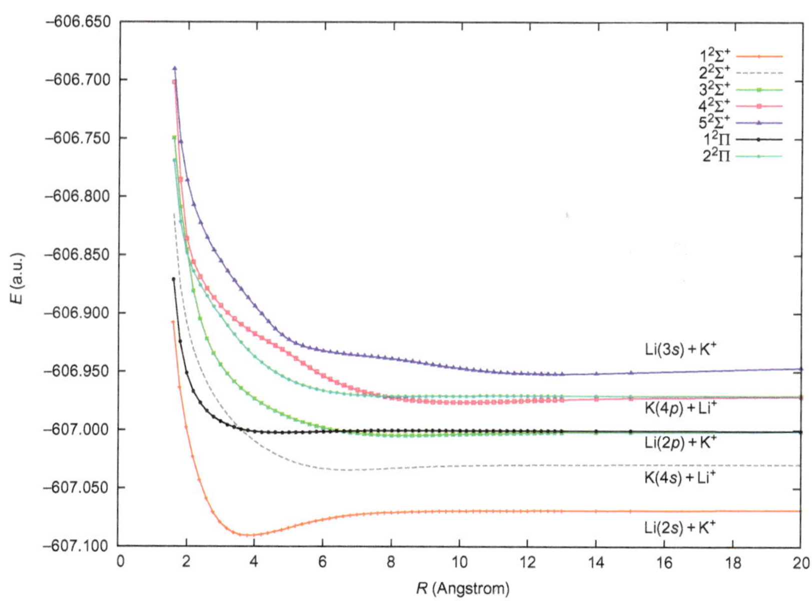

Fig. 1 Potential energy curves of the LiK^+ molecular ion with the EA-EOM-CCSD method for the five lowest dissociation limits in the unANO-RCC+ basis set.

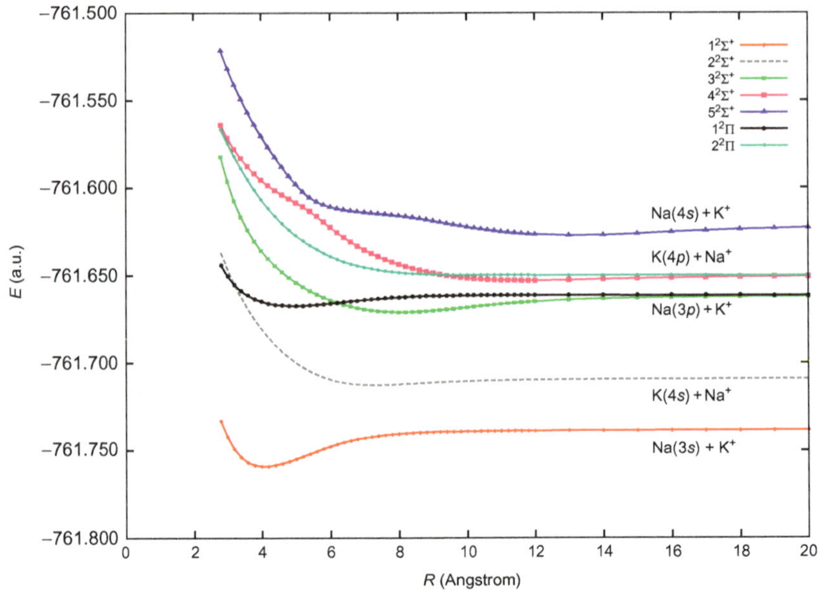

Fig. 2 Potential energy curves of the NaK^+ molecular ion with the EA-EOM-CCSD method for the five lowest dissociation limits in the unANO-RCC+ basis set.

3.1 Atomic Calculations

In our previous calculations[3,7–9] we found out that the most reliable results were obtained with the uncontracted ANO-RCC basis augmented with set of diffuse functions (unANO-RCC+). The optimum strategy would be to run the relativistic calculations for the same basis set and observe the effect of the Douglas–Kroll correction. Such a route turned out to be very difficult to realize since the failure of the DK2 calculations for the unANO-RCC+ basis set. Hence we adopted a different strategy: we performed the DK2 calculations for the best available Sapporo-type relativistic basis set and then evaluated the DK2 corrections by comparison with nonrelativistic results for the same basis set.

The DK2 corrections, found in this manner, were subsequently employed to correct the result obtained with the unANO-RCC+ basis set. This scheme may be illustrated by the results listed in Table 1. Here we have the values of the atomic energy levels for the three involved atoms: Li, Na, and K. In second column we have atomic energy levels obtained with nonrelativistic EA-EOM-CCSD calculations, while the next column provides the analogous results done with the DK2 relativistic approach. The relativistic corrections are listed in column 4. As we can see for the Li atom

Table 1 Atomic Energy Levels (eV) for Li, Na, and K (EA-EOM-CCSD Method)

Sym.	SPK-AQZP	SPKRAQZP DK2	Δ^{a}_{DK2}	Δ^{b}	unANO-RCC+	Δ^{b}	unANO-RCC++Δ_{DK2}	Δ^{b}	Exp.[c]
Li									
^2P	1.9119	1.9123	0.0004	0.0645	1.8492	0.0014	1.8496	0.0018	1.8478
^2S	3.3626	3.3629	0.0003	−0.0102	3.3707	−0.0024	3.3710	−0.0021	3.3731
^2P	4.1101	4.1105	0.0004	0.2762	3.8339	−0.0004	3.8342	−0.0001	3.8343
^2D	3.9641	3.9644	0.0003	0.0858	3.8757	−0.0029	3.8761	−0.0025	3.8786
Na									
^2P	2.0712	2.0778	0.0066	−0.0245	2.0871	−0.0152	2.0937	−0.0086	2.1023
^2S	3.1517	3.1568	0.0051	−0.0346	3.1721	−0.0193	3.1772	−0.0142	3.1914
^2D	3.6532	3.6598	0.0066	0.0428	3.5939	−0.0231	3.6005	−0.0165	3.6170
^2P	3.9753	3.9818	0.0065	0.2292	3.7311	−0.0215	3.7376	−0.0150	3.7526
K									
^2P	1.5895	1.6063	0.0168	−0.0072	1.6067	−0.0068	1.6235	0.0100	1.6135
^2S	2.5818	2.5949	0.0131	−0.0121	2.5917	−0.0153	2.6048	−0.0022	2.6070

[a]Relativistic correction.
[b]Deviation from experiment.
[c]Refs. 34–36.

the relativistic corrections are equal to 0.0003–0.0004 eV always with plus sign, i.e., increasing the energy levels. Similar behavior is observed for the Na atom but here the relativistic corrections are larger by more than order of magnitude, being equal to 0.0051–0.0066 eV. Finally for the K atom they increase to 0.0131–0.0168 eV. Note that the atomic energy levels obtained with the Sapporo basis set show rather large deviations from experimental values, of the order of a few hundredth of eV (in two cases are larger than 0.2 eV); see next column in Table 1. The situation is much improved when we go to the unANO-RCC+ results. Here agreement with the experiment is of the order of 0.001–0.003 eV for the Li atom, 0.01–0.02 eV for the Na atom and of similar accuracy for the K atom. By adding the Douglas–Kroll correction the agreement is nearly in all cases significantly improved, i.e., the DK2 correction goes to right direction.

In Table 2 we compared the results obtained with inclusion of the full T_3 operator both at the ground state (CCSDT) and at the EA-EOM level, with the standard EA-EOM-CCSD values. For the Li atom the T_3 effect is

Table 2 Atomic Energy Levels (eV) for Li, Na, and K (EA-EOM-CCSD and EA-EOM-CCSDT Methods)

Sym.	unANO-RCC+ SD	$\Delta^a_{SDT\text{-}SD}$	unANO-RCC+ SDT	Δ^bSD	Δ^bSDT	Exp.[c]
Li						
^2P	1.8492	−0.0001	1.8491	0.0014	0.0013	1.8478
^2S	3.3707	0.0002	3.3709	−0.0024	−0.0022	3.3731
^2P	3.8339	0.0004	3.8341	−0.0004	−0.0002	3.8343
^2D	3.8757	0.0004	3.8761	−0.0029	−0.0025	3.8786
Na						
^2P	2.0871	0.0092	2.0963	−0.0152	−0.0060	2.1023
^2S	3.1721	0.0105	3.1826	−0.0193	−0.0088	3.1914
^2D	3.5939	0.0128	3.6067	−0.0231	−0.0103	3.6170
^2P	3.7311	0.0120	3.7431	−0.0215	−0.0095	3.7526
K						
^2P	1.6067	−0.0028	1.6039	−0.0068	−0.0096	1.6135
^2S	2.5917	−0.0039	2.5878	−0.0153	−0.0192	2.6070

[a]Triples correction.
[b]Deviation from experiment.
[c]Refs. 34–36.

small, of the order of 0.0001–0.0002 eV; in all cases improving the results. For the Na atom the T_3 contribution in all cases but one exceeds 0.01 eV and brings the error practically below 0.01 eV. Only for the first term ^2P the T_3 correction is smaller and equal to 0.0092eV. Contrary to the expectation the T_3 effect for the K atom is smaller than for Na case, equal to 0.0038 and 0.0039 eV and moves the results away from the experimental results. This indicates that the basis set for the K atom requires improvement.

3.2 LiK$^+$ Molecular Ion

For the LiK$^+$ system we investigated seven electronic states correlating to five dissociation limits. These are (states in parenthesis) Li(2*s*) + K$^+$ ($1^2\Sigma^+$), Li$^+$ + K(4*s*) ($2^2\Sigma^+$), Li(2*p*) + K$^+$ ($3^2\Sigma^+$, $1^2\Pi$), Li$^+$ + K(4*p*) ($4^2\Sigma^+$, $2^2\Pi$), Li(3*s*) + K$^+$ ($5^2\Sigma^+$). All these PECs together with their asymptotic energy values are shown in Fig. 1.

In Table 3 we collect selected spectroscopic constants computed with two basis sets: SPKRAQZP used in the relativistic second-order Douglas–Kroll calculations and the unANO-RCC+ employed in the nonrelativistic calculations. The same computational strategy has been adopted as in the case of atomic calculations discussed above, i.e., the calculations done with the SPK-AQZP were used to evaluate the relativistic correction to the given spectroscopic constant which subsequently was added to the nonrelativistic value obtained with the unANO-RCC+ basis set. For the ground state only ($1^2\Sigma^+$) additional calculations were done with inclusion of the full triple operator both at the ground state and at the EOM level. This enabled us to evaluate the effect of more complete electronic correlation, however, for the smaller basis set (POL1). The inclusion of the full T_3 operator reduces the optimum bond length by 0.005 Å and increases the D_e value by 16 cm^{-1} with negligible effect on the harmonic and anharmonic frequencies.

For a given excited electronic state we give the value computed at the relativistic level, next the DK2 correction, then the nonrelativistic unANO-RCC+ value and, finally, the latter value with added relativistic correction. In most cases the relativistic correction reduces the computed equilibrium bond length by ca. 0.01 Å. Only for the $2^3\Sigma^+$ state the shortening of the bond is larger and equal to 0.035 Å. The effect of the DK2 on the D_e value in most cases exceeds 100 wave numbers and is comparable to the DK2 effect on the energy of the atomic levels of K atom, see Table 1. The excitation energy is affected to lesser degree by the relativistic

Table 3 EA-EOM-CCSD Results for Selected Electronic States of LiK^+

Sym.	R_e (Å)	D_e (cm^{-1})	T_e (cm^{-1})	ω_e (cm^{-1})	$\omega_e x_e$ (cm^{-1})	Ref.
Dissociation limit Li (2s) + K^+						
$1^2\Sigma^+$	3.842	4723	0	152.35	1.05	POL1
	3.837	4739	0	152.43	1.05	POL1(CCSDT)
	3.870	4899	0	152.21	0.94	DK2 SPKRAQZP
	−0.009	163	0	0.87	0.01	Δ_{DK2}^{a}
	3.861	4780	0	153.55	0.99	unANO-RCC+
	3.852	4943	0	154.42	1.00	unANO-RCC+ + Δ_{DK2}
	3.858	4865	0			Theory[20]
	3.884	4663		152.12	1.45	Theory[21]
Dissociation limit K (4s) + Li^+						
$2^2\Sigma^+$	6.783	1073	12,196	62.28	0.91	DK2 SPKRAQZP
	0.014	−31	−59	−0.57	0.02	Δ_{DK2}^{a}
	6.746	1098	12,292	63.21	0.91	unANO-RCC+
	6.760	1067	12,233	62.64	0.93	unANO-RCC+ + Δ_{DK2}
	6.694	1045	8477			Theory[20]
	6.768	1038	12,103	62.17	0.56	Theory[21]
Dissociation limit Li (2p) + K^+						
$3^2\Sigma^+$	8.264	1032	19,291	42.68	0.51	DK2 SPKRAQZP
	−0.035	133	33	0.36	−0.01	Δ_{DK2}^{a}
	8.376	776	18,918	40.58	0.57	unANO-RCC+
	8.341	909	18,951	40.940	0.560	unANO-RCC+ + Δ_{DK2}
	8.504	749	14,901			Theory[20]
	8.377	745	18,824	39.74	0.34	Theory[21]
$1^2\Pi$	4.703	608	19,714	62.47	1.41	DK2 SPKRAQZP
	−0.006	119	46	0.00	−0.01	Δ_{DK2}^{a}
	4.761	319	19,375	55.70	1.79	unANO-RCC+

Table 3 EA-EOM-CCSD Results for Selected Electronic States of LiK^+—cont'd

Sym.	R_e (Å)	D_e (cm^{-1})	T_e (cm^{-1})	ω_e (cm^{-1})	$\omega_e x_e$ (cm^{-1})	Ref.
	4.755	438	19,421	55.70	1.78	unANO-RCC+ + Δ_{DK2}
	4.831	246				Theory[20]
	4.868	50	9419	49.16	2.98	Theory[21]
Dissociation limit K (4p) + Li+						
$4^2\Sigma^+$	10.143	1163	25,061	39.86	0.31	DK2 SPKRAQZP DK
	−0.006	7	37	0.12	−0.01	Δ^a_{DK2}
	9.974	1242	25,108	41.29	0.33	unANO-RCC+
	9.968	1249	25,145	41.41	0.32	unANO-RCC+ + Δ_{DK2}
	10.123	5297	21,495			Theory[20]
	10.049	1247	24,916	38.89	0.89	Theory[21]
$2^2\Pi$	8.619	107	26,117	25.94	1.19	DK2 SPKRAQZP
	0.003	−1	45	−0.03	0.00	Δ^a_{DK2}
	8.444	79	26,271	25.47	1.29	unANO-RCC+
	8.447	78	26,316	25.44	1.29	unANO-RCC+ + Δ_{DK2}
	8.398	127	6597			Theory[20]
	8.858	96	26,067	20.76	14.64	Theory[21]
Dissociation limit Li (3s) + K+						
$5^2\Sigma^+$	12.368	1350	30,673	38.55	0.20	DK2 SPKRAQZP
	−0.002	120	45	−0.01	0.00	Δ^a_{DK2}
	12.829	1460	30,506	32.07	0.07	unANO-RCC+
	12.827	1580	30,551	32.06	0.07	unANO-RCC+ + Δ_{DK2}
	12.303	1519	27,208			Theory[20]
	12.833	1531	30,344	33.89	0.13	Theory[21]

[a]Relativistic correction.

treatment: in most cases it increases by 30–50 wave numbers. A negligible relativistic effect (below 1%) is observed for the listed harmonic and anharmonic frequencies.

Since the experimental data for the LiK^+ are unavailable the computed unANO-RCC+ values modified with DK2 correction can be confronted only with other theoretical works.[20,21] Both of them are based on the ECP approach simulating the effect of all inner shell electrons. The results obtained in the current work are generally much closer to those reported in Ref. 21. For example, T_e value for $2^2\Sigma^+$ computed in this work is equal to 12,233 cm^{-1}, in Ref. 21—12,103 cm^{-1} while in Ref. 20—8477 cm^{-1}. Similarly dissociation energy for the $4^2\Sigma^+$ state—this work: D_e = 1249 cm^{-1}; Ref. 21: D_e = 1247 cm^{-1}; Ref. 20: D_e = 5297 cm^{-1}. Excitation energy for the $2^2\Pi$ state—this work: T_e = 26,316 cm^{-1}; Ref. 21: 26,067 cm^{-1}; Ref. 20: 6597 cm^{-1}.

3.3 NaK^+ Molecular Ion

For the NaK^+ system the same electronic states were considered as in the case above. In Fig. 2 we present the PECs for seven lowest lying electronic states of NaK cation.

The selected molecular constants are listed in Table 4. As in the previous case, we report results of the relativistic calculations at the second-order Douglas–Kroll level for the Sapporo quadruple zeta basis set (rows indicated by DK2 SPKRAQZP) and evaluate the magnitude of the relativistic correction Δ_{DK2}. Next the nonrelativistic calculations were done for the relatively large basis set (unANO-RCC+) and these results were modified by adding the Δ_{DK2} correction. Thus the final values of the present calculations are given in the rows with the header unANO-RCC+ + Δ_{DK2}.

The magnitude of the relativistic corrections to the dissociation energy is of the order of 100–150 wave numbers except for the very low D_e values for which it goes down to a few cm^{-1}. The effect of the relativistic correction on the equilibrium bond length is small, below 0.01 Å (only for the ground state slightly more: 0.012 Å). The effect on the adiabatic excitation energy is very stable: in all but one case lie between 60 and 85 wave numbers. Only for the state $2^2\Sigma^+$ it goes down by 35 cm^{-1}.

Also in the NaK^+ case we do not have experimental data at our disposition hence we can confront our results only with other theoretical studies. In Table 4 we report results obtained by Ghanmi et al.[25] and Valence et al.[26] (the latter gives theoretical results only for three states). We observe

Table 4 EA-EOM-CCSD Results for Selected Electronic States of NaK^+

Sym.	R_e (Å)	D_e (cm^{-1})	T_e (cm^{-1})	ω_e (cm^{-1})	$\omega_e x_e$ (cm^{-1})	Ref.
Dissociation limit Na(3s) + K^+						
$1^2\Sigma^+$	4.104	4438	0	89.15	0.35	POL1
	4.092	4427	0	89.40	0.35	POL1(CCSDT)
	4.077	4826	0	91.67	0.34	DK2 SPKRAQZP
	−0.012	152	0	0.46	0.00	Δ^a_{DK2}
	4.077	4602	0	91.25	0.34	unANO-RCC+
	4.065	4754	0	91.71	0.34	unANO-RCC+ + Δ_{DK2}
		4597	0			Exp.[37]
	4.101	4442	0	91.00	0.36	Theory[25]
	4.022	4678	0			Theory[26]
Dissociation limit Na^+ + K(4s)						
$2^2\Sigma^+$	7.330	766	10,170	33.30	0.38	DK2 SPKRAQZP
	0.005	−17	−35	−0.26	0.01	Δ^a_{DK2}
	7.291	765	10,243	33.48	0.38	unANO-RCC+
	7.296	748	10,208	33.22	0.39	unANO-RCC+ + Δ_{DK2}
	7.308	742	10,142	32.63	0.32	Theory[25]
	6.879	1290				Theory[26]
Dissociation limit K^+ + Na(3p)						
$3^2\Sigma^+$	8.086	2261	19,323	34.50	0.09	DK2 SPKRAQZP
	−0.022	142	63	0.19	0.00	Δ^a_{DK2}
	8.093	2065	19,370	34.41	0.09	unANO-RCC+
	8.071	2207	19,433	34.60	0.09	unANO-RCC+ + Δ_{DK2}
	8.133	2080	19,330	34.01	0.08	Theory.[25]
$1^2\Pi$	4.976	1537	20,047	43.77	0.21	DK2 SPKRAQZP
	−0.002	119	86	−0.10	0.00	Δ^a_{DK2}
	4.933	1309	20,127	44.47	0.23	unANO-RCC+
	4.931	1428	20,213	44.37	0.23	unANO-RCC+ + Δ_{DK2}
	5.070	1119	20,292	39.92	0.44	Theory[25]

Continued

Table 4 EA-EOM-CCSD Results for Selected Electronic States of NaK^+—cont'd

Sym.	R_e (Å)	D_e (cm^{-1})	T_e (cm^{-1})	ω_e (cm^{-1})	$\omega_e x_e$ (cm^{-1})	Ref.
Dissociation limit $Na^+ + K(4p)$						
$4^2\Sigma^+$	11.819	643	23,250	18.31	0.15	DK2 SPKRAQZP
	0.002	4	82	0.01	0.00	Δ^a_{DK2}
	11.803	611	23,356	17.56	0.12	unANO-RCC+
	11.805	615	23,274	17.57	0.12	unANO-RCC+ + Δ_{DK2}
	11.854	682	23,226	17.77	0.89	Theory[25]
$2^2\Pi$	10.316	8	23,885	9.25	0.60	DK2 SPKRAQZP
	0.010	0	85	−0.07	−0.01	Δ^a_{DK2}
	Repulsive					unANO-RCC+
	Repulsive					unANO-RCC+ + Δ_{DK2}
	10.541	20	23,887	8.78	0.51	Theory[25]
	Repulsive					Theory[26]
Dissociation limit $Na\ (4s) + K^+$						
$5^2\Sigma^+$	12.741	1215	29,072	22.69	0.08	DK2 SPKRAQZP
	0.003	115	78	0.00	0.01	Δ^a_{DK2}
	13.175	1176	29,011	19.80	0.05	unANO-RCC+
	13.178	1291	29,089	19.80	0.06	unANO-RCC+ + Δ_{DK2}
	13.272	1371	28,827	19.76	0.10	Theory[25]

[a]Relativistic correction.

general compatibility of our results and those of Ref. 25. In particular for all $^2\Sigma^+$ states the agreement is very good both for D_e and T_e values. For example, for the $2^2\Sigma^+$ state the D_e in this work is equal to 748 cm^{-1} and in Ref. 25 742 cm^{-1} (the value reported in Ref. 26 is equal to 1290 cm^{-1}). The differences in the T_e values reported in this work and in Ref. 25 stay below 100 cm^{-1}.

4. CONCLUSIONS

Quantum chemical calculations from *first principles* are applied in the theoretical study of PECs for the lowest lying electronic states of LiK^+

and NaK^+ molecular ions. The results obtained are of comparable quality to other theoretical values available in the literature, obtained with model or effective potentials. In the whole range of interatomic distances the calculations were done for the RHF reference function with all electrons correlated and with the inclusion of scalar relativistic effects. The crucial element in the overall computational plan to obtain smooth and accurate PECs is to adopt the EA strategy relying (i) on the removal of one electron from the LiK^+ or NaK^+ molecular ions to obtain a convenient RHF reference and (ii) on using the EA scheme to recover the original structure by attaching the electron to the LiK^{+2} and NaK^{+2} ions. The reported data demonstrate that the rigorous ab initio calculations can be applied to determine both the short- and long-range interaction for ionic AB^+ molecules. The latter are of particular importance in the studies of interatomic forces, ultra-cold collisions and a formation of the ultra-cold molecules.[38]

ACKNOWLEDGMENT

This work has been supported by the National Science Centre, Poland under Grant No. 2013/11/B/ST4/02191.

REFERENCES

1. Musiał, M. Multireference Fock Space Coupled Cluster Method in the Effective and Intermediate Hamiltonian Formulation for the (2,0) Sector. *J. Chem. Phys.* **2012**, *136*, 134111.
2. Musiał, M.; Bartlett, R. J. Multi-Reference Fock Space Coupled Cluster Method in the Intermediate Hamiltonian Formulation for Potential Energy Surfaces. *J. Chem. Phys.* **2011**, *135*, 044121.
3. Musiał, M.; Kucharski, S. A. First Principle Calculations of the Potential Energy Curves for Electronic States of the Lithium Dimer. *J. Chem. Theory Comput.* 2014, *10*, 1200.
4. Musiał, M.; Kucharski, S. A.; Bartlett, R. J. Multireference Double Electron Attached Coupled Cluster Method With Full Inclusion of the Connected Triple Excitations: MR-DA-CCSDT. *J. Chem. Theory Comput.* **2011**, 7, 3088.
5. Musiał, M.; Kowalska-Szojda, K.; Lyakh, D. I.; Bartlett, R. J. Potential Energy Curves Via Double Electron Affinity Calculations: Dissociation of Alkali Metal Dimers. *J. Chem. Phys.* **2013**, *138*, 194553.
6. Lyakh, D. I.; Musiał, M.; Lotrich, V.; Bartlett, R. J. Multireference Nature of Chemistry: The Coupled-Cluster View. *Chem. Rev.* **2012**, *112*, 182; Musial, M.; Bartlett, R. J. Spin-Free Intermediate Hamiltonian Fock Space Coupled-Cluster Theory With Full Inclusion of Triple Excitations for Restricted Hartree Fock Based Triplet States. *J. Chem. Phys.* **2008**, *129*, 244111.
7. Musiał, M.; Medrek, M.; Kucharski, S. A. Potential Energy Curves of Li_2^+ From All-Electron EA-EOM-CCSD Calculations. *Mol. Phys.* **2015**, *113*, 2943.
8. Bewicz, A.; Musiał, M.; Kucharski, S. A. Potential Energy Curves of the Na_2^+ Molecular Ion From All Electron ab Initio Relativistic Calculations. *Mol. Phys.* **2017**, https://doi.org/10.1080/00268976.2017.1292010.

9. Skupin, P.; Musiał, M.; Kucharski, S. A. Potential Energy Curves for the Low-Lying Electronic States of K_2^+ From ab Initio Calculations With All Electrons Correlated. *J. Phys. Chem. A* **2017**, *121*(7), 1480–1486. https://doi.org/10.1021/acs.jpca. 6b09920.
10. Čizek, J. On the Correlation Problem in Atomic and Molecular Systems. Calculation of Wavefunction Components in Ursell-Type Expansion Using Quantum-Field Theoretical Methods. *J. Chem. Phys.* **1966**, *45*, 4256. Paldus, J.; Čizek, J.; Shavitt, I. Correlation Problems in Atomic and Molecular Systems. IV. Extended Coupled-Pair Many-Electron Theory and Its Application to the BH_3 Molecule. *Phys. Rev. A* **1972,** *5*, 50.
11. Bartlett, R. J. Coupled-Cluster Approach to Molecular Structure and Spectra: A Step toward Predictive Quantum Chemistry. *J. Phys. Chem.* **1989**, *93*, 1697. Bartlett, R. J.; Stanton, J. F. In *Reviews in Computational Chemistry*; K. B. Lipkowitz; Boyd, D. B., Eds.; VCH Publishers: New York, 1994, Vol. 5, p. 65.
12. Paldus, J.; Li, X. A Critical Assessment of Coupled Cluster Method in Quantum Chemistry. *Adv. Chem. Phys.* **1999**, *110*, 1.
13. Bartlett, R. J.; Musiał, M. Coupled-Cluster Theory in Quantum Chemistry. *Rev. Mod. Phys.* **2007**, *79*, 291.
14. Nooijen, M.; Bartlett, R. J. Equation of Motion Coupled Cluster Method for Electron Attachment. *J. Chem. Phys.* **1995**, *102*, 3629.
15. Krylov, A. I. Equation-of-Motion Coupled-Cluster Methods for Open-Shell and Electronically Excited Species: The Hitchhiker's Guide to Fock Space. *Annu. Rev. Phys. Chem.* **2008**, *59*, 433.
16. Musiał, M.; Bartlett, R. J. Equation-of-Motion Coupled Cluster Method With Full Inclusion of Connected Triple Excitations for Electron-Attached States: EA-EOM-CCSDT. *J. Chem. Phys.* **2003**, *119*, 1901.
17. Musiał, M. The Excited, Ionized and Electron Attached States Within the EOM-CC Approach With Full Inclusion of Connected Triple Excitations. *Mol. Phys.* **2010**, *108*, 2921.
18. Kamiya, M.; Hirata, S. Higher-Order Equation-of-Motion Coupled-Cluster Methods for Electron Attachment. *J. Chem. Phys.* **2007**, *126*, 134112.
19. Gour, J. R.; Piecuch, P. Efficient Formulation and Computer Implementation of the Active-Space Electron-Attached and Ionized Equation-of-Motion Coupled-Cluster Methods. *J. Chem. Phys.* **2006**, *125*, 234107.
20. Al-dossary, O. M.; Khelifi, N. Theoretical Study of LiK and LiK^+ in Adiabatic Representation. *Russ. J. Phys. Chem. A* **2014**, *88*, 73.
21. Berriche, H.; Ghanmi, C.; Ben Ouada, H. Theoretical Study of the Electronic States and Transition Dipole Moments of the LiK^+ Molecule. *J. Mol. Spectrosc.* **2005**, *230*, 161.
22. Patil, S. H.; Tang, K. T. Simple Model Potential and Model Wave Functions for (H-alkali)+ and (alkali-alkali) + Ions. *J. Chem. Phys.* **2000**, *113*, 676.
23. von Szentpály, L.; Fuentealba, P.; Preuss, H.; Stoll, H. Pseudopotential Calculations on Rb_2^+, Cs_2^+, RbH^+, CsH^+ and the Mixed Alkali Dimer Ions. *Chem. Phys. Lett.* **1982**, *93*, 555.
24. Müller, W.; Meyer, W. Ground-State Properties of Alkali Dimers and Their Cations (Including the Elements Li, Na, and K) From ab Initio Calculations With Effective Core Polarization Potentials. *J. Chem. Phys.* **1984**, *80*, 3311.
25. Ghanmi, C.; Berriche, H.; Ben Ouada, H. Theoretical Investigation of the Molecular Structure and Transition Dipole Moments of the NaK^+ Low Lying Electronic States. *J. Mol. Spectrosc.* **2006**, *235*, 158.
26. Valence, A.; Bernier, A.; El, M. Maddarsi, Σ and Π Molecular States for NaK^+, KRb^+ and $NaRb^+$ Molecular Ions. *Chem. Phys.* **1986**, *103*, 151.

27. Douglas, M.; Kroll, N. M. Quantum Electrodynamical Corrections to the Fine Structure of Helium. *Ann. Phys.* **1974**, *82*, 89; Hess, B. A. Relativistic Electronic-Structure Calculations Employing a Two-Component No-Pair Formalism With External-Field Projection Operators. *Phys. Rev. A* **1986,** *33*, 3742; Jensen, G.; Hess, B. A.; Revision of the Douglas-Kroll Transformation. *Phys. Rev. A*, **1989,** *39*, 6016; Barysz, M., Sadlej, A. J. Two-Component Methods of Relativistic Quantum Chemistry: From the Douglas-Kroll Approximation to the Exact Two-Component Formalism. *J. Mol. Struct.:THEOCHEM* **2001,** *573*, 181.
28. Schmidt, M. W.; Baldridge, K. K.; Boatz, J. A.; Elbert, S. T.; Gordon, M. S.; Jensen, J. H.; Koseki, S.; Matsunga, N.; Nguyen, K. A.; Su, S.; Windus, T. L.; Dupuis, M.; Montgomery, J. A. General Atomic and Molecular Electronic Structure System. *J. Comput. Chem.* **1993**, *14*, 1347. Piecuch, P.; Kucharski, S. A.; Kowalski, K.; Musiał, M. Efficient Computer Implementation of the Renormalized Coupled-Cluster Methods: The R-CCSD[T], R-CCSD(T), CR-CCSD[T], and CR-CCSD(T) Approaches. *Comp. Phys. Commun.* **2002**, *149*, 71; Gour, J. R.; Piecuch, P.; Wloch, M.; Active-Space Equation-of-Motion Coupled-Cluster Methods for Excited States of Radicals and Other Open-Shell Systems: EA-EOM-CCSDt and IP-EOM-CCSDt. *J. Chem. Phys.* **2005,** *123*, 134113.
29. Stanton, J. F.; Gauss, J.; Watts, J. D.; Nooijen, M.; Oliphant, N.; Perera, S. A.; Szalay, P. G.; Lauderdale, W. J.; Kucharski, S. A.; Gwaltney, S. R.; Beck, S.; Balková, A.; Musiał, M.; Bernholdt, D. E.; Baeck, K.-K.; Sekino, H.; Rozyczko, P., Huber, C.; Bartlett, R. J. ACES II Program is a Product of the Quantum Theory Project. University of Florida; Almlöf, J.; Taylor, P. Integral Packages Included are VMOL; Taylor, P. R. VPROPS; Helgaker, T. U.; Jensen, H. J. Aa.; Olsen, J.; Joergensen, P.; Taylor, P. R. A Modified Version of ABACUS Integral Derivative Package.
30. Roy, R. J. L. LEVEL 8.0: A Computer Program for Solving the Radial Schrödinger Equation for Bound and Quasibound Levels. University of Waterloo Chemical Physics Research Report CP-663 2007. See http://leroy.uwaterloo.ca/programs/.
31. Sadlej, A. J. Medium-Size Polarized Basis Sets for High-Level Correlated Calculations of Molecular Electric Properties. *Collect. Czech. Chem. Commun.* **1988**, *53*, 1995.
32. Roos, B. O.; Veryazov, V.; Widmark, P. O. Relativistic Atomic Natural Orbital Type Basis Sets for the Alkaline and Alkaline-Earth Atoms Applied to the Ground-State Potentials for the Corresponding Dimers. *Theor. Chem. Acc.* **2004**, *111*, 345.
33. Noro, T.; Sekiya, M.; Koga, T. Segmented Contracted Basis Sets for Atoms H Through Xe: Sapporo-(DK)-nZP Sets (n=D, T, Q). *Theor. Chem. Acc.* **2012**, *131*, 1124.
34. Radziemski, L. J.; Engelman, R.; Brault, J. W. Fourier-Transform-Spectroscopy Measurements in the Spectra of Neutral Lithium, ^6I and ^7I (Li I). *Phys. Rev. A* **1995**, *52*, 4462.
35. Sansonetti, J. E. Wawelengths, Transition Probabilities, and Energy Levels for the Spectra of Sodium (Na I-Na XI). *J. Phys. Chem. Ref. Data* **2008**, *37*, 1659.
36. Sugar, J.; Corliss, C. Atomic Energy Levels of the Iron-Period Elements: Potassium Through Nickel. *J. Phys. Chem Ref. Data* **1985**, *14* (Suppl. 2), 1.
37. Leutwyler, S.; Hofmann, M.; Harri, H. P.; Schumacher, E. The Adiabatic Ionization Potentials of the Alkali Dimers Na_2, NaK and K_2. *Chem. Phys. Lett.* **1981**, 77, 257.
38. Tomza, M.; Goerz, M. H.; Musial, M.; Moszynski, R.; Koch, C. P. Optimized Production of Ultracold Ground-State Molecules: Stabilization Employing Potentials With Ion-Pair Character and Strong Spin-Orbit Coupling. *Phys. Rev. A* **2012**, *86*, 043424.

INDEX

Note: Page numbers followed by "*f*" indicate figures, "*t*" indicate tables, and "*np*" indicate footnotes.

A

Absolute hardness, 62
Algebraic singularities, 55–56
Alkali metal diatomics, 334
Aluminum (Al), Pauli kinetic energy modulating factor, 65–67, 66–67*f*, 71*f*
Angular momentum, 155–158, 207–208
Angular quantum numbers, 214–216
Antisymmetrized geminal power (AGP), 256–257
Asymptotic behavior, 36
Asymptotic expansions, 37, 41–42
  classical Euler–Maclaurin (E–M) expansion, 38–39
  Gauss–Legendre quadrature, error expansions for, 46–49
  Legendre polynomials, 50–52
  Legendre series coefficients, 52–56
ATMOL, 5, 16, 28
Atomic electronic structure, 6
Atomic orbital bases, 16
Atomic self-consistent fields, 6–8
Atomic units, total energy and variance, 275*t*
Atoms, 247–251
AUTOCL, 5, 28

B

Bernoulli numbers, 50–51, 51*np*
Beryllium atom
  excitation energies, 264–266, 265*f*, 266–267*t*
  Jastrow factor, 266–267
  oscillator strength, 267, 268*t*
Bipolar harmonics, 214–215
Bohr–Schrödinger theory, 62
Broken symmetry wave function, 297–298
Brown's formula, 17–21

C

Carbon atom, 223–224
  ground-state leading configuration, 227
  low-lying states, 236*t*
Classical Euler–Maclaurin (E–M) expansion, 38–39
  computation of singular integrals, 44–45
  Navot's generalizations, 39–40
  recent generalizations, 40–43
  remarks on, 45–46
Commutation relations, 317–319
Complete basis set (CBS), 10
Complex Jastrow factor, 274–275
Configuration interaction (CI), 6, 188, 273
  ALCHEMY computer program, 224
  dynamic (hyper)polarizabilities, 300–301
  electronic wave functions, 297
  methods, 316
  ^{3}P ground state of carbon atom, 231–233*t*, 235*t*
    angular energy limits, 233*t*
    calculations, 229–234
    selection of, 227–229
  time-dependent, 299–300
Configuration interaction Monte Carlo (CIMC) method
  atoms, 247–251
  functions and recursion relations, 245–246
  molecules, 247–250
  single-particle states, construction, 246–247
  Slater determinants, 242–243
Configuration state functions (CSFs), 10–11, 297–298
Contracted Gaussian-type function (CGF), 149–150
Correlation energy, 108–112, 109*t*, 110*f*
Coulomb functions
  consecutive coefficients, 94–95, 95*f*
  definitions, 82–84
  irregular, 86–91
  Laguerre-type functions, 84–91
  notations, 82–84
  orthogonal polynomials

Coulomb functions (*Continued*)
Meixner–Pollaczek polynomials, 92–96
Pollaczek polynomials, 91–92
regular, 85–86
Coulomb interaction, 104–105, 108–111
Coulomb Schrödinger equation, 82–83
Coulomb Sturmian (CS) functions, 214–215
Coulomb Sturmians orbitals (CSO), 134–135
definitions, 134–135
radial functions, 136–138
three-dimensional orbitals, 139
Coupled cluster (CC), 19
Coupled-cluster single double and perturbative triple [CCSD(T)], 280, 284, 287
Coupled-cluster theory, 335–336
Coupled-perturbed Hartree–Fock (CPHF), 174–176
Cu(111), 272–274

D

Degenerate L-eigenfunction solutions, 228
Degenerate spaces, 13
Density functional theory (DFT), 60, 68, 212
Slater determinant, 273
Depurated inversion method (DIM), 126–127
inverted effective potentials, 122–126
magnesium atom, 123*f*, 128*f*
nodes and derivatives, 119–121, 119*f*
Diatomic molecules
correlation energy, 108–112, 109*t*, 110*f*
direct diagram, 110–111
domain decomposition algorithm, 105–107
exchange diagram, 110–111
Goldstone diagrams, 110–111, 110*f*
Hartree–Fock equations, 104–105
horizontal interface, 108, 109*t*
model problem, 107–108
orbital equations, 112–115
pair equation
first-order, 114–115, 115*t*
for two noninteracting electrons, 113–114
PDE, 106
single-electron equation, 112–113
spheroidal coordinates, 108
vertical interface, 106, 106*f*
Difference radial-density (DRD), 282
Difference total densities (DTD), 282
Diffusion Monte Carlo approaches, 224
DIM. *See* Depurated inversion method (DIM)
Dipole moments, 288, 289*t*
Dirac–Frenkel variational principle, 259–261
DMC algorithm, 275
Domain decomposition algorithm, 105–107
Double perturbation, HF, 176–177
Douglas–Kroll correction, 338, 341
Dynamic hyperpolarizabilities, 300–302
Dynamic polarizabilities, 300–302

E

EA-EOM-CC module, 336–337
EA-EOM-CCSD module, 336, 337–338*f*, 338–340, 339*t*
EA-EOM-CCSDT module, 336, 340*t*
Effective core potentials (ECP), 334
Effective inverted charge, 123, 124–125*f*
Electron-attachment equation-of-motion (EA-EOM), 340–341
Electron repulsion integrals (ERI), 139–141, 144*f*
4-center, 134–135
Equation-of-motion (EOM), 334
coupled cluster theory, 335–336
Ergonals, 14–15
Error expansions, Gauss–Legendre quadrature, 46–49
Euler angle, 195
Euler–Maclaurin (E–M) expansions, 37
Exchange–correlation functional, 212
Excitation-based methods, 316
Excitation energy, 259–261
beryllium atom, 264–266, 265*f*, 266–267*t*
Excited Slater determinants, 297–298
Exponential Correlated Gaussians (ECGs), 224
Exponential correlation
for four-body systems, 196–197
for three-body systems, 196
two-electron integrals, 194–196

Exponential-type orbitals (ETOs), 134, 139–141
Extended Hy-CI (E-Hy-CI), 189

F

First-principle calculations, 273
First principle computational scheme, 335
Fisher information entropy, 60–61
F12 methods, 147–148
Fock space (FS), 242–243, 334
Four-electron computer program, 229–230
Four-electron integrals, 149–151
algorithm, 158–162, 159*f*
recurrence relations, 154–158
unlinked, 198–201
Four-electron operators, 150–151, 151*f*
Fredholm integral equations
singular and hypersingular, 44
weakly singular, 44
Frequency-dependent hyperpolarizability, 301
Frequency-dependent polarizability, 301
Full configuration interaction (FCI), 334
Fundamental integrals, 149–150, 152–154

G

Gauge-including AOs (GIAO), 169–171
NDDO approximation, 171–172
ZDO approximation on, 173–174
Gaunt coefficients, 191–193, 199–201, 207–208
Gaussian basis functions
algorithm, 158–162, 159*f*, 160–161*t*
four-electron integrals, 149–151, 151*f*
fundamental integrals, 152–154, 152*t*
recurrence relations
horizontal recurrence, 158
transfer recurrence, 157–158
vertical recurrence, 154–157, 157*f*
Gaussian geminals, 147–148, 153–154
Gaussian integral representation
Kernels *F(t)*, 152, 152*t*
of two-electron operator, 152
Gauss–Legendre quadrature, 37, 148
error expansions for, 46–49
Graphical unitary group approach (GUGA), 26, 298–299

H

Hadamard finite part (HPF), 40
Hamiltonian matrix, 22–23, 259
Hamiltonian operator, 316–317
Hartree–Fock Roothaan (HFR)
for field and moments, 169–171
with field and moments, 171–172
INDO formalism for field and moments within London approximation, 173–174
London approximation for fields, 172–173
perturbation exchange, 177–178
second-order energy correction using CPHF, 174–176
Hartree–Fock (HF) wave functions, 121*f*, 130–131, 334
argon, 125*f*
density, 282
depurated inversion method, 126–127
inverted effective potentials, 122–126
nodes and derivatives, 119–121, 119*f*
double perturbation, 176–177
equations, 7–8
for atoms, 104
for diatomic molecule, 104–105
Kohn–Sham equations, 118
orbitals, 62
Pauli exclusion principle, 118
perturbation exchange, 177
potential field, 118
radial density, 72–73, 73–74*f*
Hartree's method, 7–8
Head–Gordon–Pople (HGP) algorithm, 158
Helium atom
angular quantum numbers, 214–216
basis set, construction, 214–216
bipolar harmonics, 214–215
Coulomb Sturmian functions, 214–215
density functional theory, 212
eigenvalue equation, 216–217
ground state energy for, 217–218, 217*t*, 219*t*
Hamiltonian matrix, 216
Hartree–Fock method, 211–212
Hilbert space of, 212–214
iterative scheme, 216–217, 220*t*
Laguerre polynomial, 214–215
nonrelativistic Hamiltonian, 212–214

Helium atom (*Continued*)
one-particle potential, 218–221, 219*f*, 220*t*
Schrödinger equation, 212–214
spatial two-particle configuration state function, 214–215
Hermiticity, 172–173, 178–179
Hilbert space, 212–214, 249
Homogeneous magnetic field, 169–171
Horizontal recurrence relations (HRRs), 158, 162
Hund's rule, 223–224
Hybrid treatments
commutation relations, 317–319
determinants and energies
doublet system $CO^+(^2\Sigma^+)$, 321–325, 322*t*
ground state linear chain $H_7(^2\Sigma^+u)$, 326–330, 327–328*t*
triplet system $CH_2(^3B_1)$, 321–325, 323–324*t*
spin-adapted hybrid methods, 319–320
Hydrogen, bond dissociation, 272
Hylleraas-CI (Hy-CI) method, 225
angular momentum formulas, 207–208
auxiliary two-body integrals G_I, 201–202
check of K_2, 203–204, 204–205*t*
exponential correlation
for four-body systems, 196–197
for three-body systems, 196
two-electron integrals, 194–196
integrals, 197
kinetic energy, 204–207
potential-energy matrix elements
exponentially correlated two-electron integrals, 194–196
one-electron integrals, 191
two-electron integrals, 191–194
unlinked three- and four-electron integrals, 198–201
wave functions, 189–196
Hylleraas functional, 147–148
Hylleraas integrals, 197
Maple coding, 203, 204*t*

I

INDO method, 173–174
Integrated density differences (IDD), 284, 285–286*f*
Interacting spaces, 13
Interelectron repulsion integrals, 142*t*
Intersection hybrid method, 326–330, 329*f*
Inverted effective potentials
Kohn–Sham equations, 122
magnesium atom, 123*f*
Schrödinger equations, 122
Ionization potential (IP)-EOM, 334
Isolated carbon atom, 223–224

J

Jastrow factor, 256, 264, 266–267
Jastrow–Slater-type wave functions, 257–258
J-matrix, 85

K

Kinetic energy, 204–206
in Hylleraas-CI, 206–207
Kohn–Sham theory, 60
modulating factor, 66*f*, 338*f*
Pauli exclusion principle, 63–64
Kohn–Sham (KS) equations, 60, 64, 118
Hartree–Fock orbitals, 122
spin-orbitals, 273

L

Laguerre polynomial, 80–81, 214–215
Laguerre-type functions
irregular Coulomb functions, 86–91
regular Coulomb function, 85–86
Laplace expansion, 193, 198–199
Legendre polynomials, 50–52, 207
Legendre series coefficients, 52–56
endpoint singularities, 53–55
interior and endpoint singularities, 55–56
Lewis pairs, 23
LiK^+, 334–335
atomic calculations, 338–341, 339–340*t*
EA-EOM-CCSD, 342–343*t*
molecular ion, 341–344
potential energy curves, 337*f*
spectroscopic constants, 337
Linear combinations of atomic orbitals (LCAO), 9, 169–171
Linear optimization method, 258–259

Linear-response theory
Dirac–Frenkel variational principle, 259–261
intermediate-normalized wave function, 259–261
time-dependent perturbation, 259–261
Linear-response variational Monte Carlo (LR-VMC)
computational details, 264
linear optimization method, 258–259
linear-response theory, 259–261
realization in, 261–263
time-dependent perturbation, 259–261
wave function parametrization, 257–258
Liu–Parr expansion, 71–73, 74*t*, 74*f*
Local effective potentials, 118–119
Local-scaling transformation, 67–71
London approximation, 172–173
Löwdin's turn over rule, 26

M

Maclaurin expansion, 174–176
Mac–Laurin–Taylor series, 168–169
Many-center interelectron repulsion integrals, 134, 139–141
Mass polarization, 206
Maximum projected noninteracting space (MAXPNI), 13
Mean absolute errors (MAE), 288
Mean absolute relative errors (MARE), 288
Meixner–Pollaczek polynomials, 92–96
Metal surface, 272, 276
Midpoint rule approximations, 45
Minimum projected interacting space (MINPI), 13
Molecular diamagnetism, perturbation exchange for, 178–179
Molecular self-consistent fields, 9
Molecular site approach, 168
Molecules, 247–250
Multiconfigurational explicit correlated wave function, 224
Multiconfigurational Hartree–Fock (MCHF) program, 8
Multireference CI, 15–16
Multireference coupled cluster (MRCC) approach, 334

N

NaK^+, 334–335
atomic calculations, 338–341, 339–340*t*
EA-EOM-CCSD, 345–346*t*
molecular ion, 344–346
potential energy curves, 338*f*
spectroscopic constants, 337
Natural orbitals (NOs), 5, 13–15
Navot's generalizations, 39–40
NDDO approximation
for GIAO, 171–172
unperturbed elements for, 172–173
N-electron determinants, 316, 319–320
excitation levels, 317–320
seniority number, 316–317, 319–320
N-electron excitation operator, 317–318
N-electron Slater determinants, 330
Neutral diatomics, 334
NMR chemical shifts, 181
Noninteracting kinetic energy, 73, 74*t*
NRHF code, 120
NWChem software package, 281

O

Offset trapezoidal rule approximations, 45
One-electron integrals, potential-energy matrix elements, 191
One-particle density, 62–63, 68, 70
One-particle orbitals, 64–67
Opposite-spin (OS), 280, 282–283
Optimized effective potential (OEP) method, 121
Orthogonal polynomials
Meixner–Pollaczek polynomials, 92–96
Pollaczek polynomials, 91–92
Oscillator strength, 267, 268*t*

P

Pacman operator, 151, 153–154
Paldus arrays, 298–299
Pauli enhancement factor, 70
Liu–Parr expansion, 71–73, 74*t*, 74*f*
local-scaling transformations, 67–71, 71–73*f*
one-particle orbitals, 64–67, 66–68*f*
Pauli potential, 63–64
shell structure, 61–63

Pauli exclusion principle, 118
Pauli kinetic energy, 71–72, 71–72*f*
Pauli potential, 63–64
Perturbation exchange
  Hartree–Fock, 177
  Hartree–Fock Roothaan, 177–178
  for molecular diamagnetism, 178–179
  theorem, 179–180
Pochhammer symbol, 191–192
Pollaczek polynomials, 91–92
Potential energy curve (PEC), 334
Potential-energy matrix elements
  connected (sub)diagrams, 190
  exponentially correlated two-electron integrals, 194–196
  linked (sub)diagrams, 190
  one-electron integrals, 191
  two-electron integrals, 191–194
Primitive Gaussian-type function (PGF), 149–150
Projection operators, 11–12
Pseudo-convexity, 62
Pt(111), 272–274

Q

Quantum Monte Carlo (QMC) methods, 224, 242, 256–257, 272, 296
  complex Jastrow factors, 274–275
  model geometries, 273
  time-step bias, DMC error sources, and specifications, 275–276

R

Real-time time-dependent density functional theory (RT-TDDFT), 296
Recurrence relations (RRs)
  horizontal recurrence, 158
  transfer recurrence, 157–158
  vertical recurrence, 154–157, 157*f*
Resolution of the identity (RI), 148–149
Restricted HF (RHF), 334, 337
Restricted open-shell Hartree–Fock (ROHF), 298, 321
Richardson extrapolation process, 39
Riemann zeta function, 40*np*
R12 method, 148–149
Rodrigues formula, 207
Romberg integration, 39
Roothaan secular equations, 171–172
Root mean square errors (RMS), 288

S

Same-spin (SS), 280, 282–283
Sapporo quadruple zeta basis set, 338, 344
Scalar relativistic effects, 335
Scaled-opposite spin MP2 (SOS-MP2), 280–281, 285
  correlation energy, 288–290
  dipole moments, 288
Schrödinger equation (SE), 7, 63–64, 225–226, 274
Second-order energy correction, using CPHF, 174–176
Selected configuration interaction (SCI)
  energy contribution of single configuration
    Brown's formulae, 19–21
    Dick Brown's original formula, 17–19
  on large systems, 26–27
  by parts, 22–23
  preselection of configurations without truncation energy error (TEE), 21–22
  selection of configurations with truncation energy error (TEE), 22
  selection process mechanics, 24–25
  tools
    atomic orbital bases, 16
    fast-converging CI expansions, 13–15
    interacting spaces, 13
    molecular, 10–11
    multireference CI, 15–16
    projection operators, 11–12
    subclasses, 12–13
  truncation energy error, 23
  valence, 23–24
Self-consistent field (SCF), 6
  atomic, 6–8
  molecular, 9
Shannon information entropy, 62–63
Shell structure, 61–63
Single-particle states, 246–247
Slater determinant, 7–8, 10, 169–171, 257–258, 319
Slater geminals, 147–148

Slater-type orbitals (STOs), 70, 134, 226–227
convergence and accuracy, checks of, 141–145
Coulomb Sturmian expansions
definitions, 134–135
radial functions, 136–138, 137–138*t*
three-dimensional orbitals, 139
electron repulsion integrals, 139–141
Spatial orbitals, 317–318
Specific reaction parameter density functional theory (SRP-DFT), 272, 276–277
Spin-adapted hybrid methods, 319–320
Spin-component scaled MP2 method (SCS-MP2), 280–281, 285
dipole moments, 288
Spin contamination, 316–317
Spin-resolved second-order Møller–Plesset method
computational details, 281
correlated density distributions, 282–283, 283*f*
dipole moments, 288, 289*t*
proportionality of densities, 282–283, 283*f*
two-dimensional scan, 284–287
Split operator approach, 299–300
Stieltjes numbers, 41
Subclass, 12–13

T

Tamm–Dancoff approximation (TDA), 257, 259–261
Taylor series expansions, 42
Tensors, 180
Three-electron integrals, 148
algorithm, 158–162, 159*f*
angular momentum, 158
recurrence relations, 149
unlinked, 198–201
Time-dependent CI, 299–300
Time-dependent configuration interaction singles (TD-CIS), 296–297, 302–304, 303*t*
Time-dependent perturbation, 259–261
Time-dependent restricted-active-space self-consistent field (TD-RASSCF) approaches, 296–297
Time-dependent Schrödinger equation, 299
Transfer recurrence relations (TRRs), 157–159, 162
Transition state (TS), 273
Trapezoidal rule approximation, 38
classical E–M expansion, 38–39
Navot's generalizations, 39–40
Truncation energy error (TEE), 5
preselection of configurations without, 21–22
selection of configurations with, 22
Two-electron integrals
exponential correlation, 194–196
potential-energy matrix elements, 191–194

U

Union hybrid method, 326–330, 329*f*

V

Valence orbitals, 24
Valence SCI (VSCI), 24
Variational Monte Carlo (VMC), 256, 274
Various valence CI (VCI), 21, 24
Vector potential, 169–171
Verlinden's theorem, 47
Vertical recurrence relations (VRRs), 153–157, 157*f*, 159, 162

W

Wave function, 188–189
first-order derivatives, 258
Hamiltonian matrix, 259
parametrization, 257–258
second-order derivatives, 259–261
Whittaker functions, 83, 87, 96–100
Wigner coefficient, 195–196

Z

ZDO approximation, 173–174
Zero-variance principle, 263

Kolibri 360 GmbH
Ferdinand-Jühlke-Straße 7
99095 Erfurt, Deutschland
produktsicherheit@kolibri360.de